Springer Series in

CHEMICAL PHYSICS 100

Springer Series in

CHEMICAL PHYSICS

The purpose of this series is to provide comprehensive up-to-date monographs in both well established disciplines and emerging research areas within the broad fields of chemical physics and physical chemistry. The books deal with both fundamental science and applications, and may have either a theoretical or an experimental emphasis. They are aimed primarily at researchers and graduate students in chemical physics and related fields.

Please view available titles in *Springer Series in Chemical Physics* on series homepage http://www.springer.com/series/676

Kaoru Yamanouchi
Dimitrios Charalambidis
Didier Normand

Editors

Progress in Ultrafast Intense Laser Science

Volume VII

With 121 Figures

Editors

Professor Kaoru Yamanouchi
The University of Tokyo
Department of Chemistry
7-3-1 Hongo, Bunkyo-ku
Tokyo 113-0033, Japan
kaoru@chem.s.u-tokyo.ac.jp

Dr. Didier Normand
IRAMIS, CEA Saclay
Bât. 524, PC No. 83
91191 Gif-sur-Yvette Cedex, France
didier.normand@cea.fr

Professor Dimitrios Charalambidis
University of Crete
Department of Physics
PO Box 2208, 71003 Heraklion, Greece
chara@iesl.forth.gr

Springer Series in Chemical Physics ISSN 0172-6218
ISBN 978-3-642-18326-3 e-ISBN 978-3-642-18327-0
DOI 10.1007/978-3-642-18327-0
Springer Heidelberg Dordrecht London New York

Library of Congress Control Number: 2006927806

Cover design: eStudio Calamar Steinen

Printed on acid-free paper

Springer is part of Springer Science+Business Media (www.springer.com)

Preface

We are pleased to present the seventh volume of Progress in Ultrafast Intense Laser Science. As the frontiers of ultrafast intense laser science rapidly expand ever outward, there continues to be a growing demand for an introduction to this interdisciplinary research field that is at once widely accessible and capable of delivering cutting-edge developments. Our series aims to respond to this call by providing a compilation of concise review-style articles written by researchers at the forefront of this research field, so that researchers with different backgrounds as well as graduate students can easily grasp the essential aspects.

As in the previous volumes of PUILS, each chapter of this book begins with an introductory part, in which a clear and concise overview of the topic and its significance is given, and moves onto a description of the authors' most recent research results. All the chapters are peer reviewed. The articles of this seventh volume cover a diverse range of the interdisciplinary research field, and the topics may be grouped into five categories: ionization of atoms and molecules (Chap. 1), ultrafast responses of protons and electrons within a molecule (Chaps. 2 and 3), molecular alignment (Chaps. 4 and 5), high-order harmonics and attosecond pulse generation (Chaps. 6–8), and acceleration of electrons and ions in laser plasmas (Chaps. 9 and 10).

From the third volume, the PUILS series has been edited in liaison with the activities of the Center for Ultrafast Intense Laser Science, the University of Tokyo, and JILS (Japan Intense Light Field Science Society), the latter of which has also been responsible for sponsoring the series and making the regular publication of its volumes possible. From the fifth volume, the Consortium on Education and Research on Advanced Laser Science, the University of Tokyo, has joined this publication activity as one of the sponsoring programs. The series has also collaborated since its inception with the annual symposium series of ISUILS (http://www.isuils.jp), which is designed to stimulate interdisciplinary discussion at the forefront of ultrafast intense laser science.

We would like to take this opportunity to thank all the authors who have kindly contributed to the PUILS series by describing their most recent work at the frontiers of ultrafast intense laser science. We also thank the reviewers who have read the submitted manuscripts carefully. One of the co-editors (KY) thanks Ms. Chie Sakuta for her help with the editing processes. Last but not least, our gratitude goes out to

Dr. Claus Ascheron, Physics Editor of SpringerVerlag at Heidelberg, for his kind support.

We hope this volume will convey the excitement of ultrafast intense laser science to the readers, and stimulate interdisciplinary interactions among researchers, thus paving the way to explorations of new frontiers.

Tokyo
Heraklion
Gif-sur-Yvette
October 2010

Kaoru Yamanouchi
Dimitrios Charalambidis
Didier Normand

Contents

Contributors

Daniel an der Brügge Institute for Theoretical Physics I, University of Duesseldorf, 40225 Duesseldorf, Germany, dadb@tp1.uni-duesseldorf.de

Sujata Bhattacharyya IACS, Kolkata-700032, India, sreebrata@yahoo.com

Alexey S. Boldarev Keldysh Institute of Applied Mathematics, Russian Academy of Sciences (RAS), Moscow, 125047, Russia, boldar@imamod.ru

Paul R. Bolton Kansai Photon Science Institute (KPSI), Atomic Energy Agency (JAEA), Kyoto, 615-0215 Japan
and
Photo-Medical Research Center (PMRC), Atomic Energy Agency (JAEA), Kyoto, 615-0215 Japan, bolton.paul@jaea.go.jp

Sergei V. Bulanov Kansai Photon Science Institute (KPSI), Atomic Energy Agency (JAEA), Kyoto, 615-0215 Japan
and
Photo-Medical Research Center (PMRC), Atomic Energy Agency (JAEA), Kyoto, 615-0215 Japan
and
A.M. Prokhorov Institute of General Physics, Russian Academy of Sciences (RAS), 119991 Moscow, Russia, bulanov.sergei@jaea.go.jp

Dimitrios Charalambidis Institute of Electronic Structure and Laser, Foundation for Research and Technology–Hellas, PO Box 1527, 71110 Heraklion, Crete, Greece
and
Department of Physics, University of Crete, PO Box 2208, 71003 Heraklion, Crete, Greece, chara@iesl.forth.gr

Frederic Chaussard Laboratoire Interdisciplinaire CARNOT de Bourgogne (ICB), UMR 5209 CNRS-Université de Bourgogne, BP 47870, 21078 Dijon Cedex, France, frederic.chaussard@u-bourgogne.fr

Hiroyuki Daido Kansai Photon Science Institute (KPSI), Atomic Energy Agency (JAEA), Kyoto, 615-0215 Japan
and
Photo-Medical Research Center (PMRC), Atomic Energy Agency (JAEA), Kyoto, 615-0215 Japan, daido.hiroyuki@jaea.go.jp

Timur Zh. Esirkepov Kansai Photon Science Institute (KPSI), Atomic Energy Agency (JAEA), Kyoto, 615-0215 Japan
and
Photo-Medical Research Center (PMRC), Atomic Energy Agency (JAEA), Kyoto, 615-0215 Japan, Timur.Esirkepov@jaea.go.jp

Anatoly Ya Faenov Kansai Photon Science Institute (KPSI), Atomic Energy Agency (JAEA), Kyoto, 615-0215 Japan
and
Photo-Medical Research Center (PMRC), Atomic Energy Agency (JAEA), Kyoto, 615-0215 Japan
and
Joint Institute of High Temperatures, Russian Academy of Sciences (RAS), Moscow, 125412, Russia, anatolyf@hotmail.com

Farhad H.M. Faisal Fakultät für Physik, Universität Bielefeld, Postfach 100131, 33501 Bielefeld, Germany, ffaisal@physik.uni-bielefeld.de

Olivier Faucher Laboratoire Interdisciplinaire CARNOT de Bourgogne (ICB), UMR 5209 CNRS-Université de Bourgogne, BP 47870, 21078 Dijon Cedex, France, olivier.faucher@u-bourgogne.fr

Yuichi Fujimura Department of Chemistry, Graduate School of Science, Tohoku University, Sendai 980-8578, Japan, fujimurayuichi@m.tohoku.ac.jp

Yuji Fukuda Kansai Photon Science Institute (KPSI), Atomic Energy Agency (JAEA), Kyoto, 615-0215 Japan
and
Photo-Medical Research Center (PMRC), Atomic Energy Agency (JAEA), Kyoto, 615-0215 Japan, fukuda.yuji@jaea.go.jp

Vladimir A. Gasilov Keldysh Institute of Applied Mathematics, Russian Academy of Sciences (RAS), Moscow, 125047, Russia, vgasilov@yandex.ru

Yukio Hayashi Kansai Photon Science Institute (KPSI), Atomic Energy Agency (JAEA), Kyoto, 615-0215 Japan
and
Photo-Medical Research Center (PMRC), Atomic Energy Agency (JAEA), Kyoto, 615-0215 Japan, hayashi.yukio@jaea.go.jp

Carlos Hernandez-García Departamento de Física Aplicada, Universidad de Salamanca, 37008 Salamanca, Spain, carloshergar@usal.es

Edouard Hertz Laboratoire Interdisciplinaire CARNOT de Bourgogne (ICB), UMR 5209 CNRS-Université de Bourgogne, BP 47870, 21078 Dijon Cedex, France, edouard.hertz@u-bourgogne.fr

Enrique Conejero Jarque Departamento de Física Aplicada, Universidad de Salamanca, 37008 Salamanca, Spain, enrikecj@usal.es

Takashi Kameshima Kansai Photon Science Institute (KPSI), Atomic Energy Agency (JAEA), Kyoto, 615-0215 Japan
and
Photo-Medical Research Center (PMRC), Atomic Energy Agency (JAEA), Kyoto, 615-0215 Japan, kameshima@spring8.or.jp

Masaki Kando Kansai Photon Science Institute (KPSI), Atomic Energy Agency (JAEA), Kyoto, 615-0215 Japan
and
Photo-Medical Research Center (PMRC), Atomic Energy Agency (JAEA), Kyoto, 615-0215 Japan, kando.masaki@jaea.go.jp

Manabu Kanno Department of Chemistry, Graduate School of Science, Tohoku University, Sendai 980-8578, Japan, kanno@m.tohoku.ac.jp

Yoshiaki Kato Kansai Photon Science Institute (KPSI), Atomic Energy Agency (JAEA), Kyoto, 615-0215 Japan
and
Photo-Medical Research Center (PMRC), Atomic Energy Agency (JAEA), Kyoto, 615-0215 Japan, y.kato@gpi.ac.jp

Shunichi Kawanishi Kansai Photon Science Institute (KPSI), Atomic Energy Agency (JAEA), Kyoto, 615-0215 Japan
and
Photo-Medical Research Center (PMRC), Atomic Energy Agency (JAEA), Kyoto, 615-0215 Japan, kawanishi.shunichi@jaea.go.jp

Keigo Kawase Kansai Photon Science Institute (KPSI), Atomic Energy Agency (JAEA), Kyoto, 615-0215 Japan
and
Photo-Medical Research Center (PMRC), Atomic Energy Agency (JAEA), Kyoto, 615-0215 Japan, kawase@sanken.osaka-u.ac.jp

Ryosuke Kodama Faculty of Engineering and Institute of Laser Engineering, Osaka University, Osaka, 565-0871 Japan, kodama@eie.eng.osaka-u.ac.jp

James Koga Kansai Photon Science Institute (KPSI), Atomic Energy Agency (JAEA), Kyoto, 615-0215 Japan
and
Photo-Medical Research Center (PMRC), Atomic Energy Agency (JAEA), Kyoto, 615-0215 Japan, koga.james@jaea.go.jp

Kiminori Kondo Kansai Photon Science Institute (KPSI), Atomic Energy Agency (JAEA), Kyoto, 615-0215 Japan
and
Photo-Medical Research Center (PMRC), Atomic Energy Agency (JAEA), Kyoto, 615-0215 Japan, kondo.kiminori@jaea.go.jp

Hirohiko Kono Department of Chemistry, Graduate School of Science, Tohoku University, Sendai 980-8578, Japan, hirohiko-kono@m.tohoku.ac.jp

Igor Kostyukov Institute for Applied Physics, Russian Academy of Sciences (RAS), 603950, Nizhni Novgorod, Russia, kost@appl.sci-nnov.ru

Jann E. Kruse Institute of Electronic Structure and Laser, Foundation for Research and Technology–Hellas, PO Box 1527, 71110 Heraklion, Crete, Greece
and
Department of Physics, University of Crete, PO Box 2208, 71003 Heraklion, Crete, Greece, jkruse@iesl.forth.gr

Bruno Lavorel Laboratoire Interdisciplinaire CARNOT de Bourgogne (ICB), UMR 5209 CNRS-Université de Bourgogne, BP 47870, 21078 Dijon Cedex, France, bruno.lavorel@u-bourgogne.fr

Ruxin Li State Key Laboratory of High Field Laser Physics, Shanghai Institute of Optics and Fine Mechanics, Chinese Academy of Sciences, Qinghe Rd 390, Jiading District, Shanghai 201800, China, ruxinli@mail.shcnc.ac.cn

Yuexun Li State Key Laboratory of High Field Laser Physics, Shanghai Institute of Optics and Fine Mechanics, Chinese Academy of Sciences, Qinghe Rd 390, Jiading District, Shanghai 201800, China
and
School of Electrical and Information Technology, Yunnan University of Nationalities, 121 Street No. 134, Kunming city, Yunnan 650031, China, yuexunli@siom.ac.cn

Peng Liu State Key Laboratory of High Field Laser Physics, Shanghai Institute of Optics and Fine Mechanics, Chinese Academy of Sciences, Qinghe Rd 390, Jiading District, Shanghai 201800, China, peng@siom.ac.cn

Alexander I. Magunov A.M. Prokhorov Institute of General Physics, Russian Academy of Sciences (RAS), 119991 Moscow, Russia, mai@mail.ru

Michiaki Mori Kansai Photon Science Institute (KPSI), Atomic Energy Agency (JAEA), Kyoto, 615-0215 Japan
and
Photo-Medical Research Center (PMRC), Atomic Energy Agency (JAEA), Kyoto, 615-0215 Japan, mori.michiaki@jaea.go.jp

Katsunori Nakai Department of Chemistry, School of Science,
The University of Tokyo, 7-3-1 Hongo, Bunkyo-ku, Tokyo, 113-0033, Japan, nakai@chem.s.u-tokyo.ac.jp

Tatsufumi Nakamura Kansai Photon Science Institute (KPSI), Atomic Energy Agency (JAEA), Kyoto, 615-0215 Japan
and
Photo-Medical Research Center (PMRC), Atomic Energy Agency (JAEA), Kyoto, 615-0215 Japan, nakamura.tatsufumi@jaea.go.jp

Koichi Ogura Kansai Photon Science Institute (KPSI), Atomic Energy Agency (JAEA), Kyoto, 615-0215 Japan
and
Photo-Medical Research Center (PMRC), Atomic Energy Agency (JAEA), Kyoto, 615-0215 Japan, ogura.koichi@jaea.go.jp

Hideki Ohmura National Institute of Advanced Industrial Science and Technology (AIST), 1-1-1, Higashi, Tsukuba, Ibaraki 305-8565, Japan, hideki-ohmura@aist.go.jp

Tomoya Okino Department of Chemistry, School of Science, the University of Tokyo, 7-3-1 Hongo, Bunkyo-ku, Tokyo, 113-0033, Japan, tokino@chem.s.u-tokyo.ac.jp

José Antonio Pérez-Hernández Centro de Láseres Pulsados CLPU, 37008 Salamanca, Spain, joseap@usal.es

Tatiana A. Pikuz Kansai Photon Science Institute (KPSI), Atomic Energy Agency (JAEA), Kyoto, 615-0215 Japan
and
Photo-Medical Research Center (PMRC), Atomic Energy Agency (JAEA), Kyoto, 615-0215 Japan
and
Joint Institute of High Temperatures, Russian Academy of Sciences (RAS), Moscow, 125412, Russia, tapikuz@yahoo.com

Alexander S. Pirozhkov Kansai Photon Science Institute (KPSI), Atomic Energy Agency (JAEA), Kyoto, 615-0215 Japan
and
Photo-Medical Research Center (PMRC), Atomic Energy Agency (JAEA), Kyoto, 615-0215 Japan, Pirozhkov.Alexander@jaea.go.jp

Luis Plaja Departamento de Física Aplicada, Universidad de Salamanca, 37008 Salamanca, Spain, lplaja@usal.es

Alexander Pukhov Institute for Theoretical Physics I, University of Duesseldorf, 40225 Duesseldorf, Germany, pukhov@tp1.uni-duesseldorf.de

Julio Ramos Departamento de Física Aplicada, Universidad de Salamanca, 37008 Salamanca, Spain, jarp@usal.es

Luis Roso Centro de Láseres Pulsados CLPU, 37008 Salamanca, Spain, roso@usal.es

Hironao Sakaki Kansai Photon Science Institute (KPSI), Atomic Energy Agency (JAEA), Kyoto, 615-0215 Japan
and
Photo-Medical Research Center (PMRC), Atomic Energy Agency (JAEA), Kyoto, 615-0215 Japan, sakaki.hironao@jaea.go.jp

Emmanouil Skantzakis Institute of Electronic Structure and Laser, Foundation for Research and Technology–Hellas, PO Box 1527, 71110 Heraklion, Crete, Greece
and
Department of Physics, University of Crete, PO Box 2208, 71003 Heraklion, Crete, Greece, skanman@iesl.forth.gr

Toshiki Tajima Kansai Photon Science Institute (KPSI), Atomic Energy Agency (JAEA), Kyoto, 615-0215 Japan
and
Photo-Medical Research Center (PMRC), Atomic Energy Agency (JAEA), Kyoto, 615-0215 Japan, Toshiki.Tajima@physik.uni-muenchen.de

Motonobu Tampo Kansai Photon Science Institute (KPSI), Atomic Energy Agency (JAEA), Kyoto, 615-0215 Japan
and
Photo-Medical Research Center (PMRC), Atomic Energy Agency (JAEA), Kyoto, 615-0215 Japan, tampo.motonobu@jaea.go.jp

Paris Tzallas Institute of Electronic Structure and Laser, Foundation for Research and Technology–Hellas, PO Box 1527, 71110 Heraklion, Crete, Greece, ptzallas@iesl.forth.gr

Pengfei Wei State Key Laboratory of High Field Laser Physics, Shanghai Institute of Optics and Fine Mechanics, Chinese Academy of Sciences, Qinghe Rd 390, Jiading District, Shanghai 201800, China
and
College of Physics and Electronic Information Engineering, Wenzhou University, Wenzhou 325035, P. R. China, personinjoy@163.com

Huailiang Xu State Key Laboratory on Integrated Optoelectronics, College of Electronic Science and Engineering, Jilin University, #2699 Qianjin Street, Changchun, 130012, China
and
Department of Chemistry, School of Science, the University of Tokyo, 7-3-1 Hongo, Bunkyo-ku, Tokyo, 113-0033, Japan, huailiang@jlu.edu.cn

Zhizhan Xu State Key Laboratory of High Field Laser Physics, Shanghai Institute of Optics and Fine Mechanics, Chinese Academy of Sciences, Qinghe Rd 390, Jiading District, Shanghai 201800, China, zzxu@mail.shcnc.ac.cn

Kaoru Yamanouchi Department of Chemistry, School of Science, the University of Tokyo, 7-3-1 Hongo, Bunkyo-ku, Tokyo, 113-0033, Japan, kaoru@chem.s.u-tokyo.ac.jp

Tomoya Yamauchi Graduate School of Maritime Sciences, Kobe University, Kobe 658-0022, Japan, yamauchi@maritime.kobe-u.ac.jp

Akifumi Yogo Kansai Photon Science Institute (KPSI), Atomic Energy Agency (JAEA), Kyoto, 615-0215 Japan
and
Photo-Medical Research Center (PMRC), Atomic Energy Agency (JAEA), Kyoto, 615-0215 Japan, yogo.akifumi@jaea.go.jp

Zhinan Zeng State Key Laboratory of High Field Laser Physics, Shanghai Institute of Optics and Fine Mechanics, Chinese Academy of Sciences, Qinghe Rd 390, Jiading District, Shanghai 201800, China, zhinan_zeng@siom.ac.cn

Shitong Zhao State Key Laboratory of High Field Laser Physics, Shanghai Institute of Optics and Fine Mechanics, Chinese Academy of Sciences, Qinghe Rd 390, Jiading District, Shanghai 201800, China
and
Department of Electrical and Computer Engineering, Cockrell School of Engineering, the University of Texas at Austin, 1 University Station C0803, Austin, TX 78712, USA, zst@siom.ac.cn

Chapter 1
Intense-Field Dirac Theory of Ionization of Hydrogen and Hydrogenic Ions in Infrared and Free-Electron Laser Fields

Farhad H.M. Faisal and Sujata Bhattacharyya

Abstract We investigate ionization of Dirac hydrogen atom and hydrogenic ions of charge states Z using the relativistic intense-field theory of spin-resolved ionization process. We discuss the energy dependence, emission angle dependence, spin dependence and charge state Z dependence of ionization rates for free-electron laser (FEL) and infrared frequencies. The results of relativistic intense-field model calculations are compared with the corresponding nonrelativistic calculations, and in specific cases also with the results of relativistic single-photon first Born approximation. Furthermore, we discuss two counterintuitive predictions made earlier: (a) an asymmetry of the up and down spin electron currents from *unpolarized* target H-atoms, and (b) an increase of the ionization probability with the increase of the ionization potential in certain domains of the charge states Z for a given laser intensity and frequency.

1.1 Introduction

Laser fields with intensities as high as $10^{21}\ \mathrm{W\,cm^{-2}}$ or more have become available recently (e.g., [1, 2]). At such intensities, even nonrelativistic bound electrons are accelerated to velocities comparable to the light velocity c, and the dynamics of electrons bound in atoms, molecules, clusters or solids could be strongly modified. Unlike in the traditional quantum electrodynamical problems, which are concerned with weak electromagnetic fields and spontaneous emission/absorption processes, where the electron motion can be treated successfully by relativistic perturbation theory [3], for highly intense laser fields the usual perturbation treatment

F.H.M. Faisal (✉)
Fakultät für Physik, Universität Bielefeld, Postfach 100131, 33501 Bielefeld, Germany
e-mail: ffaisal@physik.uni-bielefeld.de

S. Bhattacharyya
IACS, Kolkata-700032, India
e-mail: sreebrata@yahoo.com

K. Yamanouchi et al. (eds.), *Progress in Ultrafast Intense Laser Science VII*,
Springer Series in Chemical Physics 100, DOI 10.1007/978-3-642-18327-0_1,

breaks down. The breakdown intensity, e.g., for optical/near-infrared wavelengths is approximately about $10^{13}\,\mathrm{W\,cm^{-2}}$ [4].

One usually defines a super-intense laser field as one for which the mean quiver energy of a free electron in the field, known as the ponderomotive potential, U_p, becomes comparable to or greater than the rest mass energy of the electron, i.e., $U_p \equiv \frac{e^2F^2}{4m\omega^2} \geq mc^2$, where F is the peak value of the electric field strength. In such fields, in general a huge number of photons can be exchanged significantly in a transition process, and the usual non-perturbative approach of direct time-dependent simulation, e.g., of the corresponding Dirac equation of the system, or the Floquet expansion technique in the case of periodic fields, can be extremely arduous if not impossible. In the former case the space-time grid, and in the latter case the size of the Floquet matrix, would tend to be too large to be practicable. One alternative possibility of investigating relativistic dynamics in super-intense laser fields is to use the relativistic generalization of the so-called intense-field KFR (Keldysh–Faisal–Reiss)-like approximation (e.g., review [5]). In this approximation, *all* order interactions of the field with a relativistic free electron are first accounted for by using the Volkov solution of the Dirac equation of a free electron in a plane wave electromagnetic field [6]. This solution satisfies the final state plane wave boundary condition when the interaction with the field is over. In contrast, for an electron in a Coulomb potential plus the laser field, no such exact solution is known. A systematic approximation of the wavefunction of the *interacting system* satisfying such a typical bound state initial condition, nevertheless, can be developed within the so-called KFR-ansatz (e.g., [5]).

One aspect of the intense-field laser–atom interaction in the relativistic domain is the coupling of the photon field with the spin degrees of freedom. The latter gives rise to phenomena-like stimulated Mott-scattering [7–9] and other spin-dependent bound state effects (e.g., [10]) and the ubiquitous phenomenon of ionization in intense fields. In the latter context, most intense-field calculations of ionization probability in relativistic domains do not provide specific spin information since the rates are mostly obtained using the convenient tracing rules. The electrons emitted from target atoms in the initial bound states and/or in the final free states of course could be prepared and/or detected with or without spin selection. Thus for the ionization process, in strong fields the rates of spin-flips and the characters of up and down spin currents can be of much experimental interest. Intense-field 4-component Dirac analysis of such spin-dependence of the ionization rates has been reported in [11] that revealed among other things a remarkable spin *asymmetry* in the ionization currents even from spin *unpolarized* ground state H-atom. The asymmetry had been predicted to survive also for ionization by weak and long wavelength photons that possess a negligible photon momentum. Another relativistic aspect in ionization process in super-intense fields is the effect of retardation due to the finite velocity of light or the effect due to a finite photon momentum (that goes beyond the usual dipole approximation of the field). This can lead to deviations from the well-known return of the electron to the core along the polarization direction responsible for the various re-scattering phenomena in intense fields (e.g., [5]), due to the transverse effect of the Lorentz force exerted by the nonnegligible magnetic field of light. An interesting prediction of an earlier Klein–Gordon theoretical relativistic model calculation [12] has been *anomalous effect* in which the probability of ionization increases with increase of the ionization potential of the target atom and ion in certain domain of Z, keeping the intensity and the frequency of the laser field the same.

Here, we shall discuss relativistic ionization processes in Hydrogen atom and hydrogenic ions theoretically, keeping the above and related aspects of the process in mind. To begin with, in the next section, we give a systematic expansion of the strong-field Dirac relativistic wavefunction of the interacting system satisfying the initial *bound* state condition. Next, we apply the leading term of the wavefunction to obtain an analytic relativistic expression of the ionization amplitudes for a general elliptically polarized laser field of arbitrary frequency and intensity. Then the ionization rates for all combinations of the spin-specific (up or down spin) ground state of the atom or ion and the (up or down spin) free state of the ejected electron are obtained. Results of specific numerical calculations for the interesting case of a circularly polarized field with a given helicity are presented and discussed, and a summary of the results is given in the end.

1.2 Relativistic Theory of Intense-Field Processes

The Dirac equation of the system of atom (or ion) + laser field is

$$i\hbar\frac{\partial}{\partial t}\Psi(t) = H_{\mathrm{D}}(t)\Psi(t), \tag{1.1}$$

where the total Dirac Hamiltonian of the system is

$$H_{\mathrm{D}}(\boldsymbol{x},t) = c\boldsymbol{\alpha}\cdot\left(-i\hbar\frac{\partial}{\partial \boldsymbol{x}} - \frac{e_0}{c}\boldsymbol{A}(\boldsymbol{x},t)\right) + V(\boldsymbol{x}) + \beta m_0 c^2. \tag{1.2}$$

We have denoted the external electromagnetic vector potential of an elliptically polarized intense laser field (ellipticity parameter ξ $[-\pi/2,\pi/2]$) by

$$\boldsymbol{A}(\boldsymbol{x},t) = A_0(x)\left(\boldsymbol{\epsilon}_1\cos(\xi/2)\cos(\omega t - \boldsymbol{\kappa}\cdot\boldsymbol{x}) - \boldsymbol{\epsilon}_2\sin(\xi/2)\sin(\omega t - \boldsymbol{\kappa}\cdot\boldsymbol{x})\right), \tag{1.3}$$

where $A_0(x)$ is the envelope and $\boldsymbol{\epsilon}_1, \boldsymbol{\epsilon}_2$ are the orthogonal unit polarization vectors along the semi-major and the semi-minor axes, respectively, of the laser field. The corresponding scalar potential $\mathscr{A}_0(\boldsymbol{x})$ is defined by the potential energy for the hydrogenic atom (or ion) of charge Z: $V(\boldsymbol{x}) \equiv e_0\mathscr{A}_0(\boldsymbol{x}) = -\frac{Ze_0^2}{|x|}$. On introducing the Dirac–Feynman four-vector notations: $x \equiv (x_0,\boldsymbol{x}), x_0 = ct, |\boldsymbol{x}| = r$; $p \equiv (p_0,\boldsymbol{p}),\ a\cdot b = a_\mu b_\mu = a_0 b_0 - \boldsymbol{a}\cdot\boldsymbol{b},\ \not{a} = \gamma_\mu a_\mu,\ \bar{u} = u^\dagger\gamma_0,\ \gamma_\mu$ are Dirac γ-matrices, and defining the constants $e \equiv \frac{e_0}{\hbar c}$ and $m \equiv \frac{m_0 c}{\hbar}$, the Dirac equation above can be rewritten in the covariant form:

$$\left(i\gamma_\mu \frac{\partial}{\partial x_\mu} - e\gamma_\mu \mathscr{A}_\mu(x)\right) \Psi(t) = m\Psi(t). \tag{1.4}$$

Note that in the present notation only two constants e and m appear in the Dirac equation of interest, instead of the original four constants. Moreover, at the end of any calculation if desired the natural constants could be restored by simply substituting them in place of e and m. In Feynman slash notation, the Dirac equation becomes

$$(\mathrm{i}\,\partial\!\!\!/ - e\mathscr{A}\!\!\!/(x) - m)\Psi(t) = 0, \tag{1.5}$$

where $\partial\!\!\!/ \equiv \gamma_\mu \frac{\partial}{\partial x_\mu}$ and $\mathscr{A}\!\!\!/ \equiv \gamma_\mu \mathscr{A}_\mu$. For a later reference, we formally also define here the total Green's function associated with the same total Dirac Hamiltonian by the inhomogeneous equation

$$(\mathrm{i}\,\partial\!\!\!/ - e\mathscr{A}\!\!\!/(x) - m)G(x.x') = \delta^4(x - x'). \tag{1.6}$$

1.2.1 The Initial and Final Rest-Interaction Hamiltonians

The unperturbed equation satisfied by the initial bound state $\phi^{(0)}$ in the Coulomb potential of the nucleus is

$$(\mathrm{i}\,\partial\!\!\!/ - \gamma_0 U(x) - m)\phi^{(0)}(x) = 0. \tag{1.7}$$

Therefore, the initial state rest-interaction Hamiltonian is given by $V_i(x){=}e\ A\!\!\!/(x)$, where we have introduced the four-vector notation: $A(x) = (0, \boldsymbol{A}(\boldsymbol{x}))$, for the external vector potential only. We take the final-state as the plane wave solution of the Dirac–Volkov equation

$$(\mathrm{i}\,\partial\!\!\!/ - e\ A\!\!\!/(x) - m)\phi(t) = 0, \tag{1.8}$$

and, hence, the final state rest-interaction Hamiltonian is $V_f(x) = \gamma_0 U(\boldsymbol{x})$, with $U(x) \equiv -\frac{Z\alpha}{|\boldsymbol{x}|}$.

1.2.2 Relativistic Volkov Solutions

The relativistic plane wave Volkov solutions [6] of (1.8) for an electron in the four-momentum state $p = (p_0, \boldsymbol{p})$ can be conveniently expressed with the help of the quantity

$$\phi_p(x) = \sqrt{\frac{m}{E}} \mathrm{e}^{-\mathrm{i}p\cdot x - \mathrm{i}f_p(x)} \left(1 + \frac{e\ \kappa\!\!\!/\ A\!\!\!/(x)}{2\kappa \cdot p}\right) \hat{u}_p, \tag{1.9}$$

where $E \equiv p_0 = +\sqrt{(p^2 + m^2)}$, and

$$f_p(x) = \frac{1}{2\kappa \cdot p} \int^{\kappa \cdot x} [(2eA(\zeta) \cdot p - e^2 A^2(\zeta))] \mathrm{d}\zeta, \tag{1.10}$$

$$\hat{u}_p = \frac{\not{p} + m}{\sqrt{2m(m + E)}}, \tag{1.11}$$

$$u_p^{(s)} = \hat{u}_p w^{(s)}, \tag{1.12}$$

where the 4-spinors $w^{(s)}$ $(s = 1, 2, 3, 4)$ are defined by

$$w^{(1=\uparrow)} = (1000)^\dagger \tag{1.13}$$

$$w^{(2=\downarrow)} = (0100)^\dagger \tag{1.14}$$

$$w^{(3=\uparrow)} = (0010)^\dagger \tag{1.15}$$

$$w^{(4=\downarrow)} = (0001)^\dagger. \tag{1.16}$$

The two positive energy Volkov solutions are given by

$$\phi_p^{(s)}(x) = \phi_p(x) w^{(s)} \qquad (s = 1, 2), \tag{1.17}$$

and the two negative energy Volkov solutions are given by

$$\phi_p^{(s)}(x) = \phi_{-p}(x) w^{(s)} \qquad (s = 3, 4). \tag{1.18}$$

1.2.3 Volkov–Feynman Propagator

The associated relativistic Green's function of an electron interacting with the external electromagnetic field is defined by the inhomogeneous equation

$$(\mathrm{i}\,\not{\partial} - e\,\not{A}(x) - m) G_F(x, x') = \delta^4(x - x'), \qquad A(x) = (0, \mathbf{A}(x)). \tag{1.19}$$

Its solution satisfying the Feynman–Stückelberg boundary condition (that automatically incorporates the interpretation of the positron states as the negative energy electron states evolving backward in time) is given by the Volkov–Feynman propagator:

$$\begin{aligned} \mathrm{G}_F(x, x') = & -\mathrm{i}\frac{m}{E}\theta(x_0 - x_0') \sum_{s=1,2} \left[\int \frac{\mathrm{d}^3 p}{(2\pi)^3} \phi_p^{(s)}(x) \bar{\phi}_p^{(s)}(x') \right] \\ & + \mathrm{i}\frac{m}{E}\theta(x_0' - x_0) \sum_{s=3,4} \left[\int \frac{\mathrm{d}^3 p}{(2\pi)^3} \phi_p^{(s)}(x) \bar{\phi}_p^{(s)}(x') \right]. \end{aligned} \tag{1.20}$$

1.2.4 Floquet Representation of Relativistic Volkov Solutions

We note that for a vector potential of a constant envelope the phase function $f_p(x)$ reduces to

$$f_p(x) = -a_p \sin(u + \chi_p) + b_p \sin(2u) + \lambda_p u, \tag{1.21}$$

where

$$u = \kappa \cdot x$$

$$= \omega t - \boldsymbol{\kappa} \cdot \boldsymbol{x} \tag{1.22}$$

$$a_p = \frac{\Lambda(\boldsymbol{p}, \xi) e A_0}{\kappa \cdot p} \tag{1.23}$$

$$\Lambda(\boldsymbol{p}, \xi) = \boldsymbol{p} \cdot \boldsymbol{\epsilon}_1 \cos(\xi/2) \sqrt{1 + (\tan\phi_p \tan(\xi/2))^2} \tag{1.24}$$

$$\tan\phi_p = \frac{\boldsymbol{p} \cdot \boldsymbol{\epsilon}_2}{\boldsymbol{p} \cdot \boldsymbol{\epsilon}_1} \tag{1.25}$$

$$\chi_p = \tan^{-1}[\tan\phi_p \tan(\xi/2)] \tag{1.26}$$

$$b_p = \frac{(eA_0)^2}{8\kappa \cdot p} \cos\xi \tag{1.27}$$

$$\lambda_p = \frac{(eA_0)^2}{4\kappa \cdot p} \tag{1.28}$$

$$\kappa \cdot p = \kappa_0 p_0 - \boldsymbol{\kappa} \cdot \boldsymbol{p}, \tag{1.29}$$

and the light four-wavevector $\kappa = (\kappa_0, \boldsymbol{\kappa})$, $\kappa_0 = \omega/c$. Using the Jacobi–Anger relation

$$\mathrm{e}^{\mathrm{i}z \sin\theta} = \sum_{n=-\infty}^{\infty} J_n(z) \mathrm{e}^{\mathrm{i}n\theta}, \tag{1.30}$$

in $\mathrm{e}^{-\mathrm{i} f_p(x)}$ appearing above, combining it with the factors in (1.9) and shifting the index n as required we obtain the Floquet representation of the relativistic Volkov solution (cf. [13]),

$$\phi_p^{(s)}(x) = \sum_{n=-\infty}^{\infty} \sqrt{\frac{m}{E}} \mathrm{e}^{-\mathrm{i}(p+\lambda_p \kappa - n\kappa)\cdot x} Q_n(p) u_p^{(s)}, \tag{1.31}$$

where

$$Q_n(p) = \left(J_n + \frac{eA_0 \not{\kappa}}{4\kappa \cdot p} (\not{\epsilon}(\xi) J_{n-1} + \not{\epsilon}^*(\xi) J_{n+1}) \right), \tag{1.32}$$

and J_n stands for the generalized Bessel function of three arguments

$$J_n \equiv J_n(a_p, b_p, \chi_p) = \sum_{m=-\infty}^{\infty} J_{n+2m}(a_p) J_m(b_p) \mathrm{e}^{(n+2m)\chi_p}, \tag{1.33}$$

and, $u_p^{(s)} = \hat{u}_p w^{(s)}$ for ($s = 1, 2$) are the positive energy Dirac spinors. The corresponding negative energy Volkov solutions, $\phi_p^{(s)}(x)$ for ($s = 3, 4$), are defined as above but by replacing $p \to -p$ and $w^{(s=1,2)} \to w^{(s=3,4)}$.

1.2.5 Floquet Representation of Volkov–Feynman Propagator

A useful Floquet representation of the Volkov–Feynman Green's function, for a constant amplitude vector potential

$$A(x) = A_0(\epsilon_1 \cos\xi \cos(\kappa \cdot x + \delta) - \epsilon_2 \sin\xi \cos(\kappa \cdot x + \delta)), \tag{1.34}$$

where $\delta[0, 2\pi]$ is an arbitrary phase, can be derived [14] by first expanding the Floquet solution of (1.19) as

$$G_F(x, x') = \sum_{n=-\infty}^{\infty} \mathrm{e}^{\mathrm{i}n\kappa\cdot x} G_{n,0}(x, x'). \tag{1.35}$$

Thus, writing a factor of unity on the right-hand side of (1.19) as $1 = \sum_{n=-\infty}^{\infty} \mathrm{e}^{\mathrm{i}n\delta}\delta_{n,0}$, and substituting it and (1.35) in (1.19) and projecting with respect to δ, or equating the coefficients of $\mathrm{e}^{\mathrm{i}n\delta}$ on both sides, we get the Floquet–Volkov propagator equation

$$\left(\mathrm{i}\,\not{\partial} - n\,\not{\kappa} - \frac{eA_0}{2}(\not{\epsilon}(\xi)a_n^- + \not{\epsilon}^*(\xi)a_n^+) - m\right) G_{n,0}(x, x') = \delta_{n,0}\delta^4(x - x') \tag{1.36}$$

with

$$\epsilon(\xi) = \epsilon_1 \cos(\xi/2) + \mathrm{i}\epsilon_2 \sin(\xi/2), \tag{1.37}$$

where $a_n^\pm$ simply increases (+), or decreases (−), the index n of $G_{n,0}$ (or any other quantity with index n) on its right. It is interesting to note that $a_n^\pm$ are in fact the semiclassical analogs of the quantum creation and annihilation operators acting on the quantum number state $|n\rangle$, and replace the corresponding quantum operators very well for large initial occupation number of the field, i.e., in the so-called "laser approximation" ([4] Sect. 6.4).

The plane wave solutions of the homogeneous Floquet–Volkov equation (1.36) (with the right hand = 0) are (cf. [14]),

$$\psi_n^{(s)}(p; x) = \mathrm{e}^{-\mathrm{i}p\cdot x}\sqrt{\frac{m}{E}}\,Q_n(p)u_p^{(s)}, \tag{1.38}$$

which can be verified by substitution and using the properties of the generalized Bessel functions (cf. Table I, [15]). They satisfy the useful bi-orthonormal relations:

$$\sum_{n=-\infty}^{\infty}\sum_{s=1}^{4}\int d^4x\bar{\psi}^{(s)}_{n-N'}(p';x)\psi^{(s)}_{n-N}(p;x) = (2\pi)^4\delta(p-p')\delta_{N,N'} \quad (1.39)$$

$$\sum_{N=-\infty}^{\infty}\int \frac{d^4p}{(2\pi)^4}\psi^{(s)}_{n-N}(p;x)\bar{\psi}^{(s')}_{n'-N}(p;x') = \delta(x-x')\delta_{s,s'}\delta_{n,n'}, \quad (1.40)$$

which may be readily proved on using the orthonormality and completeness properties (cf. Table I, [15]) of the generalized Bessel functions J_{n-N} of the plane wave states and of the Dirac spinors. Therefore, following [14] the solution of the propagator equation (1.36), satisfying the Feynman–Stückelberg boundary condition, can be written as

$$G_{n,n'}(x,x') = \sum_{N=-\infty}^{\infty}\int \frac{d^4p}{(2\pi)^4}e^{-ip\cdot(x-x')}Q_{n-N}(p)\frac{\not{p}_N+m}{p_N^2-m^2+i0}Q^{\dagger}_{n'-N}(p), \quad (1.41)$$

where $Q_n(p)$ are given by (1.32), and the four-vector $p_N \equiv p+\lambda_p\kappa - N\kappa$. Note that in this form the relativistic Floquet–Volkov propagator strongly resembles the usual Feynman propagator [3] to which it reduces in the absence of the field (for $A_0 = 0$). It permits one to extend the usual Feynman covariant perturbative technique in a rather analogous way to processes in the presence of an intense external field. Finally, the Floquet representation of the Green's function $G_F(x,x')$ for a constant envelope time periodic field is given by the simple substitution of $G_{n,0}(x,x')$ from the above (1.41) into (1.35):

$$G_F(x,x') = \sum_{n=-\infty}^{\infty} e^{in\kappa\cdot x}\sum_{N=-\infty}^{\infty}\int \frac{d^4p}{(2\pi)^4}e^{-ip\cdot(x-x')}Q_{n-N}(p)\frac{\not{p}_N+m}{p_N^2-m^2+i0}Q^{\dagger}_{-N}(p). \quad (1.42)$$

1.2.6 Dirac Wavefunction of the Interacting System

To derive a systematic approximation of the wavefunction of the interacting system, we rewrite the Dirac equation (1.4) in the integral equation form as

$$\Psi(x) = \phi^{(0)}(x) + \int d^4x' G(x,x')V_i(x')\phi^{(0)}(x'), \quad (1.43)$$

where, $G(x,x')$ is the total Green's function of the system, $\phi^{(0)}(x)$ is the unperturbed initial state in the absence of the laser field, and $V_i(x')$ is the laser–matter interaction Hamiltonian in the initial state. The total Green's function $G(x,x')$ can be expanded in terms of the final state Volkov–Feynman propagator $G_F(x,x')$ as

$$G(x,x') = G_F(x,x') + \int \mathrm{d}^4x'' G_F(x,x'')V_f(x'')G_F(x'',x') + \ldots . \quad (1.44)$$

Substituting the above expansion in the equation for the wavefunction (1.43), we get

$$\begin{aligned}\Psi(x) &= \phi^{(0)}(x) + \int \mathrm{d}^4x' G_F(x,x')V_i(x')\phi^{(0)}(x') \\ &\quad + \int\int \mathrm{d}^4x''\mathrm{d}^4x' G_F(x,x'')V_f(x'')G_F(x'',x')V_i(x')\phi^{(0)}(x') + \cdots . \end{aligned} \quad (1.45)$$

Thus, we get the laser-modified Dirac wavefunction for the interacting system in the form:

$$\Psi(x) = \phi^{(0)}(x) + \Psi^{(1)}(x) + \Psi^{(2)}(x) + \cdots , \quad (1.46)$$

where the first-order (KFR1) wavefunction is given by

$$\Psi^{(1)}(x) = \int \mathrm{d}^4x' G_F(x,x')V_i(x')\phi_i^{(0)}(x'), \quad (1.47)$$

and the second-order (KFR2) correction is given by

$$\Psi^{(2)}(x) = \int \mathrm{d}^4x''\mathrm{d}^4x' G_F(x,x'')V_f(x'')G_F(x'',x')V_i(x')\phi_i^{(0)}(x'), \quad (1.48)$$

and so on.

1.3 Relativistic Ionization Amplitudes

To obtain the transition amplitude (or the S-matrix element) for ionization, we project the final Volkov state of a given momentum $\boldsymbol{p}$ on to the change of the wavefunction from the initial state $(\Psi(x) - \phi^{(0)}(x))$ due to the laser interaction. Thus, the amplitude for ionization is

$$\begin{aligned}\mathscr{A}_{\mathrm{fi}} &= \int \mathrm{d}^4x \bar{\phi}_p(x)(\Psi(x) - \phi^{(0)}(x)) \\ &= \int \mathrm{d}^4x \bar{\phi}_p(x)(\Psi^{(1)}(x) + \Psi^{(2)}(x) + \cdots) \\ &= -\mathrm{i}\int \mathrm{d}^4x \bar{\phi}_p(x)\, e \not{A}(x)\phi^{(0)}(x) - \mathrm{i}\int\int \mathrm{d}^4x\mathrm{d}^4x' \bar{\phi}_p(x)\gamma_0 U(x) G_F(x,x') \\ &\quad \times e \not{A}(x')\phi^{(0)}(x') + \cdots , \end{aligned} \quad (1.49)$$

where we have used the orthogonality of the Volkov wavefunctions and the relation

$$\bar{\phi}_p(x) = \phi_p^\dagger(x)\gamma_0, \tag{1.50}$$

which is the Dirac-adjoint of the Volkov wavefunction with four-momentum p. The lowest order relativistic intense-field amplitude (denoted as KFR1) is therefore given by the leading term

$$\mathscr{A}_{\rm if}^{(1)} = -{\rm i} \int {\rm d}^4 x \bar{\phi}_p(x)\, e\, \not{A}(x)\phi_i^{(0)}(x), \tag{1.51}$$

and the second-order amplitude (denoted as KFR2) is given by

$$\mathscr{A}_{\rm if}^{(2)} = -{\rm i} \int\int {\rm d}^4 x {\rm d}^4 x' \bar{\phi}_p(x)\gamma_0 U(x) G_F(x,x') e \not{A}(x')\phi_i^{(0)}(x'), \tag{1.52}$$

and so on.

In the sequel, we shall restrict ourselves to the lowest order (KFR1) amplitude for bound-free transitions, (1.51). The evaluation of the four-dimensional integrations can be carried out analytically [11]. The results are given for the ionization amplitudes in terms of simple algebraic formulas involving explicit functions. They permit us to readily calculate not only the energy and angle differential and integrated rates but also the individual *spin-resolved* ionization probabilities in intense fields [11]. The analytical formulas are given below for the *general* case of an *elliptically* polarized radiation field, which automatically reduces to the important special cases of linear and circular polarizations for the ellipticity parameter $\xi = 0$ and $\xi = \pm\pi/2$ (right, left) helicity. We use the expressions to investigate the energy spectra, angular distributions, the spin currents, and the total rates of ionization at both infrared and XUV frequencies. They are applied to study the dependence of the ionization probability of H and a sequence of hydrogenic ions. Results of specific relativistic calculations are compared with the corresponding nonrelativistic calculations.

1.4 Analytic Evaluations

In this section, we evaluate the lowest order (KFR1) ionization amplitude in greater detail and give the analytical formulas for the corresponding spin-specific ionization transition amplitudes and rates.

1.4.1 Covariant Expressions for Ground-State Dirac Wavefunction and Its Fourier Transform

To evaluate the relativistic ionization amplitudes, it is found very useful first to express the Dirac hydrogenic bound state wavefunction of charge state Z in a covariant form as ([16] p. 391):

$$\phi_{1s}^{(s)}(\boldsymbol{r}) = N_{1s} r^{\gamma'-1} \mathrm{e}^{-p_B r} \, \not{n}_\mu(\hat{\boldsymbol{r}}) w^{\uparrow,\downarrow} \qquad (s = 1, 2), \tag{1.53}$$

$$N_{1s} = (2p_B)^{\gamma'+\frac{1}{2}} \left(\frac{1+\gamma'}{8\pi \Gamma(1+2\gamma')} \right)^{\frac{1}{2}}$$

$$p_B = mZ\alpha$$

$$\gamma' = \sqrt{1-(Z\alpha)^2}\,, \tag{1.54}$$

$$\beta' = (1-\gamma')/(Z\alpha)\,, \tag{1.55}$$

$$\not{n}_\mu = \gamma_\mu n_\mu$$

$$n_\mu = (n_0, \boldsymbol{n}) \equiv (1, \mathrm{i}\beta' \hat{\boldsymbol{r}}). \tag{1.56}$$

Next, it is useful to derive the Fourier transform (FT) of the bound state also in the covariant form. To this end, we obtain the following two integrals:

$$c_0(p) = \int \mathrm{d}^3 r \mathrm{e}^{-\mathrm{i}\boldsymbol{p}\cdot\boldsymbol{r}} r^{\gamma'-1} \mathrm{e}^{-p_B r}$$

$$= \frac{4\pi}{p} \frac{\Gamma(\gamma'+1)}{(p_B^2+p^2)^{\frac{\gamma'+1}{2}}} s_0(p), \tag{1.57}$$

$$s_0(p) = \sin\left((\gamma'+1) \tan^{-1} \frac{p}{p_B} \right), \tag{1.58}$$

and

$$g_1(\boldsymbol{p}) = \int \mathrm{d}^3 r \mathrm{e}^{-\mathrm{i}\boldsymbol{p}\cdot\boldsymbol{r}} r^{\gamma'-1} \mathrm{e}^{-p_B r} \boldsymbol{\gamma}\cdot\hat{\boldsymbol{r}}$$

$$= \mathrm{i} c_1(p) \boldsymbol{\gamma}\cdot\hat{\boldsymbol{p}}, \tag{1.59}$$

$$c_1(p) = \frac{4\pi}{p} \frac{\Gamma(1+\gamma')}{(p^2+p_B^2)^{\frac{\gamma'+1}{2}}} \left((p_B/p) s_0(p) - \frac{\gamma'+1}{\gamma'} (1+(p_B/p)^2)^{\frac{1}{2}} s_1(p) \right)$$

$$s_1(p) = \sin\left(\gamma' \tan^{-1} \frac{p}{p_B} \right). \tag{1.60}$$

Thus, the FT of Dirac H atom ground state:

$$\tilde{\phi}_{1s}^{(s)}(\boldsymbol{p}) = \int \mathrm{d}^3 r \mathrm{e}^{-\mathrm{i}\boldsymbol{p}\cdot\boldsymbol{r}} \phi_{1s}^{(s)}(\boldsymbol{r})$$

$$= N_{1s} \left(c_0(p) + \beta' c_1(p) \boldsymbol{\gamma}\cdot\hat{\boldsymbol{p}} \right) w^{\uparrow,\downarrow}$$

$$= N_{1s} c_0(p) \not{b}_\mu(\boldsymbol{p}) w^{\uparrow,\downarrow} \qquad (s=1,2), \tag{1.61}$$

where

$$\begin{aligned}
b_\mu(\boldsymbol{p}) &= (1, \boldsymbol{b}(\boldsymbol{p})) \\
\not{b} &= \gamma_\mu b_\mu \\
\boldsymbol{b}(\boldsymbol{p}) &= -g(p)\hat{\boldsymbol{p}}, \\
g(p) &= \beta' \left(\frac{p_B}{p} - \frac{\gamma'+1}{\gamma'} (1 + (p_B/p)^2)^{\frac{1}{2}} \frac{\sin(\gamma' \tan^{-1}(p/p_B))}{\sin((\gamma'+1)\tan^{-1}(p/p_B))} \right) \\
w^\uparrow &= w^{(s=1)} \\
w^\downarrow &= w^{(s=2)}.
\end{aligned} \tag{1.62}$$

1.4.2 Explicit Formulas: Spin-Specific Ionization Amplitudes and Rates

We may now evaluate the transition amplitude for ionization (1.51) and the probability per unit time (or rates) of ionization.

1.4.2.1 Explicit Ionization Amplitudes

The amplitudes are obtained by substituting the ground-state wavefunction (1.53) and the Volkov solution (1.31) in (1.51), performing the $\mathrm{d}x_0$ integration in terms of the delta-function and the spatial integration using the FT of the ground state, (1.61). Thus, for the transition between the spin states $s \to s'$, one gets:

$$\mathscr{A}^{(1)}_{s\to s'} = -2\pi \mathrm{i} \sum_n \delta(p_0 + p_0^B + \lambda_p \kappa_0 - n\kappa_0) \times T^{(n)}_{s\to s'}(\boldsymbol{q}), \tag{1.63}$$

where $p_0^B \equiv \sqrt{m^2 - p_B^2}$, $p_B = mZ\alpha$, and

$$T^{(n)}_{s\to s'}(\boldsymbol{q}) = \frac{eA_0}{2} N(E) \bar{u}_p^{(s')} \not{B}_n^* \tilde{\phi}^{(s)}_{1s}(\boldsymbol{q}), \qquad s, s' = (u, d) \tag{1.64}$$

$$= \frac{eA_0}{2} N(E) N_{1s} c_0(\boldsymbol{q}) t^{(n)}_{s\to s'}(\boldsymbol{q}), \tag{1.65}$$

$$t^{(n)}_{s\to s'}(\boldsymbol{q}) = \bar{u}_p^{(s')} \not{B}_n^* \not{b}\, w^{(s)}. \tag{1.66}$$

The four vectors $B_n = (B_n^0, \boldsymbol{B}_n)$ are defined in terms of the generalized Bessel functions J_n, (1.33),

$$B_n^0 = \frac{eA_0\kappa_0}{4\kappa\cdot p}\left(2J_n + \cos\xi\,(J_{n+2}+J_{n-2})\right) , \tag{1.67}$$

$$\boldsymbol{B}_n = \boldsymbol{\epsilon}(\xi)J_{n-1} + \boldsymbol{\epsilon}^*(\xi)J_{n+1} + \hat{\boldsymbol{\kappa}}B_n^0 \tag{1.68}$$

$$\boldsymbol{\epsilon}(\xi) = \boldsymbol{\epsilon}_1\cos(\xi/2) + \mathrm{i}\boldsymbol{\epsilon}_2\sin(\xi/2) \tag{1.69}$$

$$\kappa\cdot p = \kappa_0 p_0 - \boldsymbol{\kappa}\cdot\boldsymbol{p}. \tag{1.70}$$

We list the other symbols together:

$$\boldsymbol{q} \equiv q\hat{\boldsymbol{q}} = \boldsymbol{p} + (\lambda_p - n)\,\boldsymbol{\kappa} \tag{1.71}$$

$$\gamma' = \sqrt{1-(Z\alpha)^2} , \tag{1.72}$$

$$\beta' = (1-\gamma')/(Z\alpha) , \tag{1.73}$$

$$N_{1s} = (2p_B)^{\gamma'+\frac{1}{2}}\left(\frac{1+\gamma'}{8\pi\Gamma(1+2\gamma')}\right)^{\frac{1}{2}} , \tag{1.74}$$

$$m_1 = \sqrt{(E+m)/(2m)} , \tag{1.75}$$

$$m_2 = -\sqrt{(E-m)/(2m)} , \tag{1.76}$$

$$N_E = \sqrt{\frac{m}{E}} , \tag{1.77}$$

$$E = +\sqrt{\boldsymbol{p}^2 + m^2} ; \tag{1.78}$$

$$\kappa_0 = \frac{\omega}{c}$$

$$\boldsymbol{\kappa} = \kappa_0\hat{\boldsymbol{\kappa}} \tag{1.79}$$

$$\kappa = (\kappa_0, \boldsymbol{\kappa}) . \tag{1.80}$$

Finally, evaluating the matrix elements explicitly (with respect to $w^{(s)}, w^{(s')}$) for $s, s' = (u, d)$, we get the algebraic expressions for the four reduced matrix elements $t^{(n)}_{s\to s'}$ [11]:

$$\begin{aligned} t^{(n)}_{u\to u} &= B_n^{0*}\left(m_1 + m_2 g(q)\left(\hat{\boldsymbol{p}}\cdot\hat{\boldsymbol{q}} + \mathrm{i}\,(\hat{\boldsymbol{p}}\times\hat{\boldsymbol{q}})_z\right)\right) \\ &+ \boldsymbol{B}_n^*\cdot(m_2\hat{\boldsymbol{p}} + m_1 g(q)\hat{\boldsymbol{q}}) \\ &- \mathrm{i}\left(\boldsymbol{B}_n^*\times(m_2\hat{\boldsymbol{p}} - m_1 g(q)\hat{\boldsymbol{q}})\right)_z \end{aligned} \tag{1.81}$$

$$\begin{aligned} t^{(n)}_{u\to d} &= m_2 g(q) B_n^{0*}\left(\mathrm{i}\,(\hat{\boldsymbol{p}}\times\hat{\boldsymbol{q}}))_x - (\hat{\boldsymbol{p}}\times\hat{\boldsymbol{q}})_y\right) \\ &- \mathrm{i}\left(\boldsymbol{B}_n^*\times(m_2\hat{\boldsymbol{p}} - m_1 g(q)\hat{\boldsymbol{q}})\right)_x \\ &+ \left(\boldsymbol{B}_n^*\times(m_2\hat{\boldsymbol{p}} - m_1 g(q)\hat{\boldsymbol{q}})\right)_y \end{aligned} \tag{1.82}$$

$$\begin{aligned} t^{(n)}_{d\to u} &= m_2 g(q) B_n^{0*}\left(\mathrm{i}\,(\hat{\boldsymbol{p}}\times\hat{\boldsymbol{q}})_x + (\hat{\boldsymbol{p}}\times\hat{\boldsymbol{q}})_y\right) \\ &- \mathrm{i}\left(\boldsymbol{B}_n^*\times(m_2\hat{\boldsymbol{p}} - m_1 g(q)\hat{\boldsymbol{q}})\right)_x \\ &- \left(\boldsymbol{B}_n^*\times(m_2\hat{\boldsymbol{p}} - m_1 g(q)\hat{\boldsymbol{q}})\right)_y , \end{aligned} \tag{1.83}$$

and

$$t_{d\to d}^{(n)} = B_n^{0*}\left(m_1 + m_2 g(q)\left(\hat{\boldsymbol{p}}\cdot\hat{\boldsymbol{q}} - \mathrm{i}\,(\hat{\boldsymbol{p}}\times\hat{\boldsymbol{q}})_z\right)\right) + \boldsymbol{B}_n^{*}\cdot\left(m_2\hat{\boldsymbol{p}} + m_1 g(q)\hat{\boldsymbol{q}}\right) + \mathrm{i}\left(\boldsymbol{B}_n^{*}\times\left(m_2\hat{\boldsymbol{k}} - m_1 g(q)\hat{\boldsymbol{q}}\right)\right)_z . \qquad (1.84)$$

1.4.3 Spin-Specific Ionization Rates

The differential rates of ionization by absorption of n laser photons from the bound spin states $s = (u = \uparrow, d = \downarrow)$ to the continuum spin states $s' = (u, d)$ can be easily expressed in terms of the absolute square of the transition amplitude (1.63), divided by the effectively long interaction time τ, and summed over the final wavenumber states. To this end, we use a convenient representation of the square of the 1-dimensional delta-function

$$\delta^2(k-k') = \delta(k-k')\times\frac{1}{2\pi}\lim_{\tau\to\infty}\int_{-c\tau/2}^{c\tau/2}\mathrm{e}^{\mathrm{i}(k-k')x_0}\mathrm{d}x_0 = \delta(k-k')\times\lim_{\tau\to\infty}\frac{1}{2\pi}(c\tau). \qquad (1.85)$$

Thus, we get the differential rate $\mathrm{d}W^{(n)}$ of ionization:

$$\mathrm{d}W_{s\to s'}^{(n)} = \left(\frac{eA_0}{2}N(E)N_{1s}c_0(q)\right)^2\left|t_{s\to s'}^{(n)}(\boldsymbol{q})\right|^2 \times(2\pi)c\delta(p_0 + p_0^B + \lambda_p\kappa_0 - n\kappa_0)\frac{1}{(2\pi)^3}\mathrm{d}^3p. \qquad (1.86)$$

Finally, noting $\mathrm{d}^3p = \boldsymbol{p}^2\mathrm{d}|\boldsymbol{p}|\mathrm{d}\Omega_p$, and performing the $\mathrm{d}|\boldsymbol{p}|$ integration $[0,\infty]$, the rate of ionization for the final electron momentum $\boldsymbol{p}$ in the solid angle element $\mathrm{d}\Omega_p$ takes the simple form [11]:

$$\frac{\mathrm{d}W_{s\to s'}}{\mathrm{d}\Omega_p}(\boldsymbol{p}) = \sum_{n\geq n_0}\left(\frac{eA_0}{2}N(E)N_{1s}c_0(q)\right)^2\left|t_{s\to s'}^{(n)}(\boldsymbol{q})\right|^2 c\frac{p_0|\boldsymbol{p}|}{(2\pi)^2}. \qquad (1.87)$$

In the above, the minimum integer value of $n = n_0$ is determined by the energy-momentum conservation relation,

$$n\omega = \epsilon_B + \epsilon_{\mathrm{kin}} + \lambda_p\omega, \qquad (1.88)$$

where $\epsilon_B = m_0c^2\left(1 - \sqrt{1-(\frac{p_B}{m})^2}\right)$ is the binding energy (or ionization potential), and $\epsilon_{\text{Kin}} = m_0c^2\left(\sqrt{1+(\frac{p}{m})^2} - 1\right)$ is the kinetic energy. We note here parenthetically that because of the implicit p-dependence of λ_p, (1.88) is an implicit equation for n that requires to be solved for its root before the contributions of all allowed $n \geq n_0$, where n_0 is the minimum allowed value of n, can be calculated. The other parameters are

$$p_0 \equiv \sqrt{m^2 + \boldsymbol{p}^2} = (n - \lambda_p)\,\kappa_0 + \sqrt{m^2 - p_B^2} \tag{1.89}$$

$$\boldsymbol{q} = |\boldsymbol{q}|\hat{\boldsymbol{q}} \equiv \boldsymbol{p} + (\lambda_p - n)\,\boldsymbol{\kappa}. \tag{1.90}$$

1.4.3.1 Special Polarizations

The above results hold for the general case of elliptical polarization. The results for the two most common cases of linear and circular polarization of the field are readily obtained from above by simply setting the ellipticity parameter $\xi = 0$ and $\xi = \frac{\pi}{2}$ (right helicity), $-\frac{\pi}{2}$ (left helicity), respectively.

Linear Polarization

The same rate formulas for the reduced amplitudes (1.81)–(1.84) hold, except that, since in this case $\xi = 0$, therefore

$$a_p(\xi = 0) = \frac{\boldsymbol{p}\cdot\boldsymbol{\epsilon}_1 e A_0}{\kappa \cdot p} \tag{1.91}$$

$$\chi_p(\xi = 0) = 0 \tag{1.92}$$

$$b_p = \frac{eA_0^2}{8\kappa \cdot p}. \tag{1.93}$$

Thus, the generalized Bessel function of three arguments (1.33) now simplifies to the generalized Bessel function of two arguments, i.e.,

$$J_n = J_n\left(\frac{\boldsymbol{p}\cdot\boldsymbol{\epsilon}_1 A_0}{c\kappa \cdot p}, \frac{eA_0^2}{8\kappa \cdot p}\right). \tag{1.94}$$

We note that in this case the quantities B_n^0 and $\boldsymbol{B}_n$ appearing in the corresponding reduced transition matrix elements are real and hence the amplitudes $t^{(n)}_{u\to u}$ and $t^{(n)}_{d\to d}$ are complex conjugate of each other. Thus, we can readily conclude that for linear polarization of the laser field, the symmetric $u \to u$ and $d \to d$ ionization rates must be equal. Similar conclusions hold for the spin-flip transitions $u \to d$ and $d \to u$.

Circular Polarization

The same formulas (1.81)–(1.84) for the reduced amplitudes hold except that, since in this case $\xi = \pm\pi/2$,

$$a_p(\xi = \pm\pi/2) = \frac{|\boldsymbol{p}| \sin\theta_p A_0}{c\kappa \cdot p}, b_p = 0 \tag{1.95}$$

$$\chi(\xi = \pm\pi/2) = \pm\phi_p, \tag{1.96}$$

therefore the generalized Bessel function of three arguments now simplifies to an ordinary Bessel function with a phase factor, i.e.,

$$J_n = J_n\left(\frac{|\boldsymbol{p}| \sin\theta_p A_0}{c\kappa \cdot p}\right) \mathrm{e}^{\pm in\phi_p}. \tag{1.97}$$

In this case, the corresponding expressions of the spin-flip reduced matrix elements $|t^{(n)}_{u\to d}|^2 \neq |t^{(n)}_{d\to u}|^2$, which implies an interesting asymmetry for the spin-flip ionization rates with respect to the photon propagation direction.

We may note also that the influence of relativity in the rates of ionization given above arises, first, from the appearance of the relativistic energy-momentum conservation, second, from the nonnegligible momentum of the photon and, third, from the spin degrees of freedom. The first two factors affect the arguments of the Bessel functions in B^0_n and $\boldsymbol{B}_n$ that determine the reduced amplitudes, $t^{(n)}_{s,s'}$, and in the dressed-momentum $\boldsymbol{q} = \boldsymbol{p} + (\lambda_p - n)\boldsymbol{\kappa}$, which is "dressed" by the field strength via the parameter λ_p. The influence of the spin degrees of freedom arises from the 4-component spinor character of both the free-state spinor and the bound-state spinor, and more interestingly depends on the parameter m_2 of the free-electron Dirac spinor, and the parameter β' (that multiply the term $g(p)$) that appears in the "weak" component of the ground-state spinor. They are small in the low velocity nonrelativistic limits but become important in high Z ions and for the inner-shells of heavy atoms.

1.4.4 The Unpolarized Rate of Ionization

The spin-averaged total ionization rate from an unpolarized target atom can be easily obtained by simply adding the four spin-specific rates given above and dividing by 2 (for the average with respect to the two equally probable initially occupied spin states):

$$\frac{\mathrm{d}W^{(\mathrm{ion})}}{\mathrm{d}\Omega_p} = \sum_{n\geq n_0} \left(\frac{eA_0}{2} N(E) N_{1s} c_0(q)\right)^2 c \frac{p_0 |\boldsymbol{p}|}{(2\pi)^2} \times \frac{1}{2} \sum_{(s,s')=u,d} \left|t^{(n)}_{s\to s'}(\boldsymbol{q})\right|^2. \tag{1.98}$$

We may note in passing that for the nonrelativistic ground 1*s*-state of H atom, $\gamma' = 1$ and for the ground *s*-state of a zero-range potential, $\gamma' = 0$. Before ending this section, we should point out that in the nonrelativistic limit (and in the dipole approximation), (1.98) reduces to the corresponding nonrelativistic KFR1 rates (e.g., Equation (44) in [5]).

1.5 Applications

We apply the formulas given above to study the angle and energy differential ionization rates, the total ionization rates, the spin-specific ionization currents, spin-flip asymmetry, spin-averaged asymmetry parameter, and the charge-state (Z) dependence of intense-field ionization process, in the case a circularly polarized incident laser field. This case is particularly interesting from the point of view of the interplay between the spin polarization and the laser helicity. To be specific and to bring out the mutual dependence of helicity of the laser photons and the helicity of the electron spin, we choose the incident laser field to be circularly polarized (cf. Fig. 1.1) with the laser propagation vector, $\boldsymbol{\kappa}$, along the z-axis and the unit polarization vector $\boldsymbol{\epsilon}$ with orthogonal components along the x and y directions ($\boldsymbol{\epsilon}_1 = \boldsymbol{\epsilon}_x, \boldsymbol{\epsilon}_2 = \boldsymbol{\epsilon}_y$); the momentum $\boldsymbol{p}$, and the two possible spin states up (u $=\uparrow$) and down (d $=\downarrow$) are also indicated. We choose the laser frequency to be in the infrared, UV and XUV or free-electron laser (FEL) domains or more specifically, a typical Ti-sapphire laser frequency $\omega = 1.55\,\text{eV}$, as well as a set of higher frequencies, $\omega = 5$, 10, 27.2, 54.4, and 200 eV.

For the purpose of numerical calculations, in this section we conveniently use the atomic units (a.u.): $e_0 = m_0 = \hbar = \alpha c = 1$ and hence in a.u. $e = 1/c = \alpha = 137.036\cdots$, and $m = c$. The peak amplitude of the vector potential A_0 (in a.u.) is related to the peak electric field strength F_0 (in a.u.), and the laser intensity I (in W cm^{-2}) by:

$$A_0 = \frac{c}{\omega} F_0 \tag{1.99}$$

$$F_0 = \sqrt{I/I_a}, \tag{1.100}$$

$$I_a = (3.51\cdots) \times 10^{16}\,\text{W cm}^{-2}. \tag{1.101}$$

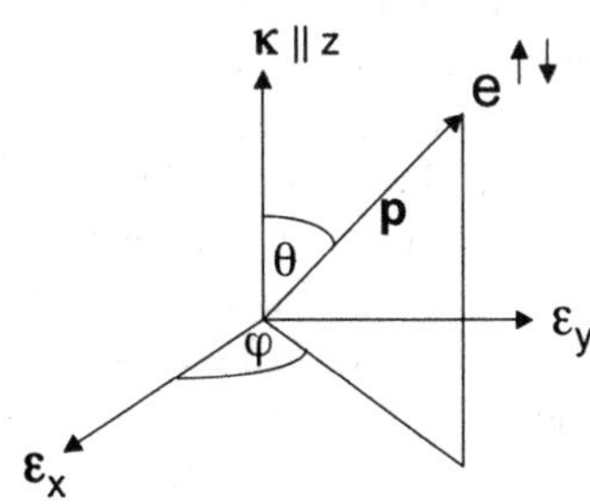

Fig. 1.1 Geometry indicating the propagation vector $\boldsymbol{\kappa}$ and the polarization components $\boldsymbol{\epsilon}_x, \boldsymbol{\epsilon}_y$ of a circularly polarized laser field, as well as the momentum $\boldsymbol{p}$ and spin states $(\uparrow, \downarrow)$ of the emitted electron

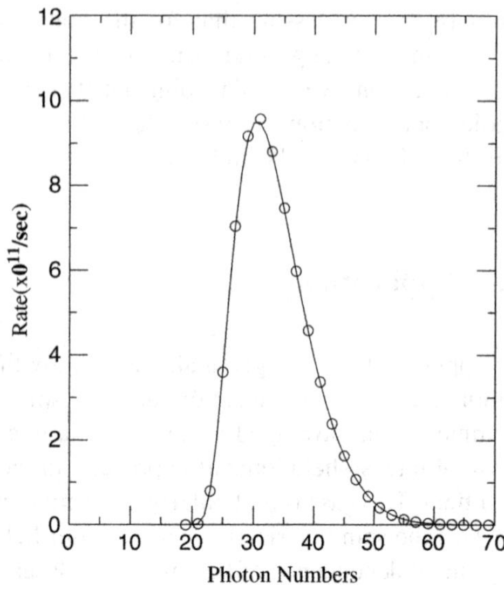

Fig. 1.2 Ionization rate: emitted electron energy spectrum (in units of photon energy) at $\omega = 200$ eV, $I = 10^{21}$ W cm^{-2}: intense-field relativistic case (*solid line*), and nonrelativistic case (*open circles*)

1.5.1 Ionization Rate: Energy Distribution

In Fig. 1.2, we show the results of calculations for the energy distribution of the electrons from ionization of Hydrogen atom at a free-electron laser frequency $\omega = 200$ eV and an intensity $I = 10^{21}$ W cm^{-2}. It can be seen (photoelectron energy in units of the photon energy) that the probability of ionization per seccond. reaches its maximum at an energy far above the value of $n = 1$ – far above the allowed (relativistic ionization threshold $\approx$13.6 eV) single-photon transition energy. This is a typical characteristic of intense-field ionization in circularly polarized fields, which is accompanied by absorption of a large number of photons *and* angular momenta. In the figure, we have compared the results of the present intense-field relativistic model (solid line) and the corresponding nonrelativistic Schrödinger calculations (open circles) using (44) given in [5]. It is interesting to observe that despite the "high" intensity, the nonrelativistic calculation at this XUV frequency is virtually indistinguishable (in the scale of the figure) from that of the relativistic theory. This implies that effectively this combination of nominally high intensity (higher than most current high intensity infrared or optical laser field) the electron motion throughout is essentially nonrelativistic in nature. This is due to the inverse frequency-square dependence of the characteristic ponderomotive energy $U_p = I/4\omega^2$(a.u.) (that corresponds to the mean kinetic energy of a free electron in a laser field). The result at a higher intensity of $I = 10^{22}$ W cm^{-2} is shown in Fig. 1.3. It can be seen from the figure that although still the relativistic and the nonrelativistic distributions peak near the same energy, the maximum height is now visibly lower in the relativistic case (solid line) than in the nonrelativistic calculation (open circles). A systematic variation of the peak of the energy distributions as a function of intensity can be seen very clearly in Fig. 1.4 that shows the results of relativistic

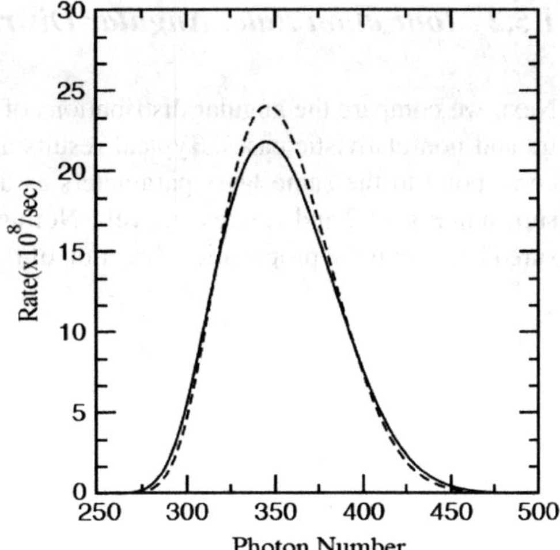

Fig. 1.3 Ionization rate: emitted electron energy spectrum (in units of photon energy) at $\omega = 200\,\text{eV}$, $I = 10^{22}\,\text{W}\,\text{cm}^{-2}$: intense-field relativistic case (*solid line*), and nonrelativistic case (*dashed line*)

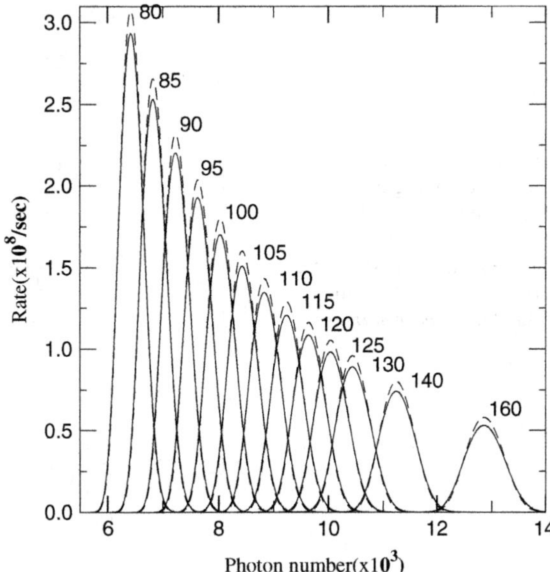

Fig. 1.4 Ionization rate: electron energy spectra (in units of photon energy) at $\omega = 5\,\text{eV}$ and a sequence of intensities $I = (80-160)$ a.u. $= (80-160) \times 3.51 \times 10^{16}\,\text{W}\,\text{cm}^{-2}$. Note how the peak values decrease systematically with increasing intensity for both relativistic (*solid line*) and nonrelativistic cases (*dashed line*)

(solid lines) and nonrelativistic calculations (open circles) for a whole sequence of intensities in the range $I = 80-160$ a.u. $= (80-160) \times 3.51 \times 10^{16}\,\text{W}\,\text{cm}^{-2}$, for a fixed laser frequency $\omega = 5\,\text{eV}$. We note that the sequence of peaks of the distributions occur at an energy of the order of U_p. This is a typical characteristic of electron energy spectrum produced by a circular polarization of the field.

1.5.2 Ionization Rate: Angular Distribution

Next, we compare the angular distributions of the emitted electron in the relativistic and nonrelativistic cases. Typical results are shown in Figs. 1.5 and 1.6. They correspond to the same laser parameters as assumed for the energy distributions shown in Figs. 1.2 and 1.3, respectively. Noting that the angles of emission are measured here from the propagation direction of the laser field, a shift of the relativistic

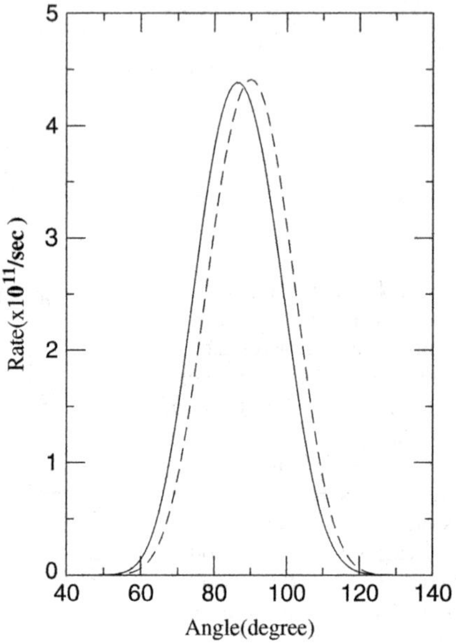

Fig. 1.5 Ionization rate: comparison of angular distributions at $\omega = 200\,\mathrm{eV}$, $I = 10^{21}\,\mathrm{W\,cm^{-2}}$: intense-field relativistic Dirac (*solid line*), nonrelativistic Schrödinger (*dashed line*)

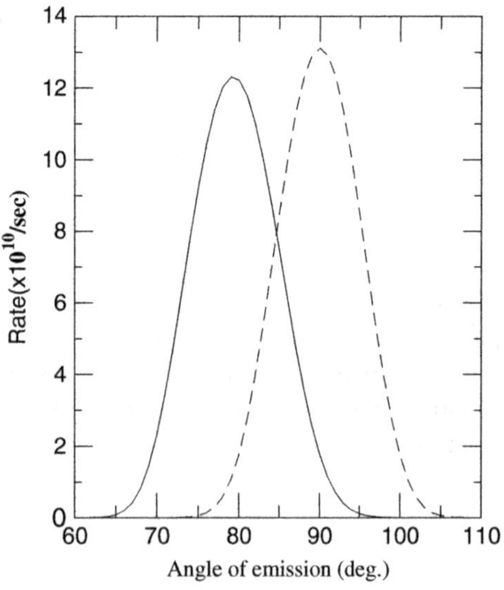

Fig. 1.6 Ionization rate: comparison of angular distributions at $\omega = 200\,\mathrm{eV}$, $I = 10^{22}\,\mathrm{W\,cm^{-2}}$: intense-field relativistic Dirac (*solid line*), nonrelativistic Schrödinger case (*dashed line*). Note the focusing effect toward the smaller angle

distribution (solid line) toward a smaller angle compared to the nonrelativistic case (dashed line) can be seen to occur in Fig. 1.5. This is a manifestation of the general relativistic *focusing* effect toward the propagation direction (cf., e.g., [5]), that in fact becomes more significant with the increase of the field intensity. This can be seen more prominently from the distribution at a higher intensity (10^{22} W cm^{-2}) shown in Fig. 1.6. Again one finds that the highest rate is lower in the relativistic case than in the nonrelativistic case.

In Fig. 1.7, we compare the angular distributions calculated at $\omega = 1$ a.u. $=$ 27.2 eV and intensity $I = 0.01$ a.u. $= 3.51 \times 10^{14}$ W cm^{-2} calculated in three different ways. It can be seen that there is virtually no distinction, at this relatively low intensity, between the intense-field relativistic (solid line) and the nonrelativistic (dashed line) calculations. Moreover, they are essentially equal to the rates given by the single-photon relativistic Born approximation (crosses). The results for the same frequency but for a two orders of magnitude higher intensity (I = 3.51×10^{16} W cm^{-2}) are shown in Fig. 1.8. It can be seen that the intense-field relativistic (solid line) and the nonrelativistic (dots) distributions are still essentially equal to each other. However, now, they both differ significantly from the results of the single-photon relativistic Born approximation (crosses) and overestimates the peak rate considerably indicating the failure of the perturbation theory.

1.5.3 *Total Ionization Rates: Intensity Dependence*

We next consider the total ionization probability per unit time (the total rate), which is obtained by integrating the distributions over all angles of emission at a

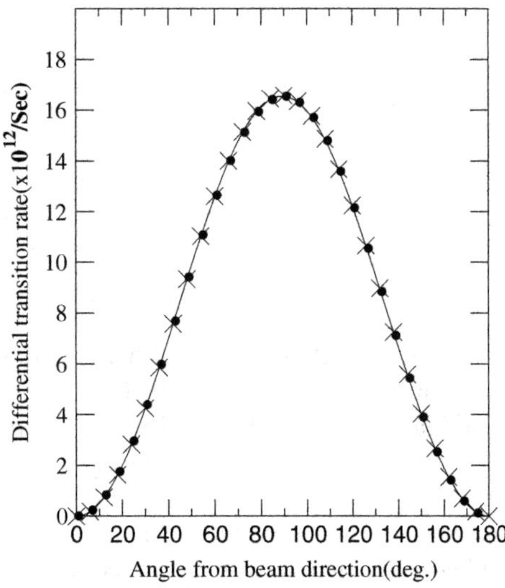

Fig. 1.7 Ionization rate: comparison of angular distributions at $\omega = 27.2$ eV, $I = 3.51 \times 10^{14}$ W cm^{-2}: intense-field Dirac (*solid line*), nonrelativistic Schrödinger (*dots*), and single-photon relativistic Born approximation (*crosses*)

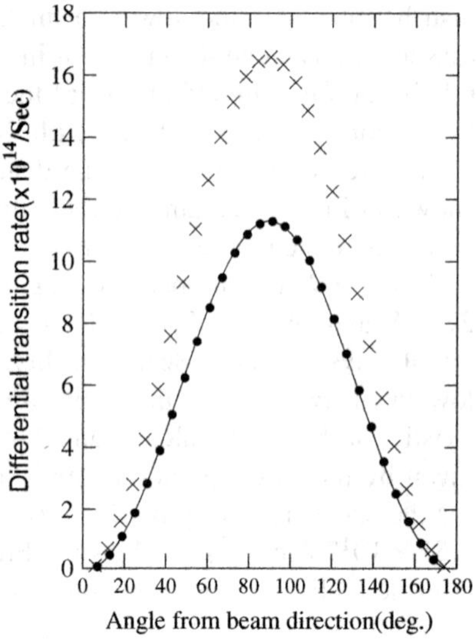

Fig. 1.8 Ionization rate: comparison of angular distributions at $\omega = 1$ a.u. $= 27.2$ eV and $I = 1$ a.u. $= 3.51 \times 10^{16}$ W cm^{-2}: intense-field Dirac (*solid line*), nonrelativistic Schrödinger (*dots*), and single-photon relativistic Born approximation (*crosses*)

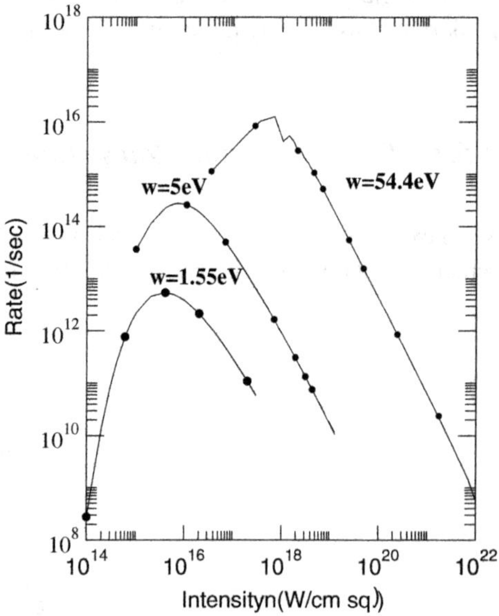

Fig. 1.9 Total ionization rate for H atom at $\omega = 1.55$, 5, and 54.4 eV, as a function of intensity in the range 10^{14} W cm^{-2}–10^{22} W cm^{-2}. Note the counter intuitive decrease of ionization probability with increasing intensity above the maxima

given energy followed by a further integration over all the emission energies of the electron, and averaging over the two initial spin states, as well as summing over the two final spin states. In Fig. 1.9, we compare the *total* rates of ionization at three different laser frequencies $\omega = 1.55$, 5, and 54.4 eV, as a function of the laser intensity from 10^{14} to 10^{22} W cm^{-2}. In all cases, the ionization rates show at first an increasing probability with increasing intensity. Then they go through a maximum and *decrease* generally with further increase of intensity. We note that this decrease of ionization rate with the increase of intensity is reminiscent of (but not identical to) the phenomenon of "adiabatic stabilization" of atoms against ionization in the high frequency limit [17]. In the latter case, the single-photon ionization threshold is supposed to lie much below the laser photon energy. In contrast, in this case, the decreasing rate occurs as well at low frequencies (i.e., below the "high frequency" limit). We note that the maximum rates are higher in strength for the higher frequencies, and they occur at higher intensities. Finally, a modulation of the total ionization rates can be detected below the maximum rate in the top curve – this behavior indicates a *channel closing* effect or the closing of the lowest n-photon ionization channels that can occur with the upward shift of the ionization threshold by the ponderomotive energy $U_p = \frac{I}{4\omega^2}$ (a.u.) with the increase of the intensity (see, e.g., [5]).

1.5.4 Spin Dependence of Ionization Currents

In the cases considered so far we have not resolved the spin degrees of freedom of the electron either in the ground state or in the free state after the ionization. In fact, the distributions shown in the above figures correspond to the usual measurement of the unpolarized ionization signal that is spin-averaged over the initial spin states of the target atom and spin-summed over the final states of the free electron. The present four-component relativistic analysis provides explicit analytic expressions for the individual spin-dependent transition amplitudes (first obtained in [11]). This allows us to further examine the *spin-resolved* electron current, its dependence on polarized or un-polarized target atoms, the spin-flip ionization probabilities, and any asymmetry in the spin currents. We recall, as indicated already in Fig. 1.1, that the field propagation direction (z-axis) is the spin quantization axis and the spin "up" orientation is defined to be along the positive z-direction.

1.5.4.1 Spin-Symmetric Transitions: Angular Distributions

We first consider the spin unchanging (spin symmetric) angular distributions of the ejected electron for $\omega = 27.2\,\mathrm{eV}$ and $I = 3.51 \times 10^{15}\,\mathrm{W\,cm^{-2}}$. Note that at this frequency the single photon ionization channel is open (since the ionization threshold energy of Dirac H-atom $\approx 13.6\,\mathrm{eV}$). In Fig. 1.10 we show the results of intense-field relativistic calculation for the symmetric $u \equiv\uparrow$ to $u \equiv\uparrow$ (solid line) and $d \equiv\downarrow$ to $d \equiv\downarrow$ (crosses) ionization rates. For the sake of comparison, we also show the results of the single-photon relativistic first Born approximation: $\uparrow$ to $\uparrow$ (open circles) and $\downarrow$ to $\downarrow$ (dots). Clearly, there is no significant distinction between the intense-field relativistic and the single-photon Born approximation. Moreover, the rates of $\uparrow$ to $\uparrow$ and $\downarrow$ to $\downarrow$ symmetric transitions are equal as we have discussed analytically in Sect. 4.2. In Fig. 1.11, we present the corresponding results for a higher intensity $I = 3.51 \times 10^{16}\,\mathrm{W\,cm^{-2}}$. Again, the spin-symmetric

Fig. 1.10 Spin symmetric ionization rate: angular distributions at $\omega = 1$ a.u. $= 27.2$ eV, $I = 3.51 \times 10^{15}$ W cm^{-2} – intense-field relativistic Dirac: ↑ to ↑ (*solid line*), ↓ to ↓ (*crosses*); single-photon relativistic Born approximation: ↑ to ↑ (*open circle*), ↓ to ↓ (*thick dots*)

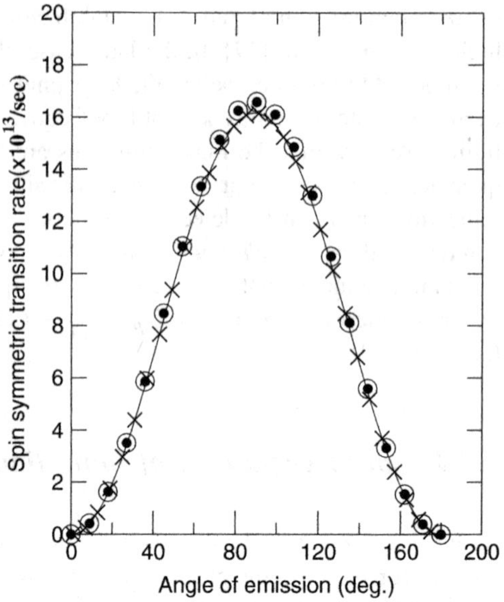

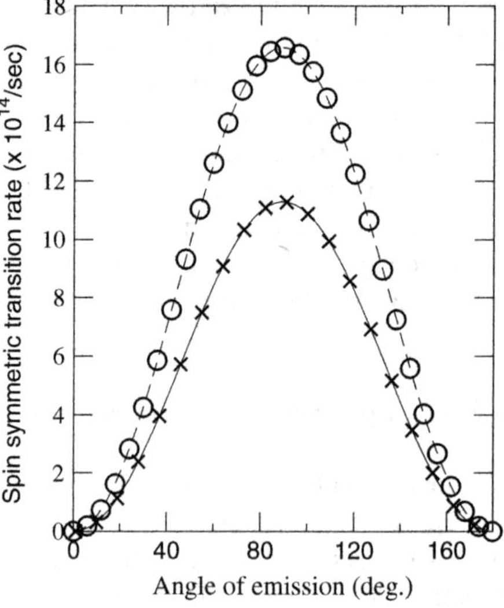

Fig. 1.11 Spin symmetric ionization rate: angular distributions at $\omega = 1$ a.u. $= 27.2$ eV, $I = 3.51 \times 10^{16}$ W cm^{-2} – intense-field relativistic Dirac: ↑ to ↑ (*solid line*), ↓ to ↓ (*crosses*); single-photon relativistic Born approximation: ↑ to ↑ (*open circle*), ↓ to ↓ (*dashed line*)

transition rates are equal, however, at the higher intensity the Born approximation results clearly overestimate the corresponding results of intense-field calculations.

1.5.4.2 Spin-Flip Transitions: Angular Distributions

In Fig. 1.12, we show the *spin-flip* ionization rates for an FEL frequency $\omega = 200$ eV, at a high intensity $I = 10^{22}$ W cm^{-2}, calculated using the present

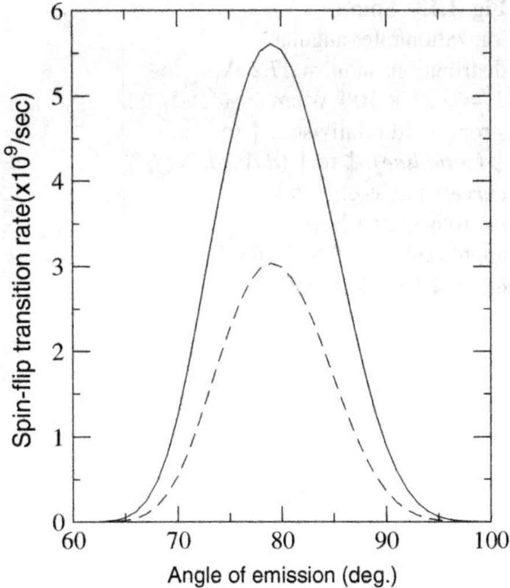

Fig. 1.12 Spin-flip ionization rate: angular distributions for $\omega = 200\,\mathrm{eV}$, $I = 3.5 \times 10^{22}\,\mathrm{W\,cm^{-2}}$: intense-field relativistic: $\uparrow$ to $\downarrow$ (*solid line*), $\downarrow$ to $\uparrow$ (*dashed line*)

intense-field theory. The $\uparrow$ to $\downarrow$ transition rates (solid line) and the $\downarrow$ to $\uparrow$ u transition rates (dashed line) are compared. It can be seen that the $\uparrow$ to $\downarrow$ rates are larger than the $\downarrow$ to $\uparrow$ rates, at all angles of emission. They are, nevertheless, of a comparable *order* of magnitude. This, as we shall see next, is in stark contrast to the *zero* probability for the $\downarrow$ to $\uparrow$ transition given by the relativistic single-photon Born approximation. In Fig. 1.13, we show the results of the angular distributions of the spin-flip ionization rates calculated using the intense-field relativistic approximation ($\uparrow$ to $\downarrow$ (solid line), $\downarrow$ to $\uparrow$ (dashed line)) and, the single-photon relativistic first Born approximation ($\uparrow$ to $\downarrow$ (dots), $\downarrow$ to $\uparrow$ $\equiv 0$ (not shown)). We note that the intense-field $\uparrow$ to $\downarrow$ and $\downarrow$ to $\uparrow$ rates are both finite but the latter is significantly weaker in strength. It is remarkable that the relativistic first Born approximation gives qualitatively different result since (as opposed to the finite $\uparrow$ to $\downarrow$ transition rates) the $\downarrow$ to $\uparrow$ transitions rates vanish (not shown) at all angles.

1.5.5 Spin Asymmetry

We may characterize the asymmetry between the spin-flip rates $u \to d$ and $d \to u$ quantitatively by an "asymmetry parameter". To this end, we define the differential up and down spin *currents* (to be calculated from (1.87) and (1.81–1.84)) as:

$$\frac{\mathrm{d}W^{\mathrm{up}}}{\mathrm{d}\Omega} = \frac{1}{2}\left(\frac{\mathrm{d}W_{u\to u}}{\mathrm{d}\Omega} + \frac{\mathrm{d}W_{d\to u}}{\mathrm{d}\Omega}\right), \tag{1.102}$$

$$\frac{\mathrm{d}W^{\mathrm{down}}}{\mathrm{d}\Omega} = \frac{1}{2}\left(\frac{\mathrm{d}W_{d\to d}}{\mathrm{d}\Omega} + \frac{\mathrm{d}W_{u\to d}}{\mathrm{d}\Omega}\right). \tag{1.103}$$

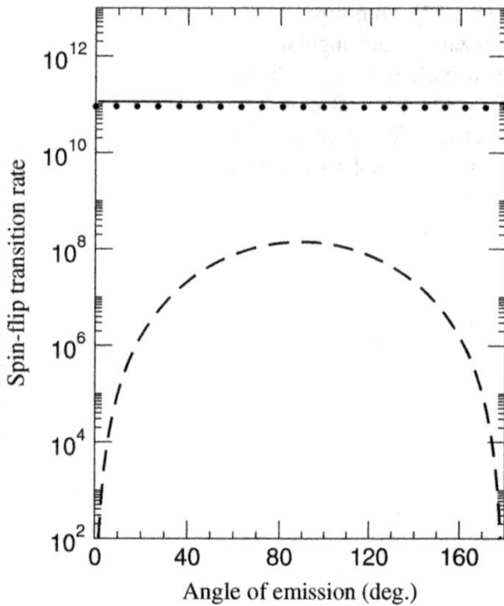

Fig. 1.13 Spin-flip ionization rate: angular distributions at $\omega = 27.2$ eV, $I = 3.51 \times 10^{16}$ W cm^{-2} – strong-field relativistic: $\uparrow$ to $\downarrow$ (*solid line*), $\downarrow$ to $\uparrow$ (*dashed curve*); single-photon relativistic first Born approximation: $\uparrow$ to $\downarrow$ (*thick dots*), $\downarrow$ to $\uparrow\equiv 0$ (not shown)

We define the *spin-flip* asymmetry parameter:

$$A(\theta) = \left(\frac{\mathrm{d}W_{u\to d}}{\mathrm{d}\Omega} - \frac{\mathrm{d}W_{d\to u}}{\mathrm{d}\Omega}\right) / \left(\frac{\mathrm{d}W^{\mathrm{up}}}{\mathrm{d}\Omega} + \frac{\mathrm{d}W^{\mathrm{down}}}{\mathrm{d}\Omega}\right). \tag{1.104}$$

Similarly, we define the asymmetry parameter for a spin *unpolarized* target as the averaged $< A > (\theta)$:

$$< A > (\theta) = \left(\frac{\mathrm{d}W^{\mathrm{up}}}{\mathrm{d}\Omega} - \frac{\mathrm{d}W^{\mathrm{down}}}{\mathrm{d}\Omega}\right) / \left(\frac{\mathrm{d}W^{\mathrm{up}}}{\mathrm{d}\Omega} + \frac{\mathrm{d}W^{\mathrm{down}}}{\mathrm{d}\Omega}\right). \tag{1.105}$$

Figure 1.14 shows the results of angle dependence of the spin-flip asymmetry A(θ) for a Ti-sapphire laser with $\omega = 1.55$ eV, $I = 10^{16}$ W cm^{-2} (outer curve) and $I = 10^{17}$ W cm^{-2} (inner curve). $A(\theta)$ is positive implying clearly that the $\uparrow$ to $\downarrow$ flip rate is greater than the $\downarrow$ to $\uparrow$ flip rate. We may remark in passing that unlike the spin asymmetry for Fano-effect [18], see also [19] observed in ordinary photoionization in the presence of spin–orbit interaction, the two curves in the figure reveal a strong dependence of $A(\theta)$ on the field intensity, at all angles. The peak value of the asymmetry is seen to lie not on the plane of polarization ($\theta = 90°$) but at a somewhat smaller angle away from it. The peak also moves further away from the polarization plane with increasing intensity. This behavior reveals the change of the electron momentum in an intense field caused by the combined effect of retardation and field intensity. This is seen analytically in the "dressed" momentum $\boldsymbol{q}$ of the electron in the presence of the field (1.90) where the extra term that adds to $\boldsymbol{p}$ depends on the photon momentum transfer $n\boldsymbol{\kappa}$, as well as on the intensity through λ_p. Figure 1.15 shows the spin-flip asymmetry $A(\theta)$ for an FEL frequency $\omega = 20$ eV and intensity $I = 10^{20}$ W cm^{-2}. Again, a strong angular dependence

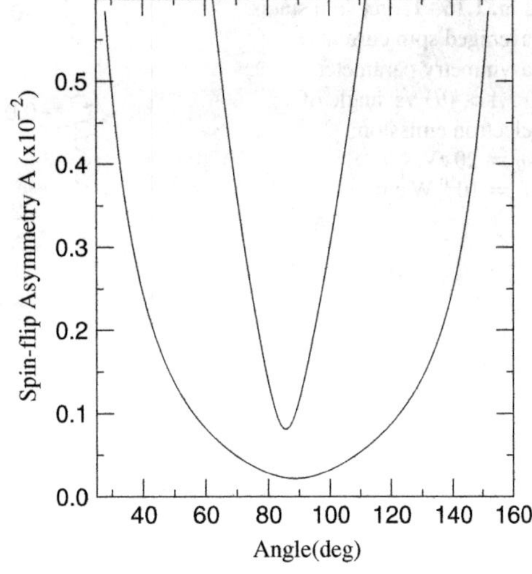

Fig. 1.14 Spin-flip asymmetry parameter $A(\theta)$ vs. angle of electron emission: $\omega = 1.55\,\text{eV}$, $I = 10^{16}\,\text{W cm}^{-2}$ (*outer curve*) and $I = 10^{17}\,\text{W cm}^{-2}$ (*inner curve*)

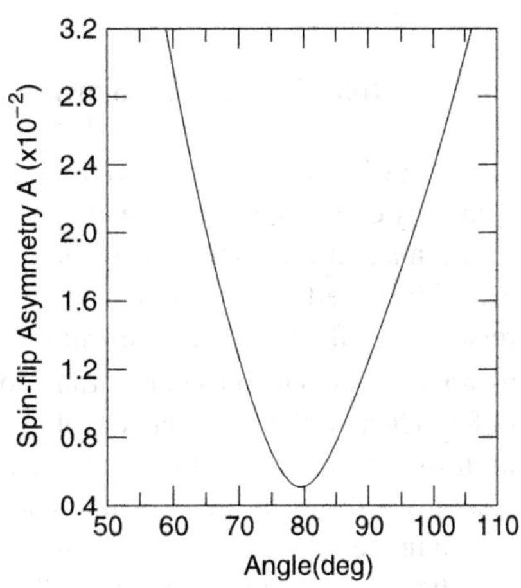

Fig. 1.15 Spin-flip asymmetry parameter $A(\theta)$ vs. angle of electron emission: $\omega = 20\,\text{eV}$, $I = 10^{20}\,\text{W cm}^{-2}$

of the asymmetry is evident at this higher intensity. One may wonder whether this asymmetry actually would survive for ionization from an *unpolarized* target, which requires a 50/50 weighted average of the up and down spin sub-states of the ground state. In Fig. 1.16 we show the results for the spin current asymmetry parameter $< A > (\theta)$ averaged over the ground spin states, as a function of the polar angle of the emitted electrons (measured from the laser field propagation direction) for an $\omega = 20\,\text{eV}$ and an intensity $I = 10^{20}\,\text{W cm}^{-2}$ as in Fig. 1.15. Aside from the change of sign (compare the definitions of the numerators of $A(\theta)$ and $< A > (\theta)$ defined by (1.104) and (1.105)), the asymmetry is seen to survive the spin averaging of the target states; it also shows a strong angular dependence as seen already.

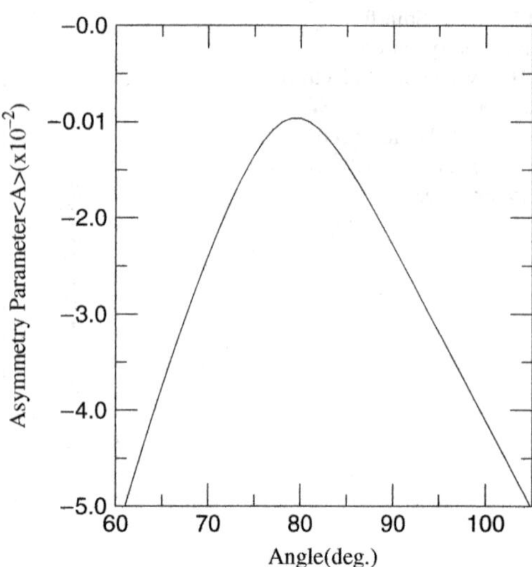

Fig. 1.16 Target spin states averaged spin current asymmetry parameter $< A > (\theta)$ vs. angle of electron emission: $\omega = 20\,\text{eV}$, $I = 10^{20}\,\text{W cm}^{-2}$

1.5.5.1 Role of Retardation or Finite Photon Momentum

Further insights into the nature of spin–photon interaction in intense fields can be gained by considering the influence of the retardation effect, and hence the influence of the magnetic component of the incident electromagnetic field on the spin-flip probability. Perhaps, the simplest way to examine the necessity or otherwise of the retardation effect for the spin-flip process is to put the light propagation vector $\boldsymbol{\kappa}$ identically equal to zero in the transition matrix elements $t^{(n)}_{u\to d}$ and $t^{(n)}_{d\to u}$ (1.82) and (1.83). Clearly, the argument a_p of the Bessel function $J_n(a_p)$ defined above in the limit of zero retardation ($\boldsymbol{\kappa} = 0$) remains finite and the field-dressed momentum $\boldsymbol{q}$ simply reduces to the free momentum $\boldsymbol{p}$. Hence, the spin-flip amplitudes do *not* vanish in the limit of zero retardation, i.e., if the magnetic component of the incident laser field is neglected. Note that in the present case there is no spin–orbit interaction in the initial state (ground s-state). So one may rightly enquire: what is the mechanism for the finite spin-flip transition probability in intense fields, in the absence of retardation and the spin–orbit interaction? We first note that (1.82) and (1.83) for the spin-flip amplitudes depend on the parameters m_2 and $g(q)$, which relate to the 'weak' components of the Dirac spinor of the free electron and the ground state of Dirac H-atom. Hence, they certainly go beyond the usual Pauli-mechanism of direct coupling of a given external magnetic field to the spin (magnetic moment) of the electron. In fact, further examination of the two equations shows that the non-vanishing coupling occurs through the vector product of the polarization $\boldsymbol{\epsilon}$ and the electron momentum $\boldsymbol{p}$. This along with the outer factor $eA_0 \equiv eF_0/\kappa_0$ (where F_0 is the electric peak field strength of the laser field in the laboratory) in the transition matrix element (see (1.64)) shows that the coupling depends on factors of the form $\boldsymbol{F_0} \times \boldsymbol{p}/c$ (a.u.) $\approx \boldsymbol{B}'$ (a.u.), where $\boldsymbol{B}'$ is in fact an effective magnetic field seen by the electron in its own frame of reference. It arises from the Lorentz transformation (e.g., [20]) of the electric field F_0 of the laser field in the laboratory, contributing to an effective (or 'motional') magnetic field in the rest frame of the electron moving with a momentum $\boldsymbol{p}$ in the laboratory. Therefore, the dominant mechanism that leads to the spin-flip transition in *intense* laser fields is the coupling of the "motional" magnetic field $\vec{B}'$ with the spin $\boldsymbol{\sigma}$ or magnetic moment $= -\frac{1}{4c^2}\boldsymbol{\sigma}$ (a.u.) of the electron. This is rather analogous but *not* identical to the Lorentz transformation of the electric field associated with the static *atomic* potential $V(r)$ into an effective magnetic field and the resulting spin–orbit interaction responsible for the well-known Fano-effect [18], see also [19] observed in ordinary photoionization in the perturbative domain of intensity. Therefore, we may also add that for a comparatively weak laser field and when the spin–orbit interaction is not negligible an additional contribution to $\boldsymbol{B}'$, of the order of $-\frac{1}{r}\frac{dV(r)}{dr}\boldsymbol{r} \times \boldsymbol{p}/2c$ (a.u.), ought to be taken into account for quantitative accuracy.

1.5.5.2 External Control of the Spin Currents

It is worth noting that the relative dominance of up or down spin electron currents can be *controlled* from outside, for example, completely reversed by changing the *helicity* of the incident light. This can be seen analytically by replacing the right circular polarization vector $\boldsymbol{\epsilon}(\xi = +\pi/2)$ by the left circular polarization vector $\boldsymbol{\epsilon}(\xi = -\pi/2)$ in (1.81)–(1.84) and observing that the following transformations of the amplitudes hold:

$$t^{(n)}{}_{u\to u} \to t^{(n)*}{}_{d\to d}, \tag{1.106}$$

$$t^{(n)}{}_{d\to d} \to t^{(n)*}{}_{u\to u}, \tag{1.107}$$

$$t^{(n)}{}_{u\to d} \to -t^{(n)*}{}_{d\to u}, \tag{1.108}$$

$$t^{(n)}{}_{d\to u} \to -t^{(n)*}{}_{u\to d}. \tag{1.109}$$

Hence, the spin-flip rates would exchange their magnitudes and the asymmetries $< A > (\theta)$ would change their signs on changing the helicity of the photons from the right circular to the left circular polarization. We may recall that the magnitude of the asymmetry parameters $< A > (\theta)$ and $A(\theta)$, in the cases explicitly considered above, are of the orders of 10^{-3} for the near-infrared wavelength and 10^{-2} for the VUV wavelength. These values lie well above the threshold efficiency $\approx 2.4 \times 10^{-4}$ of currently available spin analyzers in the laboratory (e.g., [21, 22]).

We point out that since the spin-flip asymmetry as shown here remains finite and large for weak and long wavelength fields (see, e.g., Fig. 1.13) and can be controlled from outside, it would be useful also for the new science of *spintronics*.

1.5.6 Charge-State Z Dependence and an Anomalous Effect

In this last section, we consider the dependence of the total ionization probability per unit time and its energy and emission angle distributions, on the charge states Z of the target ions, and consider an anomalous behavior of ionization probability vs. ionization potential.

1.5.6.1 Total Ionization Rates Vs. Z

It is known ([4], p. 96) that in the perturbative region of intensities the rate of ionization decreases very rapidly with inverse powers of the ionic charge Z due to the strong increase in the ionization potential of the target ion, $I_p(Z) = \frac{Z^2}{2}$ (a.u.). Thus, it is usually expected that it would be harder to ionize a system (at the same intensity and frequency of the field) if the electron is bound more tightly. However, as noted above, an anomalous *increase* of the ionization probability with the increase of the binding energy of the electron has been predicted [12] to occur, from a model atom investigation based on the relativistic Klein–Gordon equation. This was seen to occur in a certain range of the binding energy. We shall also confirm the anomalous effect of increase of ionization probability of hydrogenic ions with increasing ionization potential in certain domain of the charge states Z. In Fig. 1.17, we show the results of total ionization rates as a function of the charge state Z, from calculations using intense-field KFR1 approximation, for both Dirac relativistic (solid lines) and Schrödinger nonrelativistic (dashed lines) cases, at $\omega = 10\,\text{eV}$, and $I = 10^{18}$ and $10^{19}\,\text{W cm}^{-2}$. It can be seen that at an intensity $I = 10^{18}\,\text{W cm}^{-2}$, there is already an indication of the counterintuitive result in which the ionization probability increases for $Z = 2$ than at $Z = 1$ as the ionization potential, $I_p = Z^2/2$, increases from 1/2 to 2 (a.u.). At a still higher intensity, $I = 10^{19}\,\text{W cm}^{-2}$, the anomalous behavior is clearly exhibited for both the charge states $Z = 2$ and $Z = 3$. We note, in fact, that at this higher intensity the effect is more pronounced in the relativistic calculation than in the nonrelativistic case. It is, therefore, not exclusively a relativistic effect but appears at present to be a manifestation of highly nonlinear dependence of the ionization dynamics on the field intensity, whose *physical* origin remains to be understood.

1.5.6.2 Energy and Angular Distributions Vs. Z

To investigate this anomalous behavior further, we have computed also the energy spectrum as well as the angular distribution of the emitted electrons from different target ions. In Fig. 1.18, the results are shown for $\omega = 10\,\text{eV}$, at $\text{I} = 10^{19}\,\text{W cm}^{-2}$, for a sequence of charge states $Z = 1, 2, 3, 4, 5$. It is clearly seen that the probability of ionization for $Z = 3$ with a higher binding energy $I_p(3) = 9/2$ (a.u.) is *higher* than for the ions with $Z = 2$ that is more weakly bound ($I_p(2) = 2$ (a.u.) $< 9/2$ (a.u.) $= I_p(3)$). This holds essentially throughout the energy distribution. Finally,

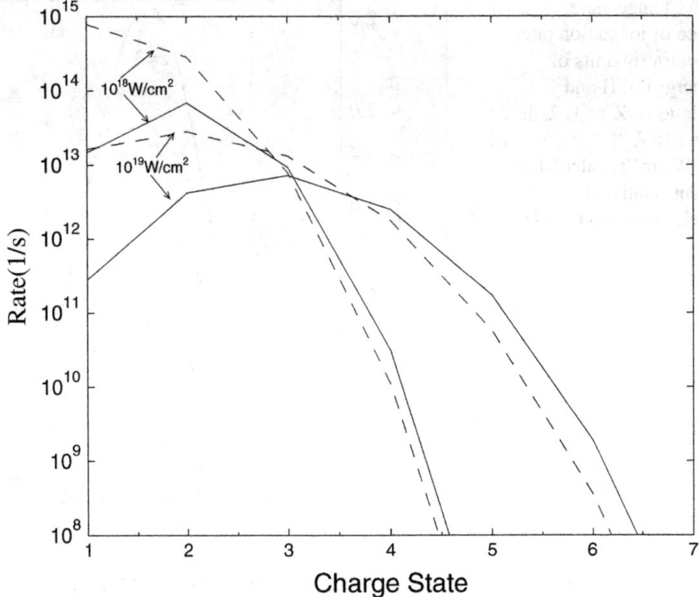

Fig. 1.17 Total ionization rates vs. charge state Z of target ion: $\omega = 10\,\text{eV}$, $I = 10^{18}$ and $10^{19}\,\text{W}\,\text{cm}^{-2}$ – intense-field relativistic Dirac case (*solid lines*), nonrelativistic Schrödinger case (*dashed lines*). Note the anomalous behavior of *larger* rates for higher ionization potentials for the charge states $Z = 2$, and $Z = 2, 3$, than for $Z = 1$, respectively, seen clearly in the relativistic results at the two intensities

in Fig. 1.19, we show the results of calculations for the angular distributions for the ionization of ions with charge states $Z = 1, 2, 3, 4, 5$. It is seen again that the electron emission rates in essentially all directions from the more tightly bound target ions with $Z = 3$ is larger than the rates of emission from the less tightly bound electron in target ions with $Z = 2$ and $Z = 1$. We note that the maximum rate of ionization occurs at about 83° from the direction of propagation of the incident laser beam.

1.6 Summary

To summarize, we have discussed the process of ionization of Dirac Hydrogen atom and hydrogenic ions of increasing charge states, using the four-component relativistic intense-field KFR1 approximation. The energy spectra, the angle dependence, the spin dependence, and the charge-state dependence of the intense-field ionization process are analyzed and discussed in detail. Specific illustrations for the special case of circularly polarized fields allow us to further investigate the helicity dependence of the intense-field ionization process and compare it with ones in

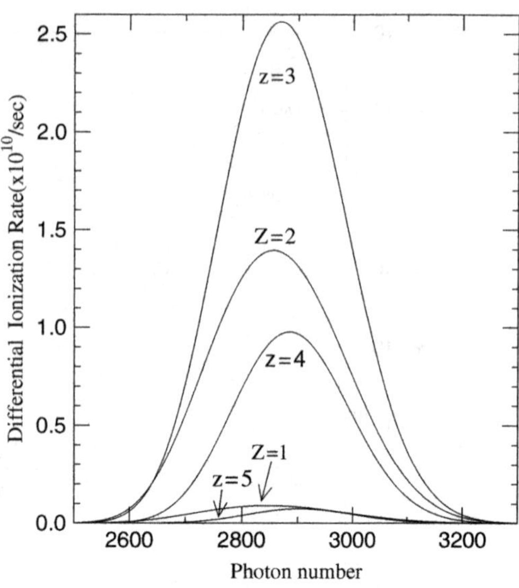

Fig. 1.18 Change-state dependence of ionization rate: energy spectra (in units of photon energy) of H and hydrogenic ions: Z = 1, 2, 3, 4, 5; $\omega = 10\,\text{eV}$, $I = 10^{19}\,\text{W}\,\text{cm}^{-2}$; calculated with present relativistic intense-field ionization model

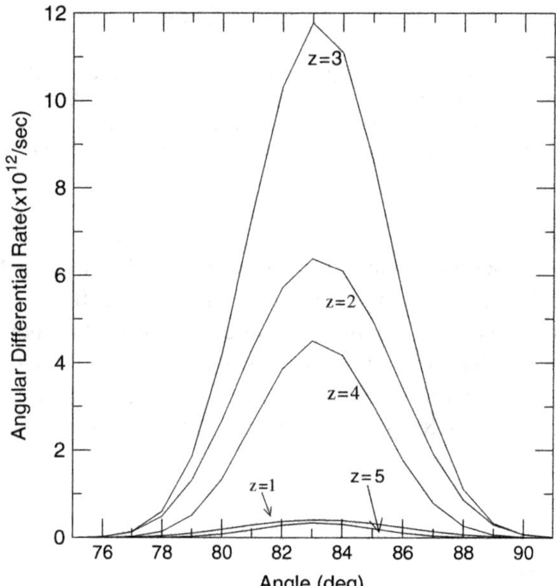

Fig. 1.19 Charge-state dependence of ionization rate: angular distribution for H and hydrogenic ions: $Z = 1, 2, 3, 4, 5$; calculated with present relativistic intense-field ionization model; $\omega = 10\,\text{eV}$, $I = 10^{19}\,\text{W}\,\text{cm}^{-2}$

the weak and long wavelength fields. To this end, results of calculations from the relativistic and the nonrelativistic intense-field approximations (and in some cases also from the relativistic first Born approximation) are compared. The frequency range is chosen to correspond to the available intense infrared and free-electron laser fields. Two counterintuitive characteristics of the ionization process predicted earlier: (a) an asymmetry of spin-up and spin-down currents from the unpolarized target and (b) an anomalous increase of ionization probability with the increase of the ionization potential are also confirmed. The strong spin-asymmetry in the case of weak and long wavelength fields and its simple control analyzed here should be of interest for the new field of spintronics.

References

1. T. Brabec, F. Krausz, Rev. Mod. Phys. **72**, 545 (2000)
2. Y.I. Salamin, S.X. Hu, K.Z. Hatsagortsyan, C. Keitel, Phys. Reports **427**, 41 (2006)
3. R.P. Feynmann, *Quantum Electrodynamics* (Benjamn, New York, 1962)
4. F.H.M. Faisal, *Theory of Multiphoton Processes* (Plenum, New York, 1987)
5. A. Becker, F.H.M. Faisal, J. Phys. B **38**, R1 (2005)
6. D.M. Wolkow, Z. Physik **94**, 250 (1935)
7. C. Szymanowski, V. Veniard, R. Taïeb, A. Maquet, Phys. Rev. A **56**, 3846–3859 (1997)
8. P. Panek, J.Z. Kaminski, F. Ehlotzky, Phys. Rev. A **64**, 033408 (2002)
9. F.B. Bunkin, A.E. Kazakov, M.V. Fedorov, Sov. Phys. Usp. **15**, 416 (1973)
10. M.W. Walser, D.J. Urbach, K.Z. Hatsagortsyan, S.X. Hu, C.H. Keitel, Phys. Rev. A **65**, 043410 (2002)
11. F.H.M. Faisal, S. Bhattacharyya, Phys. Rev. Lett. **93**, 053002 (2004)
12. T. Radozycki, F.H.M. Faisal, Phys. Rev. A. **48**, 2407 (1993)
13. F.H.M. Faisal: *Radiation Effects and Defects in Solids*, vol. 122-123, pp. 27–50 (Gordon and Breach, London 1991)
14. F.H.M. Faisal, *High-Energy Physics Directory* SPIRES-HEP: R BI-TP-90-30 (1990)
15. F.H.M. Faisal, Comput. Phys. Rep. **9**, 56 (1989)
16. F.H.M. Faisal, in *Lectures in Strong-Field Physics*, ed. by T. Brabec (Springer, Berlin 2008)
17. M. Pont, M. Gavrila, Phys. Rev. Lett. **65**, 2362 (1990)
18. U. Fano, Phys. Rev. **178**, 131 (1969)
19. P. Lambropoulos, M.R. Teague, J. Phys. B. **9**, 587 (1976)
20. J.S. Jackson, *Classical Electrodynamics*, 2nd edn. (Wiley, New York, 1975)
21. M. Getzlaff, B. Heidemann, J. Bansmann, C. Westphal, G. Schönhense, Rev. Sci. Inst. **69**, 3913 (1998)
22. J. Kessler, *Polarized Electrons*, 2nd edn. (Springer, Berlin, 1985)

Chapter 2
Ultrafast Hydrogen Migration in Hydrocarbon Molecules Driven by Intense Laser Fields

Huailiang Xu, Tomoya Okino, Katsunori Nakai, and Kaoru Yamanouchi

Abstract By referring to our recent studies conducted using the coincidence momentum imaging method, experimental evidences of the ultrafast hydrogen migration are shown. For allene ($CH_2{=}C{=}CH_2$) the momentum correlation maps and proton distribution maps of triply charged allene constructed from the observed momentum vectors of fragment ions revealed that the spatial distribution of the migrating hydrogen atom (or proton) covers the entire range of an allene molecule. It was also revealed that the hydrogen migration plays a decisive role in breaking selectively chemical bonds within molecules, showing its potential applications for chemical reaction controls. For methanol, it was shown that there are two distinctively different stages in the hydrogen migration processes in singly charged methanol, i.e., ultrafast hydrogen migration occurring within the intense laser field (~38 fs), and slower postlaser pulse hydrogen migration (~150 fs), showing quantum mechanical nature of light protons.

2.1 Introduction

Molecules exposed to an intense laser field exhibit a variety of characteristic dynamical processes such as molecular alignment, laser-induced rearrangement of chemical bonds, and multiple bond breaking called Coulomb explosion [1–7]. The hydrogen migration, in which hydrogen atoms or protons migrate from one site

T. Okino, K. Nakai, and K. Yamanouchi (✉)
Department of Chemistry, School of Science, The University of Tokyo, 7-3-1 Hongo, Bunkyo-ku, Tokyo, 113-0033, Japan
e-mail: tokino@chem.s.u-tokyo.ac.jp, nakai@chem.s.u-tokyo.ac.jp, kaoru@chem.s.u-tokyo.ac.jp

H. Xu
Department of Chemistry, School of Science, The University of Tokyo, 7-3-1 Hongo, Bunkyo-ku, Tokyo, 113-0033, Japan
e-mail: huailiang@jlu.edu.cn
and
State Key Laboratory on Integrated Optoelectronics, College of Electronic Science and Engineering, Jilin University, Changchun, 130012, China

K. Yamanouchi et al. (eds.), *Progress in Ultrafast Intense Laser Science VII*, Springer Series in Chemical Physics 100, DOI 10.1007/978-3-642-18327-0_2,

to another within a molecule, is a chemical-bond rearrangement process that has been observed in hydrocarbon molecules in intense laser fields [8]. Since the hydrogen migration processes leading to chemical bond rearrangement may open new reaction pathways that could not be realized from the initial geometry of molecules, the understanding of the motion of hydrogen atoms or protons within a molecule will provide new strategies for controlling chemical reactions by means of the manipulation of hydrogen migration processes.

The hydrogen migration is found to proceed extremely rapidly, even within the sub-10 fs time domain, when it is induced by an intense laser field. This leads to a difficulty in tracing ultrafast hydrogen migration within a molecule in time domain even when pump-and-probe measurements are achieved using sub-10 fs laser pulses. It was also noted that ultrafast hydrogen migration leading to molecular isomerization by an intense laser field could not be treated theoretically by conventional approaches based on the Born–Oppenheimer (BO) approximation, and a new theoretical approach beyond the BO approximation was recently proposed [9].

It should be noted that recent advances in laser science and technology have made it possible to investigate how far and how fast hydrogen atoms or protons migrate within a molecule. In this article, we introduce our recent experimental studies on ultrafast hydrogen migration in several kinds of hydrocarbon molecules induced by intense laser fields with the laser peak intensity in the range of 10^{13}–10^{15} W/cm^2. By using Coulomb explosion coincidence momentum imaging (CMI) method [10], the fragment ions originated from the Coulomb explosion of a single parent ion are detected in momentum coincidence, so that the charge number of the parent ions and the dissociation pathways can be identified unambiguously.

In the CMI measurements, the three-dimensional momentum vectors of each fragment ion are determined in the laboratory frame for every single event of the Coulomb explosion. The correlation among the momentum vectors of the fragment ions carries information on the nuclear dynamics of molecules, since the momentum vectors vary sensitively to the change in the geometrical structure of molecules before molecules undergo the Coulomb explosion. Therefore, the instantaneous geometrical structure of a parent molecule just before the Coulomb explosion can be constructed from the momentum vectors of the resultant fragment ions. In particular, when a proton is ejected as one of the three fragment ions in the three-body Coulomb explosion process, the spatial position of a migrating proton can be mapped within the molecule in a straightforward manner.

This article is organized as follows. In Sect. 2.2, we present a typical experimental setup used for the Coulomb explosion CMI measurements. In Sect. 2.3, by referring to our studies on allene we show that ultrafast hydrogen migration can be studied by the CMI measurements of two-body Coulomb explosion processes In Sect. 2.4, we show for allene that a momentum correlation map and a proton map representing the spreaded probability distribution of a proton within a molecule can be constructed from the observed three-dimensional momentum vectors of the fragment ions ejected in three-body Coulomb explosion processes. In Sect. 2.5, we introduce our recent studies on the hydrogen migration in methanol probed in real time using the CMI method combined with a pump-probe technique. Finally, the summary and conclusions are presented in Sect. 2.6.

2.2 Coincidence Momentum Imaging

The CMI technique was developed to derive information on geometrical molecular structures of multiply charged polyatomic molecules generated by the irradiation of short wavelength light [11], thc beamfoil charge stripping [12], and the charge exchange collision with highly charged atomic ions [13]. This technique was introduced first by our group in 2001 for studying the decomposition processes of molecules in intense laser fields [10].

In Fig. 2.1, the schematic diagram of a typical experimental apparatus used for the Coulomb explosion CMI measurements in our laboratory is shown [14]. The light source was a Ti:Sapphire femtosecond laser system (Pulsar 5000, Amplitude Technologies) composed of a regenerative amplifier, a two-pass preamplifier and a cryogenically cooled four-pass amplifier, and finally, compressed by a two-grating compressor. In the laser system, the output pulses of 25 fs, and 780 nm at a repetition of 75 MHz from a Ti: Sapphire oscillator (Femtosource S20, Femtolasers) were positively chirped to about 100 ps in an aberration-free stretcher, and then amplified at a high-repetition-rate (5 kHz) amplification stage. To shorten the pulse duration, an

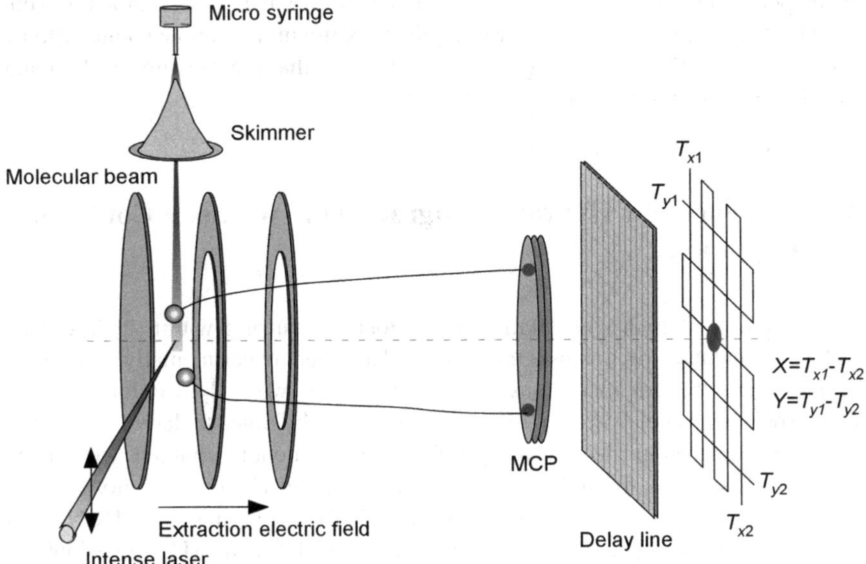

Fig. 2.1 Schematic diagrams of the experimental setup. The position $(X, Y = (T_{x1}-T_{x2}), (T_{y1}-T_{y2})$ of the fragment ions projected on the detector can be obtained, where $x1, x2, y1, y2$ represent the two ends of two wire layers x and y, respectively

acousto-optic programmable dispersive filter (Dazzler, Fastlite) was placed between the stretcher and the regenerative amplifier to control simultaneously the spectral phase and amplitude of the pulses, whose central wavelength is ~790 nm. The ultrashort laser pulses ($\Delta t \sim 40$ fs, $I = 10$–$100\,\mu$J/pulse) were introduced into an ultrahigh-vacuum chamber through a quartz lens ($f = 15$ cm) to achieve the laser field intensity of 10^{13}–10^{14} W/cm^2 at the focal spot, which was estimated from the pulse duration measured by a SPIDER and the radius of the focal spot measured by a CCD camera.

The sample gas was introduced into the sample vacuum chamber through a micro-syringe (0.51 mmϕ), and skimmed by a skimmer (0.48 mmϕ) to form a molecular beam in the main ultrahigh vacuum chamber pumped differentially, whose base pressure was about 3×10^{-11}–8×10^{-11} Torr. The molecular beam and the laser beam crossed at right angles, and the ions generated at the laser focal spot in the molecular beam were projected onto a position-sensitive detector (PSD) with delay-line anodes readout (RoentDek DLD 80) by three equally spaced ($d = 15$ mm) parallel-plate electrodes in the velocity mapping configuration [15]. The laser polarization direction, electrode plates, and the surface of the detector are all set to be parallel to the plane formed by the molecular beam and laser beam axes. The spatial and temporal resolutions of the PSD detector were about 0.255 mm and 0.5 ns, respectively. The three-dimensional momentum vectors of the respective fragment ions were determined by their positions and arrival time on the detector. To detect securely all the fragment ions from a single parent ion, it was necessary to keep the number of events per laser shot less than unity. In most of the measurements performed in our laboratory, the number of events per laser shot was kept to be 0.25–0.55 events/pulse by lowering the pressure in the sample chamber to be 1×10^{-7}–4×10^{-7} Torr during the experiment, so that the pressure in the main chamber became 2×10^{-10}–8×10^{-10} Torr.

2.3 Evidences for Hydrogen Migration in Two-Body Coulomb Explosion

The migration of hydrogen atom(s) or proton(s) occurring within hydrocarbon molecules induced by intense laser fields has been experimentally confirmed by observing the fragment ions ejected from the two-body Coulomb explosion processes. The hydrogen migration induced by intense laser fields was found first in acetonitrile (CH_3CN) [16], in which doubly charged acetonitrile molecule dissociates through three different two-body Coulomb explosion processes, $CH_3CN^{2+} \rightarrow CH_{3-n}{}^+ + H_nCN^+ (n = 0\text{–}2)$. The formation of HCN^+ and H_2CN^+, recorded respectively in coincidence with $CH_2{}^+$ and CH^+ showed unambiguously that the migration of hydrogen atom(s) or proton(s) proceeded within an acetonitrile molecule. The intense laser-induced hydrogen migration processes were also observed in the two-body Coulomb explosion processes of several other small hydrocarbon molecular species, such as acetylene [17], methanol [18], allene

[14] and 1,3-butadiene [19]. The hydrogen migration was also confirmed by the H/D exchange in deuterated methanol [18, 20]. Furthermore, the investigation of two-body Coulomb explosion of methanol under the excitations of laser pulses with pulse durations of sub-10 fs (~7 fs) and 21 fs was carried out, which exhibits different angular distributions of the recoil vectors of the fragment ions through the direct C–O bond breaking pathway, $CH_3OH^{2+} \rightarrow CH_3{}^+ + OH^+$, and the migration pathway $CH_3OH^{2+} \rightarrow CH_2{}^+ + OH_2^+$ [21].

In the latter part of this section, we will focus our attention on allene (CH_2CCH_2), the simplest member of the alkadiene compounds, to show how the ultrafast hydrogen migration process can be investigated by the CMI maps of the two-body Coloumb explosion processes [14].

Figure 2.2 shows the momentum imaging maps of $CH_m{}^+(m = 1\text{–}3)$ appearing in coincidence with $C_2H_{4-m}{}^+(m = 1\text{–}3)$, which represents the following three two-body Coulomb explosion pathways, in which one of the two C–C bonds is broken,

$$C_3H_4^{2+} \rightarrow CH_m^+ + C_2H_{4-m}^+ (m = 1 - 3). \tag{2.1}$$

For $m = 2$, and 3, the CMI maps exhibit a pair of clear crescent-like patterns, while the extent of the anisotropy for $m = 1$ is less pronounced.

From the three-dimensional momentum distributions of the fragment ions, the angular distributions, $I(\theta)$, of the fragment ions $CH_m{}^+(m = 1\text{–}3)$ can be obtained, where θ is the ejection angle of the fragment ions with respect to the laser polarization direction. For evaluating of the extent of anisotropy in the explosion process, the expectation value of the squared cosine [22] defined as

$$<\cos^2\theta> = \frac{\int I(\theta)\cos^2\theta \sin\theta d\theta}{\int I(\theta)\sin\theta d\theta} \tag{2.2}$$

is calculated. A larger number of $<\cos^2\theta>$ represents the spatial distribution of the fragmentation more anisotropic along the laser polarization direction, and

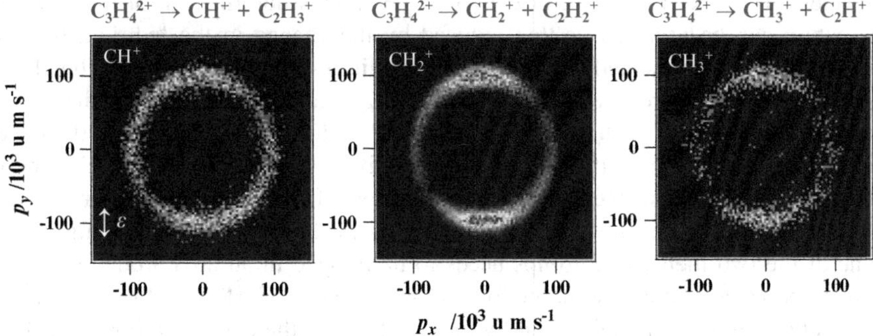

Fig. 2.2 The observed two-dimensional coincidence momentum maps of $CH^+, CH_2{}^+, CH_3{}^+$, recorded in coincidence, respectively, with $C_2H_3{}^+, C_2H_2{}^+, C_2H^+$, through the two-body Coulomb explosion processes of $C_3H_4^{2+}$. The laser polarization direction (ε) is set to be parallel to the p_y-axis as indicated by the *arrow*

$<\cos^2\theta> = 1/3$ corresponds to an isotropic distribution. For these three C–C bond breaking pathways of (2.1), $CH_m{}^+ + C_2H_{4-m}{}^+$ ($m = 1$–3), $<\cos^2\theta> = 0.41$ ($m = 1$), 0.54($m = 2$), and 0.49 ($m = 3$) are obtained.

The relatively large $<\cos^2\theta>$ values for the C–C bond breaking pathway without hydrogen atom migration ($m = 2$) and that with the migration of one hydrogen atom to the other end ($m = 3$) indicate that, for these channels, the allene molecules whose C = C = C skeletal axis is directed along the laser polarization direction are doubly ionized with larger probability, and that the precursor species $CH_2{}^+ \cdots C_2H_2{}^+$ and $CH_3{}^+ \cdots C_2H^+$ are prepared on the repulsive Coulombic potential energy surfaces so that the two-body Coulomb explosion proceeds more rapidly than the overall molecular rotation. After the formation of $CH_3{}^+ \cdots C_2H^+$ within the intense laser field, the probability that a hydrogen atom jumps between the two distant moieties, $CH_3{}^+$ and C_2H^+, may be significantly small. This means that the hydrogen migration is expected to be finished prior to the C···C bond breaking. Since the doubly charged precursor species, $CH_m{}^+ \cdots C_2H_{4-m}{}^+$, are considered to be produced by the most intense part of the laser pulse, the hydrogen migration processes are expected to proceed within the half-duration of the ultrafast laser pulse, i.e. ~20 fs.

On the contrary, the $<\cos^2\theta>$ value for the $m = 1$ pathway is smaller ($<\cos^2\theta> = 0.41$) than that for the pathways of $m = 2$($<\cos^2\theta> = 0.54$) and $m = 3$($<\cos^2\theta> = 0.49$). Since all the three precursor species $CH_m{}^+ \cdots C_2H_{4-m}{}^+$ ($m = 1$–3) prepared on the Coulombic repulsive potential after the interaction with the ultrashort intense laser field are expected to dissociate immediately into two fragment ions the differences among the $<\cos^2\theta>$ values may not be ascribed to the differences in the lifetimes of these precursor species, but rather reflect the geometrical structures of the precursor species. The migration of the hydrogen atom (or proton) in $C_3H_4^{2+}$ is expected to induce the structural deformation of the C–C–C skeleton, leading to the deflection of the ejection direction of CH^+ and $CH_3{}^+$ with respect to the molecular principal a-axis whose direction is expected to be along the laser polarization. The smaller $<\cos^2\theta>$ value for the $m = 1$ pathway, in which CH^+ is ejected, may reflect the larger bending angle, ∠C–C–C, in the precursor species $CH^+ \cdots C_2H_3{}^+$ than in $CH_3^+ \cdots C_2H^+$. This deformation of the skeletal structure can be ascribed as that induced by the change in the hybridization of chemical bonds associated with the hydrogen migration along the skeletal bonds.

Furthermore, the relative yields of the two migration pathways, that is, the $m = 1$ and $m = 3$ pathways with respect to the $m = 2$ pathway are found to be 0.08 and 0.03, respectively. The differences in the relative yields of CH^+ and $CH_3{}^+$ may be related to the distance for a proton to move. To form the precursor species $CH^+ \cdots C_2H_3{}^+$, resulting in the $m = 1$ pathway, the hydrogen atom migrating from one of the two methylene groups needs to be trapped in an area around the central carbon site. In contrast, in the formation of $CH_3{}^+ \cdots C_2H^+$, the hydrogen atom needs to migrate to the other end by passing through the central carbon area. The lower yield of $CH_3{}^+$ than CH^+ is consistent with the picture that a hydrogen atom (or a proton) migrates first into the central carbon area, and then, proceeds to reach the other end.

2.4 Momentum Correlation Maps and Proton Maps for Tracing Ultrafast Hydrogen Migration

As already described in Sect. 2.1, the ultrafast hydrogen migration within a hydrocarbon molecule induced by intense laser fields can be investigated by the analysis of three-body Coulomb explosion processes from the momentum correlation maps and the proton maps constructed from the observed momentum vectors of three fragment ions.

For acetylene, the isomerization to vinylidene type geometry induced by an intense laser field was studied in [23] by the three-body Coulomb explosion CMI of $C_2H_2^{3+} \rightarrow H^+ + C^+ + CH^+$. Through the distributions of the angle between the momenta of H^+ and C^+ or H^+ and CH^+ obtained from the two laser pulse durations of 9 fs and 35 fs, a sharp distribution peaked at ~20° and a significantly broad distribution extending to ~120° were obtained, respectively, with the excitations of 9 fs and 35 fs, indicating that the hydrogen migration process prefers to proceed when the long 35 fs laser pulses are used, while the hydrogen atom remains near the original carbon site in the acetylene configuration when 9 fs laser pulses are used.

For visualizing the spatial distribution of a migrating proton, our group introduced a proton distribution maps, called a proton map, from the CMI maps of a three-body Coulomb explosion process in which a proton is ejected as one of the three fragment ions. The proton maps have been used for investigating the hydrogen migration processes for allene [24] and 1,3-butadiene [25].

In this section, by referring our recent studies on the hydrogen migration in allene, we are going to show how the momentum correlation map and the proton map are constructed from the CMI data and how the dynamical information of the hydrogen migration induced by an ultrashort intense laser field and associated skeletal chemical-bond breaking processes are extracted from these maps.

2.4.1 *Momentum Correlation Maps*

Figure 2.3 shows the CMI map of $CH_2{}^+$, recorded in coincidence with H^+ and C_2H^+, and that of CH^+, recorded in coincidence with H^+ and $C_2H_2{}^+$ for an allene molecule. From these two CMI maps, the existence of the two three-body Coulomb explosion pathways from triply charged allene, $C_3H_4^{3+}$, that is,

$$\textit{Pathway I} : C_3H_4^{3+} \rightarrow H^+ + C_2H^+ + CH_2^+, \tag{2.3}$$

$$\textit{Pathway II} : C_3H_4^{3+} \rightarrow H^+ + CH^+ + C_2H_2^+ \tag{2.4}$$

is securely identified.

Since a triply charged molecule is considered to be decomposed into three fragment ions through the three-body Coulomb explosion immediately after its formation, the momentum vectors of the three fragment ions projected onto a

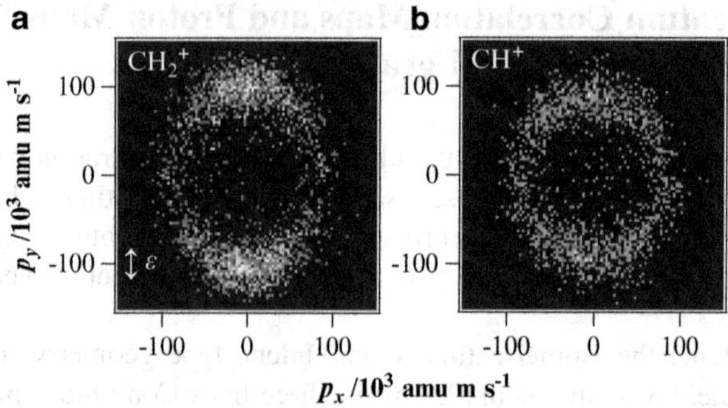

Fig. 2.3 The observed two-dimensional coincidence momentum maps of (**a**) $CH_2{}^+$, and (**b**) CH^+ produced through the three-body Coulomb explosion processes of $C_3H_4^{3+}$. The laser polarization direction (ε) is set to be parallel with the p_y-axis as indicated by the *arrow*

three-dimensional (3D) momentum space can be used to construct the geometrical structure of molecules just before the Coulomb explosion, and consequently, to map the position of the migrating proton within an allene molecule.

We first introduce the momentum correlation maps of the three fragment ions obtained for the two pathways. The correlation among the three fragment ions of each pathway is plotted in terms of the two angle variables, χ and θ [26], defined as

$$\chi_i = \cos^{-1}\left[\left(\frac{\Delta\mathbf{p}_i}{\Delta p_i}\right)\cdot\left(\frac{\mathbf{p}_{3i}}{p_{3i}}\right)\right] \tag{2.5}$$

and

$$\theta_i = \cos^{-1}\left[\left(\frac{\mathbf{p}_{1i}}{p_{1i}}\right)\cdot\left(\frac{\mathbf{p}_{2i}}{p_{2i}}\right)\right], \tag{2.6}$$

where $i = 1$ and 2 represent, respectively, *Pathway I* (2.3) and *Pathway II* (2.4), $\mathbf{p}_{1i}, \mathbf{p}_{2i}$, and $\mathbf{p}_{3i}$ are the momentum vectors of the three fragment ions produced in *Pathway*(i), and the momentum difference $\Delta\mathbf{p}_i$ is defined as $\Delta\mathbf{p}_i = \mathbf{p}_{1i} - \mathbf{p}_{2i}$. The variables p_{1i}, p_{2i}, and p_{3i} and Δp_i are the absolute values of $\mathbf{p}_{1i}, \mathbf{p}_{2i}, \mathbf{p}_{3i}$, and $\Delta\mathbf{p}_i$, respectively. The parameter χ_i is the angle between $\Delta\mathbf{p}_i$ and $\mathbf{p}_{3i}$, and θ_i is the angle between $\mathbf{p}_{1i}$ and $\mathbf{p}_{2i}$.

In Fig. 2.4a, the $\chi_1 - \theta_1$ plot of *Pathway I* is shown, where χ_1 represents the extent of hydrogen migration from the moiety C_2H^+ to $CH_2{}^+$, and θ_1 represents the angle between the ejection directions of the two moieties C_2H^+ and $CH_2{}^+$. When $\chi_1 \sim 0°$, the proton is located at around its original position, and when $\chi_1 \sim 180°$, the proton is located at the other end of the allene molecule to form the propyne (HC≡C–CH_3) configuration. In this way, by constructing the $\chi_1 - \theta_1$ correlation map, the position of the migrating proton within the $C_3H_4^{3+}$ molecule just before the Coulomb explosion can roughly be estimated. When $\theta_1 \sim 0°$ or 180°, the two moieties C_2H^+ and $CH_2{}^+$ are considered to be ejected along the linear C–C–C skeletal structure. When θ_1 is off from 0° or from 180°, the ejection directions of

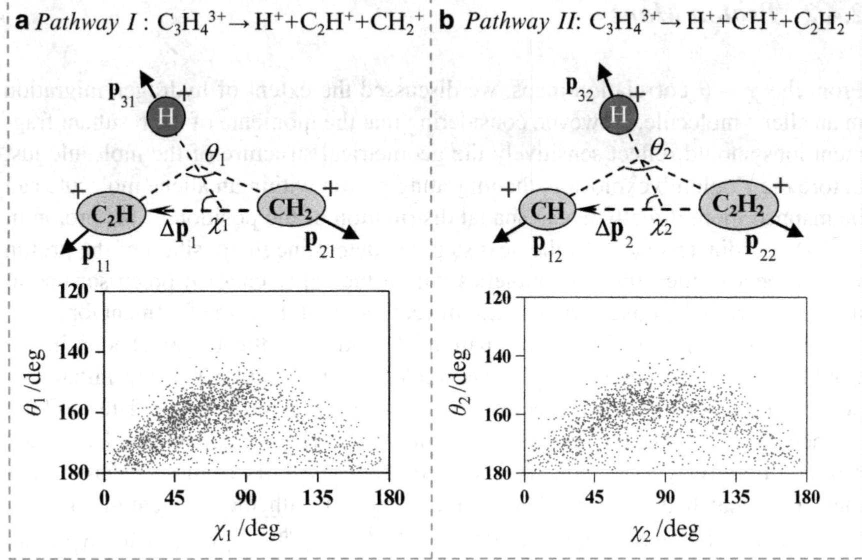

Fig. 2.4 The schematic diagrams of the two angle parameters, χ_i and θ_i, and the $\chi_i - \theta_i$ correlation maps for the three-body Coulomb explosion pathways of (**a**) *Pathway I*: $C_3H_4^{3+} \rightarrow H^+ + C_2H^+ + CH_2{}^+$ and (**b**) *Pathway II*: $C_3H_4^{3+} \rightarrow H^+ + CH^+ + C_2H_2{}^+$

the two moieties C_2H^+ and $CH_2{}^+$ are no longer along the direction of the initially linear C–C–C molecular principal axis.

The resultant $\chi_1 - \theta_1$ correlation map of $C_3H_4^{3+}$ for *Pathway I* is plotted as shown in the lower half of Fig. 2.4a. It can be seen that the distribution spreads in the wide range of $\chi_1 = 0-180°$, whereas the distribution along the θ_1 direction is in the range of $\theta_1 = 150-180°$. The broad distribution of the proton in the full χ_1 angle range can be regarded as direct evidence of the ultrafast hydrogen migration process in which the proton moves from one end of the molecule toward the other end in the intense laser field. When the proton is located in the area around its original position ($\chi_1 \sim 0°$) or at the other end of the molecule ($\chi_1 \sim 180°$), the distribution along θ_1 is found to be around $\theta_1 = 180°$, representing that the two moieties C_2H^+ and $CH_2{}^+$ are ejected along the C–C–C axis from the allene ($H^+ \cdots HCC^+ \cdots CH_2^+$) or propyne ($HCC^+ \cdots CH_2{}^+ \cdots H^+$) type geometrical configurations.

In Fig. 2.4b, a schematic diagram of the definition of two angle parameters, χ_2 and θ_2, for *Pathway II* is shown, and the corresponding $\chi_2 - \theta_2$ correlation map is plotted. In this case, χ_2 denotes the extent of hydrogen migration from the moiety CH^+ to $C_2H_2{}^+$, and θ_2 represents the angle between the ejection directions of the two moieties CH^+ and $C_2H_2{}^+$. It can be found that the event distribution spreads, similarly to those in Fig. 2.4a, in the entire χ_2 range of $\chi_2 = 0 - 180°$, and in the θ_2 range of $\theta_2 = 150–180°$, showing that the ultrafast hydrogen migration proceeds within an allene molecule.

2.4.2 Proton Maps

From the $\chi - \theta$ correlation maps, we discussed the extent of hydrogen migration in an allene molecule. However, considering that the momenta of the resultant fragment ions should reflect sensitively the geometrical structure of the molecule just before the Coulomb explosion, the migrating proton within an allene molecule can be mapped more directly as the spatial distribution of the position of the proton in the 3D coordinate space. As the next step, we determine the position of the proton with respect to the other two moieties within the triply charged precursor parent molecules from the observed momentum vectors of all the three fragment ions.

In the analysis, the classical equation of motion of the fragment ions in the Coulombic field is numerically solved under the assumption that the initial values of the velocity of the respective fragment ions are zero, to calculate the final momentum vectors of the respective fragment ions. The numerical calculations are repeated iteratively until the calculated momentum vectors of the respective fragment ions match the observed momentum vectors with the momentum error of $|\delta \mathrm{P}| \leq 1.4 \times 10^3$ amu m/s, where $|\delta \mathrm{P}| = \sqrt{\sum_{j=1}^{3} (|\mathrm{p}_j^{\mathrm{obs}}| - |\mathrm{p}_j^{\mathrm{cal}}|)^2}$ with $|\mathrm{p}_j^{\mathrm{obs}}|$ and $|\mathrm{p}_j^{\mathrm{cal}}|$ being the absolute values of the calculated and observed momentum vectors of the j-th fragment ion ($j = 1$–3).

The geometrical structures of the triply charged allene molecule constructed from the measured momentum vectors of the fragment ions for *Pathway I* and *Pathway II* are plotted in the form of the proton map as shown in Fig. 2.5, exhibiting that the proton has a very broad distribution covering the wide areas around the two ion moieties C_2H^+ and ${CH_2}^+$ in *Pathway I*, and those around the two ion moieties, CH^+ and ${C_2H_2}^+$ in *Pathway II*. It is noted that the migrating proton exhibits a denser distribution in the area around the ion moiety to which it is initially bonded, that is, C_2H^+ in *Pathway I* and CH^+ in *Pathway II*. The distribution in the area around CH_2^+ in *Pathway I* and that in the area around ${C_2H_2}^+$ in *Pathway II* indeed indicate that the migrating proton reaches the other end of the allene molecule prior to the C–C bond breaking. It can also be seen in both Fig. 2.5a, b that the proton density in the spatial region between those two relatively dense areas is much smaller, showing that the proton stays with a much lower probability in the spatial region between the two heavy moieties.

The ratio of the density of the proton distribution in the area of $r \cos\theta < 0$ with respect to that in $r \cos\theta > 0$ is much larger in Fig. 2.5a than in Fig. 2.5b. To examine this difference quantitatively, the $r \cos\theta$ distributions of the proton for *Pathway I* (open bar) and *Pathway II* (solid bar) are plotted in Fig. 2.6a. For comparison, the total event numbers are normalized for both pathways. It is seen in Fig. 2.6a that the relative intensity of *Pathway I* is larger in the region of $r \cos\theta < 0$, while the relative intensity of *Pathway II* is larger in the region of $r \cos\theta > 0$. This may be interpreted as follows by referring to Fig. 2.6b.

When a proton is trapped in the region of $r \cos\theta < 0$ in the course of the migration from one end to the other, as shown in the upper-left side of Fig. 2.6b, the precursor species $H^+ \cdots C_2H^+ \cdots {CH_2}^+$ may be formed preferentially, that is, the

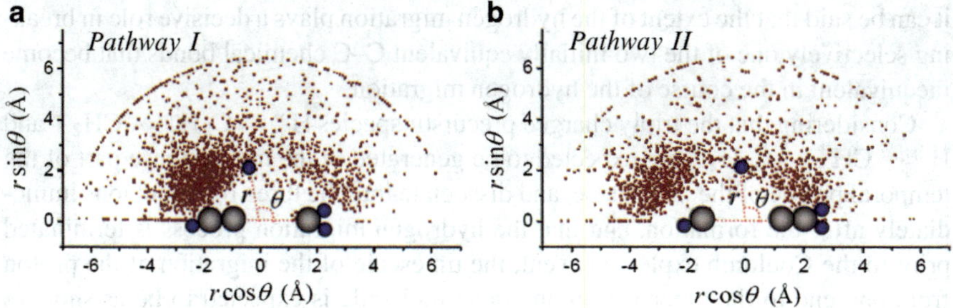

Fig. 2.5 The geometrical structure of $C_3H_4^{3+}$ reconstructed directly from the observed momentum vectors of the respective fragment ions ejected through the Coulomb explosion pathways of (**a**) *Pathway I*: $C_3H_4^{3+} \rightarrow H^+ + C_2H^+ + CH_2{}^+$ and (**b**) *Pathway II*: $C_3H4^{3+} \rightarrow H^+ + CH^+ + C_2H_2{}^+$. The parameter *r* is the distance between the proton and the center of mass of "C_2H^+ and CH_2^+" in *Pathway I* or that of "CH^+ and $C_2H_2^+$" in *Pathway II*, and θ is the angle between *r* and the line connecting the two heavy moieties in each pathway. The dots distributed along the circular outer boundary ($r = 6.4$ Å) of the proton maps result from those events with $r \geq 6.4$ Å

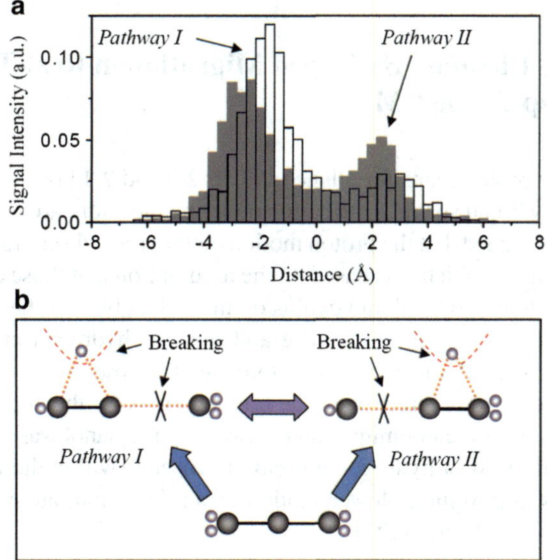

Fig. 2.6 (**a**) The $r \cos\theta$ distributions for *Pathway I* and *Pathway II* (2.4), and (**b**) the schematic diagram for the position of the C–C chemical bond broken preferentially depending on the spatial position of the migrating proton

C–C bond between C_2H^+ and $CH_2{}^+$ is broken with the larger probability. This is *Pathway I*. In contrast, if the position of the migrating proton is in the region of $r \cos\theta > 0$, as shown in the upper-right side of Fig. 2.6b, the precursor species $H^+ \cdots CH^+ \cdots C_2H_2{}^+$ may be formed preferentially, that is, the C–C bond between CH^+ and $C_2H_2{}^+$ is broken with the larger probability. This is *Pathway II*. Therefore, it can be said that the extent of the hydrogen migration plays a decisive role in breaking selectively one of the two initially equivalent C–C chemical bonds that become inequivalent in the course of the hydrogen migration.

Considering that the triply charged precursor species $H^+ \cdots C_2H^+ \cdots CH_2{}^+$ and $H^+ \cdots CH^+ \cdots C_2H_2{}^+$ are expected to be generated in the most intense part of the temporal profile of the laser pulse, and dissociate into the three fragment ions immediately after the formation, and that the hydrogen migration process is terminated prior to the Coulomb explosion event, the timescale of the migration of the proton from one end to the other within an allene molecule is expected to be as short as around a half of the laser pulse duration, that is, $\sim$20 fs.

In our more recent report [25], we constructed the proton maps of the two three-body Coulomb explosion pathways, $C_4H_6^{3+} \rightarrow H^+ + CH_3{}^+ + C_3H_2{}^+$ and $C_4H_6^{3+} \rightarrow H^+ + C_2H^+ + C_2H_4{}^+$ of 1,3-butadiene induced by an ultrashort intense laser field ($\Delta t = 40$ fs, $\lambda = 795$ nm and $I = 4.5 \times 10^{14}$ W/cm^2), and showed that two protons migrate within a 1,3-butadiene molecule prior to the three-body decomposition.

2.5 Tracing Ultrafast Hydrogen Migration in Real-Time by Pump-Probe CMI

As shown in the proton maps for allene in Sects 2.3 and 2.4 [14, 24], the migrating proton has a very broad spatial distribution covering the entire area around an allene molecule. It was argued that the proton motion could proceed very rapidly occurring within the light field, which was based on the assumption that those doubly or triply charged species undergo Coulomb explosion immediately after they are formed by the most intense part of the laser pulse and that the hydrogen migration should proceed before the Coulomb explosion. Moreover, the observation that the extent of anisotropy in the ejection direction of the fragment ions for the migration pathways is the same as that for the nonmigration pathways in methanol was considered to be an evidence that ultrafast hydrogen migration proceeds within the laser field [18]. On the contrary, a postpulse slower motion of an deuterium atom was shown for deuterated acetylene dication [27].

To show more explicitly how rapidly hydrogen migration proceeds, and to distinguish the hydrogen (or a proton) migration within the laser field from the post laser-pulse hydrogen migration, we investigated recently the hydrogen migration in methanol in real time by the pump-and-probe technique [28]. It was revealed that the hydrogen migration proceeds within the laser pulse, as well as after the laser–molecule interaction as the postlaser pulse hydrogen migration.

The pump-and-probe experiment was carried out by simultaneously monitoring two types of two-body Coulomb explosion processes in methanol, that is, the pathway in which the C–O bond is broken without the hydrogen migration,

$$CH_3OH^{2+} \rightarrow CH_3{}^+ + OH^+, \tag{2.7}$$

and the pathway in which the C–O bond is broken after the migration of one hydrogen atom from the methyl group to the hydroxyl group,

$$CH_3OH^{2+} \rightarrow CH_2^+ + OH_2^+. \tag{2.8}$$

In this experiment, a pair of linearly polarized laser pulses with the same pulse energies (40 μJ/pulse) was prepared through a Michelson-type interferometer with a variable time delay Δt. The two laser pulses were then focused onto an effusive molecular beam of methanol in an ultrahigh vacuum chamber with a base pressure of $\sim 3 \times 10^{-11}$ Torr. The field intensity at the focal spot was $\sim 2 \times 10^{14}$ W/cm^2 for both pump and probe pulses. The time delay Δt was varied from 100 to 800 fs. The minimum time delay of $\Delta t = 100$ fs was chosen to avoid the optical interference of the two laser pulses. The increment of the time delay was set to be 50 fs with the uncertainty of 0.5 fs.

In Fig. 2.7a–d, the CMI maps of $CH_3{}^+$, appearing in coincidence with OH^+, are shown, which were obtained with the pump only (a), and with the pump and probe pulses whose time delays are $\Delta t = 200$ fs (b), 500 fs (c) and 800 fs (d). It can be clearly seen in Fig. 2.7b–d that as Δt increases, a new circular momentum component emerges and its peak momentum value decreases gradually as Δt increases. In Fig. 2.7e–h, the CMI maps of $CH_2{}^+$ recorded in coincidence with $OH_2{}^+$ are shown.

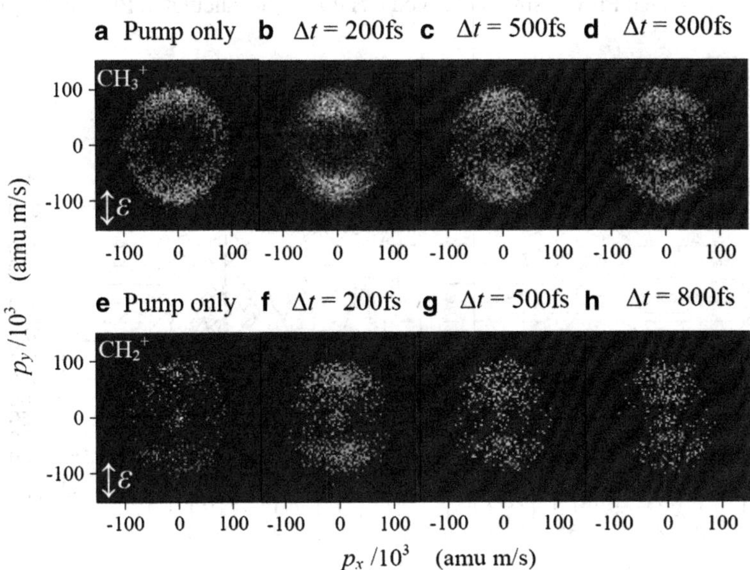

Fig. 2.7 The recorded CMI maps of $CH_3{}^+$ and $CH_2{}^+$, appearing in coincidence, respectively, with OH^+ and $OH_2{}^+$. The laser polarization directions (ε) of both the pump and probe pulses were set to be parallel to the p_y-axis as indicated by the *arrow*. The signals appearing in the central areas of Fig. 2.7e–h are the accidental false coincidence events from residual H_2O in the vacuum chamber

The new components appearing in Fig. 2.7b–d, and those in Fig. 2.7f–h show that the dissociating singly charged molecular ions, $(CH_3 \cdots OH)^+$ and $(CH_2 \cdots OH_2)^+$, prepared by the pump pulse are ionized further by the second pulse into the doubly charged molecular ions $CH_3{}^+ \cdots OH^+$ and $CH_2{}^+ \cdots OH_2{}^+$, leading to the Coulomb explosion into $CH_3{}^+ + OH^+$ and $CH_2{}^+ + OH_2{}^+$, respectively.

From the three-dimensional momentum distributions of the fragment ions, the sum of the kinetic energy released from a pair of the fragment ions, E_{kin}, was obtained for both pathways as shown in Fig. 2.8. The kinetic energy distributions for both pathways can be classified into two parts: the high-energy component, where the kinetic energy distributions are independent of Δt, and the low-energy component, where the peak position of the kinetic energy distributions shift toward lower energies as Δt increases, exhibiting a time-dependent behavior.

The time-dependent low-energy component in Fig. 2.8 is considered to reflect the temporal evolution of a wavepacket of the dissociating $(CH_3 \cdots OH)^+$ and that of $(CH_2 \cdots OH_2)^+$. It was found that the kinetic energy distribution of the time-independent high-energy component in Fig. 2.8a is centered at ~5.9 eV, while the one in Fig. 2.8b is centered at ~5.2 eV. In contrast, when the nonmigration pathway through (2.7) and the migration pathway through (2.8) are induced only by the pump laser, the kinetic energy distributions of the fragment ions ejected from both the migration and the nonmigration pathways are peaked at ~5.9 eV.

The observation above can be interpreted in terms of the landscape of the theoretically obtained PES of singly charged CH_4O^+. The calculated PES of the ground

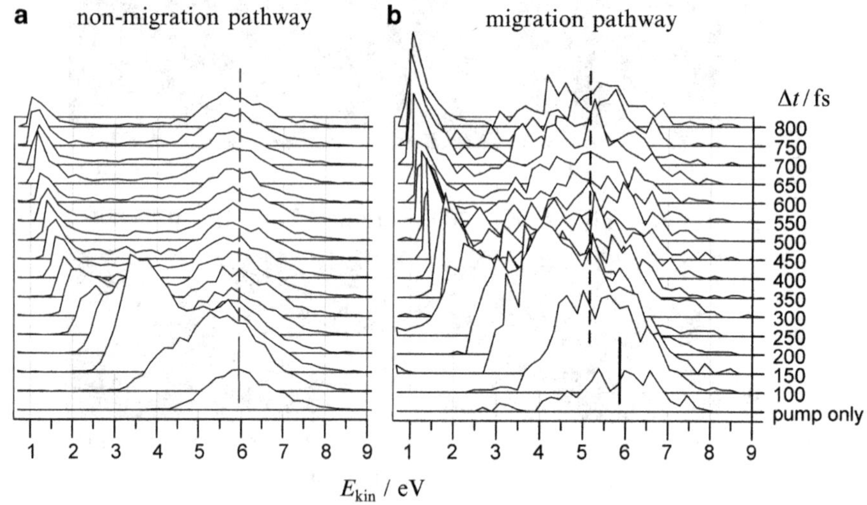

Fig. 2.8 The kinetic energy distributions of the fragment ions released from the pathways (**a**) $CH_3OH^{2+} \rightarrow CH_3{}^+ + OH^+$ and (**b**) $CH_3OH^{2+} \rightarrow CH_2{}^+ + OH_2^+$. The energy distributions for these two pathways after the irradiation with only one laser pulse excitation are both peaked at ~5.9 eV (*solid lines*). The *black dash line* shows the peak positions at 5.2 eV for the migration pathway; while the *red dash line* shows the peak positions at 5.9 eV for the nonmigration pathway

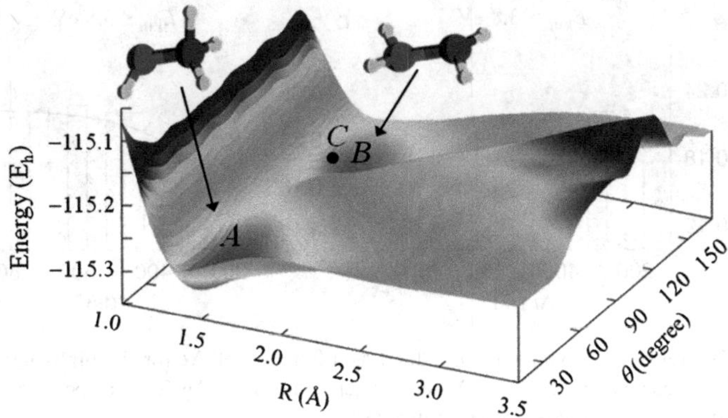

Fig. 2.9 The ground-state potential energy surface of CH_4O^+. R is the C–O bond distance, and θ is the angle between the C–O bond and the vector that connects the migrating hydrogen atom to the center of the C–O bond. All of other degrees of freedom are optimized

states of CH_4O^+ obtained by the density functional theory at the UB3LYP/6-31G(d) level with the Gaussian 03 program [29] are shown in Fig. 2.9. There are two minima at around $A(1.37\,Å, 45°)$ and $B(1.46\,Å, 145°)$ on the PES of CH_4O^+, representing the two geometrical structures, CH_3OH^+ and $CH_2OH_2{}^+$, respectively.

Therefore, the difference in the kinetic energy distributions shown in the high-energy component of Fig. 2.8a and that of Fig. 2.8b can be ascribed to the difference in the geometrical structures of the CH_3OH^+ and $CH_2OH_2{}^+$ molecular ions. When the pump laser pulse prepares nuclear wave packets of CH_3OH^+ and $CH_2OH_2{}^+$ on the PES around A and B shown in Fig. 2.9, the probe laser pulse, after a certain temporal delay, ionizes CH_3OH^+ from the well A into $(CH_3 \cdots OH)^{2+}$, and $CH_2OH_2^+$ from the well B into $(CH_2 \cdots OH_2)^{2+}$, which explode into the two moieties in each pathway by the C–O bond breaking. Since the C–O bond length in the well A (1.37 Å) is shorter than that in the well B (1.46 Å), the kinetic energy released from the dissociation of $(CH_3 \cdots OH)^{2+}$ is expected to be larger than that from the dissociation of $(CH_2 \cdots OH_2)^{2+}$, resulting in the observation that the peak position of the kinetic energy distribution at 5.9 eV for the high-energy component in Fig. 2.8a is larger than 5.2 eV shown in Fig. 2.8b.

The calculated energy of the transition state located between CH_3OH^+ and $CH_2OH_2{}^+$ is ~1.51 and ~1.48 eV higher than the two minima of the potential wells around A and B, respectively. Therefore, when methanol molecules are prepared on the PES in the wells of A and B, the hydrogen migration would not proceed after the pump laser pulse, and thus leads to the relative ion yield obtained from the migration pathways (mig.) with respect to the total ion yield (total = non-mig.+mig.) to be constant, as shown in Fig. 2.10a.

The calculated threshold energy of the C–O bond dissociation is ~3.2 eV, higher than the energy (~1.5 eV) of the transition state for the isomerization reaction, i.e., the hydrogen migration between CH_3OH^+ and $CH_2OH_2^+$. When methanol molecules ionized by the first pulse are prepared on those PESs of vibrationally

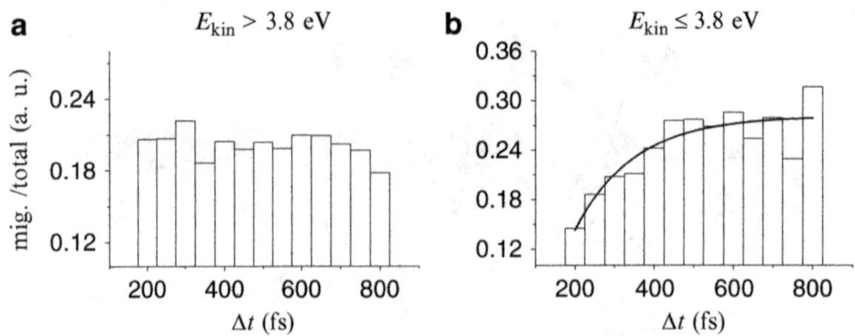

Fig. 2.10 The relative ion yields of mig./total as a function of Δt for the high-energy ($E_{kin} > 3.8$ eV) and low-energy ($E_{kin} \leq 3.8$ eV) components shown in Fig. 2.8b, respectively. The *solid line* shows the exponential fit to the experimental data

excited states of singly charged state with the energy higher than the transition energy, it is expected that the postpulse hydrogen migration can proceed after the light–molecule interaction. In this case, as Δt increases, the relative ion yield obtained from the migration pathway (mig.) with respect to the total ion yield (total = non-mig.+mig.) increases with an exponential curve of ~150 fs, as shown in Fig. 2.10b As a result, the observed temporal evolution of the kinetic energy spectra reveals that there are two distinctively different stages in the hydrogen migration processes in the singly charged methanol, that is, (a) ultrafast hydrogen migration occurring within the intense laser field (~38 fs) and (b) the slower postlaser pulse hydrogen migration (~150 fs) [28].

In addition, the observation that the peak position of 5.2 eV for the migration pathway shown in the upper strip exhibits a large difference from the pump-only value further reveals the dynamics of the hydrogen migration within a singly charged methanol molecule. The observed pump-only peak positions at ~5.9 eV for both the nonmigration and the migration pathways indicate that the distance between the two dissociating moieties in the precursor species $CH_3{}^+ \cdots OH^+$ and that in $CH_2{}^+ \cdots OH_2{}^+$ does not change so much during the double ionization processes from neutral methanol by a 38-fs laser pulse. Since the Coulomb explosion is considered to occur immediately after the formation of doubly charged $CH_3{}^+ \cdots OH^+$ and $CH_2{}^+ \cdots OH_2{}^+$ species, the hydrogen migration should proceed in the singly charged manifold during the period of light–molecule interaction. Therefore, the singly charged $(CH_2 \cdots OH_2)^+$ can be prepared in an area C, where the C–O distance is close to that in the area *A* for $(CH_3 \cdots OH)^+$, rather than in the area around the bound well *B*.

The observation that the peak position of the kinetic energy distribution in Fig. 2.8b at $\Delta t = 100$ fs decreases from 5.9 to 5.2 eV shows that the C$\cdots$O distance in $CH_2OH_2^+$ is stretched from *C* to *B* after being irradiated with the pump pulse.

2.6 Conclusion

Using the Coulomb explosion CMI method, various types of the hydrogen migration processes were identified for hydrocarbon molecules. It was shown from the momentum correlation maps and proton maps in allene that the proton can migrate from one end of an allene molecule to the other within the short laser period of time, and that the extent of the hydrogen migration can play a decisive role in determining which one of the two initially equivalent C=C chemical bonds is more preferentially broken. Using the pump-probe CMI method, it was demonstrated in singly charged methanol that the hydrogen migration processes can proceed both within the intense laser field and after the interaction of molecule with the intense laser field.

These new findings may be regarded as evidences of quantum mechanical nature of light hydrogen atoms or protons appearing when hydrocarbon molecules interacting with an ultrashort intense light field. On the contrary, the effect of the hydrogen migration on the breaking of chemical bonds opens up a new possibility of controlling chemical bond breaking by manipulating the motion of protons by an intense laser field.

Acknowledgements We thank the groups of S.L. Chin (Laval University, Quebec, Canada), and A. Baltuska, M. Kitzler (Vienna University of Technology, Vienna, Austria) for fruitful collaboration. The present research was supported by two grants from the Ministry of Education, Culture, Sports, Science and Technology (MEXT), Japan (the Grant-in-Aid for Specially Promoted Research on Ultrafast Hydrogen Migration (#19002006), and the Grant-in-Aid for Global COE Program for Chemistry Innovation). HLX would also like to acknowledge the financial support from NSFC 11074098.

References

1. K. Yamanouchi, Science **295**, 1659 (2002)
2. J.H. Posthumus, Rep. Prog. Phys. **67**, 623 (2004)
3. E. Baldit, S. Saugout, C. Cornaggia, Phys. Rev. A **71**, 021403 (2005)
4. W. Fuß, W.E. Schmid, S.A. Trushin, Chem. Phys. **316**, 225 (2005)
5. P. Liu, T. Okino, Y. Furukawa, T. Ichikawa, R. Itakura, K. Hoshina, K. Yamanouchi, H. Nakano, Chem. Phys. Lett. **423**, 187 (2006)
6. A.N. Markevitch, D.A. Romanov, S.M. Smith, R.J. Levis, Phys. Rev. Lett. **96**, 163002 (2006)
7. H. Yazawa, T. Shioyama, Y. Suda, M. Yamanaka, F. Kannari, R. Itakura, K. Yamanouchi, J. Chem. Phys. **127**, 124312 (2007)
8. K. Hoshina, Y. Furukawa, T. Okino, K. Yamanouchi, J. Chem. Phys. **129**, 104302 (2008)
9. T. Kato, K. Yamanouchi, J. Chem. Phys. **131**, 164118 (2009)
10. H. Hasegawa, A. Hishikawa, K. Yamanouchi, Chem. Phys. Lett. **349**, 57 (2001)
11. J. Ullrich, R. Moshammer, A. Dorn, R. Dörner, L.Ph.H. Schmidt, H. Schimidt-Böcking, Rep. Prog. Phys. **66**, 1463 (2003)
12. Z. Vager, D. Zajfman, T. Graber E.P. Kanter Phys. Rev. Lett. **71**, 4319 (1993)
13. F.A. Rajgara, M. Krishnamurhy, D. Mathur, T. Nishide, T. Kitamura, H. Shiromaru, Y. Achiba, N. Kobayashi, Phys. Rev. A **64**, 032712 (2001)
14. H.L. Xu, T. Okino, K. Yamanouchi, Chem. Phys. Lett. **349**, 255 (2009)
15. A.T.J.B. Eppink, D.H. Parker, Rev. Sci. Instrum. **68**, 3477 (1997)

16. A. Hishikawa, H. Hasegawa, K. Yamanouchi, J. Electron Spectrosc. Relat. Phenom. **141**, 195 (2004)
17. A.S. Alnaser, I. Litvinyuk, T. Osipov, B. Ulrich, A. Landers, E. Wells, C.M. Maharjan, P. Ranitovic, I. Bochareva, D. Ray, C.L. Cocke, J. Phys. B **39**, S485 (2006)
18. T. Okino, Y. Furukawa, P. Liu, T. Ichikawa, R. Itakura, K. Hoshina, K. Yamanouchi, H. Nakano, Chem. Phys. Lett. **423**, 220 (2006)
19. H.L. Xu, T. Okino, K. Nakai, K. Yamanouchi, S. Roither, X. Xie, D. Kartashov, M. Schöffler, A. Baltuska, M. Kitzler, Chem. Phys. Lett. **484**, 119 (2010)
20. T. Okino, Y. Furukawa, P. Liu, T. Ichikawa, R. Itakura, K. Hoshina, K. Yamanouchi, H. Nakano, J. Phys. B **39**, S515 (2006)
21. R. Itakura, P. Liu, Y. Furukawa, T. Okino, K. Yamanouchi, H. Nakano J. Chem. Phys. **127**, 104306 (2007)
22. T. Okino, Y. Furukawa, P. Liu, T. Ichikawa, R. Itakura, K. Hoshina, K. Yamanouchi, H. Nakano, Chem. Phys. Lett. **419**, 223 (2006)
23. A. Hishikawa, A. Matsuda, E.J. Takahashi, and M. Fushitani, J. Chem. Phys. **128**, 084302 (2008)
24. H.L. Xu, T. Okino, K. Yamanouchi, J. Chem. Phys. 131, 151102 (2009)
25. H.L. Xu, T. Okino, K. Nakai, K. Yamanouchi, S. Roither, X. Xie[3], D. Kartashov, L. Zhang, A. Baltuska, M. Kitzler, Phys. Chem. Chem. Phys. (2010), DOI: 10.1039/c0cp00628a
26. A. Hishikawa, H. Hasegawa, K. Yamanouchi, Chem. Phys. Lett. **361**, 245 (2002)
27. A. Hishikawa, A. Matsuda, M. Fushitani, E.J. Takahashi, Phys. Rev. Lett. **99**, 258302 (2007)
28. H.L. Xu, C. Marceau, K. Nakai, T. Okino, S.L. Chin, K. Yamanouchi, J. Chem. Phys. **133**, 071103 (2010)
29. M.J. Frisch et al., *GAUSSIAN 03, Revision C.02* (Gaussian Inc., Wallingford CT, 2004)

Chapter 3
Control of π-Electron Rotations in Chiral Aromatic Molecules Using Intense Laser Pulses

Manabu Kanno, Hirohiko Kono, and Yuichi Fujimura

Abstract Our recent theoretical studies on laser-induced π-electron rotations in chiral aromatic molecules are reviewed. π electrons of a chiral aromatic molecule can be rotated along its aromatic ring by a nonhelical, linearly polarized laser pulse. An ansa aromatic molecule with a six-membered ring, 2,5-dichloro[n](3,6) pyrazinophane, which belongs to a planar-chiral molecule group, and its simplified molecule 2,5-dichloropyrazine are taken as model molecules. Electron wavepacket simulations in the frozen-molecular-vibration approximation show that the initial direction of π-electron rotation depends on the polarization direction of a linearly polarized laser pulse applied. Consecutive unidirectional rotation can be achieved by applying a sequence of linearly polarized pump and dump pulses to prevent reverse rotation. Optimal control simulations of π-electron rotation show that another controlling factor for unidirectional rotation is the relative optical phase between the different frequency components of an incident pulse in addition to photon polarization direction. Effects of nonadiabatic coupling between π-electron rotation and molecular vibrations are also presented, where the constraints of the frozen approximation are removed. The angular momentum gradually decays mainly owing to nonadiabatic coupling, while the vibrational amplitudes greatly depend on their rotation direction. This suggests that the direction of π-electron rotation on an attosecond timescale can be identified by detecting femtosecond molecular vibrations.

3.1 Introduction

Recent progress in laser technology has opened up a new research area on photo-induced ultrafast electron dynamics [1]. Laser control of electron motions is a fascinating target in optical and molecular sciences [2–10]. Electrons treated so

---

M. Kanno (✉), H. Kono, and Y. Fujimura
Department of Chemistry, Graduate School of Science, Tohoku University, Sendai 980-8578, Japan
e-mail: kanno@m.tohoku.ac.jp, hirohiko-kono@m.tohoku.ac.jp, fujimurayuichi@m.tohoku.ac.jp

K. Yamanouchi et al. (eds.), *Progress in Ultrafast Intense Laser Science VII*, Springer Series in Chemical Physics 100, DOI 10.1007/978-3-642-18327-0_3,

far can be basically classified into two groups. One involves σ electrons of small molecules, which are localized at their chemical bond. σ electrons have been primarily treated for investigation of high harmonic generation and ultrafast imaging [11–13]. The other group involves π electrons of aromatic molecules, which are delocalized in their aromatic ring. Investigation of the rotations of delocalized electrons has provided a basic design for ultrafast switching devices. For example, in quantum ring systems, quantum dynamical calculations have shown that photo-induced charge currents can be controlled by varying the time delay and strength of the pulses [14]. Optimal control of ring currents by terahertz laser pulses has been reported as a realistic approach to construct a laser-driven single-gate qubit [15]. In molecular systems, Barth et al. carried out a quantum simulation of attosecond electron dynamics in Mg porphyrin, which is one of the medium-sized aromatic molecules [16–18]. They have shown that π electrons can be rotated along the ring of the aromatic molecule by producing optically active degenerate excited states using a circularly polarized laser pulse.

In this review article, we present the theoretical results on ultrafast intense-laser-induced π-electron rotations in chiral aromatic molecules. Electrons in chiral molecules have peculiar properties originating from asymmetric potentials. For example, π electrons of a chiral aromatic molecule can be rotated along its aromatic ring by a nonhelical, linearly polarized laser pulse [19–22], and the rotation direction is intrinsic to the chiral molecule of interest because nonhelical photons have no angular momentum. Therefore, its electron dynamics directly reflects the asymmetry of the molecule. This suggests that molecular chirality can be identified by observing the rotation direction of π electrons.

In the next section, we briefly summarize the concept of electronic angular momentum of aromatic molecules with degenerate electronic states in terms of molecular orbitals (MOs) to introduce approximate angular momentum eigenstates in chiral aromatic molecules, which have no degenerate electronic states but a pair of quasidegenerate electronic states.

In Sect 3.3, we show that the initial direction of π-electron rotation in a chiral aromatic molecule depends on the polarization direction of a linearly polarized laser pulse and then π electrons continue to rotate clockwise and counterclockwise (or counterclockwise and clockwise). π electrons in chiral aromatic molecules can be rotated using a circularly polarized laser pulse as well. However, the rotation direction of π electrons is predetermined by that of the polarization plane of the applied circularly polarized pulse as in the case of achiral aromatic molecules. We also present a scenario for a consecutive unidirectional rotation of π electrons. This can be realized by applying a sequence of linearly polarized pump and dump pulses to prevent reverse rotation.

In Sect 3.4, we show the results of optimal control simulations of unidirectional π-electron rotation. The results show that another controlling factor for unidirectional rotation is the relative optical phase between the different frequency components of an incident pulse in addition to photon polarization direction.

In Sect 3.5, we present the results on nonadiabatic effects of π-electron rotations. Most of the theoretical treatments of π-electron rotations have been performed in the frozen-nuclear-motion approximation since time constants of π-electron rotations are in attoseconds, while those of nuclear motions are in femtoseconds. However, the frozen approximation breaks down when the duration of π-electron rotations becomes close to the period of molecular vibrations, and we should take into account nonadiabatic coupling effects. Nuclear wave packets (WPs) on the potential energy surfaces (PESs) of quasidegenerate electronic states created by an ultrashort pulse interfere with each other in nonadiabatic transition. The initial rotation direction of π electrons controlled by the polarization direction of the linearly polarized laser pulse determines whether the interference between the two WPs is constructive or destructive, and the interference varies the amplitudes of vibrational modes. This suggests that an enantiomer can be identified by observing the amplitudes of vibrational modes in a transient spectrum.

In the final section, we present a summary and perspectives of π-electron rotations of chiral aromatic molecules and their control.

3.2 Molecular Symmetry and Angular Momentum Eigenstates

The concept of angular momentum eigenstates is the key to understanding the mechanism of optically induced π-electron rotation in aromatic molecules. As a preparation for the following sections, we summarize in this section how photogeneration of an angular momentum eigenstate is linked with molecular symmetry and photon polarization.

3.2.1 *Angular Momentum Eigenstates: Complex and Real Orbitals*

First of all, let us begin with a description of angular momentum eigenstates of π electrons in an aromatic molecule of $D_{N\mathrm{h}}$ symmetry. The z-axis is taken to be the C_N axis. According to MO theory, complex MOs $\{|\pi_m\rangle\}$ of the molecule are given as linear combinations of atomic orbitals (LCAO-MOs) in the form [23]

$$|\pi_m\rangle = \frac{1}{N^{1/2}} \sum_{j=1}^{N} \exp\left(im\frac{2j\pi}{N}\right)|\mathrm{p}_{zj}\rangle = \frac{1}{N^{1/2}} \sum_{j=1}^{N} \exp\left(im\phi_j\right)|\mathrm{p}_{zj}\rangle, \qquad (3.1)$$

where ϕ_j and $|\mathrm{p}_{zj}\rangle$ are the azimuth and p_z orbital at the jth atom in the aromatic ring, respectively. The integer m reads $m = -N/2+1,\ldots,0,\ldots,N/2$. The energy levels of $\{|\pi_m\rangle\}$ are well known as a Frost circle [24]: $|\pi_0\rangle$ and $|\pi_{N/2}\rangle$ are the lowest and highest MOs, respectively, and for the other values of m, $|\pi_m\rangle$ and $|\pi_{-m}\rangle$ are degenerate. By approximating a molecular polygon with a complete cylindrical ring, the symmetry of the molecule becomes $D_{\infty\mathrm{h}}$ and the z component of electronic angular momentum is quantized. Note that the expansion coefficients $N^{-1/2}\exp(im\phi_j)$ in (3.1) have the same mathematical form as the eigenfunctions of the angular momentum operator $\hat{\ell}_z = -i\hbar\partial/\partial\phi$, $(2\pi)^{-1/2}\exp(im\phi)$, except for the normalization constant. Hence, the complex MO $|\pi_m\rangle$ can be regarded as an angular momentum eigenstate and its eigenvalue of $\hat{\ell}_z$ is $m\hbar$ for - $N/2+1 \leq m \leq N/2-1$ or zero for $m = N/2$. Here, real MOs $|\pi_{mx}\rangle$ and $|\pi_{my}\rangle$ are defined as linear combinations of the complex ones $|\pi_m\rangle$ and $|\pi_{-m}\rangle$:

$$|\pi_{mx}\rangle = 2^{-1/2}(|\pi_{+m}\rangle + |\pi_{-m}\rangle), \quad |\pi_{my}\rangle = -2^{-1/2}\mathrm{i}(|\pi_{+m}\rangle - |\pi_{-m}\rangle). \quad (3.2)$$

From (3.2), one readily obtains

$$|\pi_{\pm m}\rangle = 2^{-1/2}(|\pi_{mx}\rangle \pm \mathrm{i}|\pi_{my}\rangle). \quad (3.3)$$

This relation between complex and real MOs is similar to that between complex AOs $|2p_{+1}\rangle$ and $|2p_{-1}\rangle$, which are angular momentum eigenstates of an electron in a hydrogen atom, and real ones $|2p_x\rangle$ and $|2p_y\rangle$.

3.2.2 Molecular Symmetry and Angular Momentum Eigenstates

We next consider the mechanism of π-electron rotation in Mg porphyrin interacting with a circularly polarized laser pulse [16–18] as an example. Mg porphyrin belongs to the D_{4h} point group, and its highest occupied and lowest unoccupied MOs (HOMO and LUMO) are nondegenerate a_{1u} and doubly degenerate e_g orbitals, respectively [25, 26]. The degenerate LUMOs are one-electron angular momentum eigenstates with $m = \pm 1$. As for multielectron states, Mg porphyrin has doubly degenerate ^1E_u excited states, whose major components are single excitations from nondegenerate MOs such as the HOMO to the LUMOs. The degenerate excited states are viewed as the eigenstates of the multielectron angular momentum operator $\hat{L}_z$ with the quantum number $M = \pm 1$. As in the case of MOs, the multielectron angular momentum eigenstates $|^1E_{u\pm}\rangle$ with $M = \pm 1$ can be expressed as linear combinations of real excited states $|^1E_{ux}\rangle$ and $|^1E_{uy}\rangle$:

$$|^1E_{u\pm}\rangle = 2^{-1/2}\left(|^1E_{ux}\rangle \pm \mathrm{i}\,|^1E_{uy}\rangle\right). \quad (3.4)$$

When a circularly polarized laser pulse is applied to Mg porphyrin, the spin angular momentum of a photon determines to which angular momentum eigenstate an excitation occurs. This is the origin of the unique correspondence between the rotation direction of π electrons and that of the polarization plane of a circularly polarized laser pulse.

On the contrary, how can π-electron rotation be induced in a chiral aromatic molecule by a linearly polarized laser pulse? Lowering the molecular symmetry allows no two-dimensional irreducible representation E for a chiral aromatic molecule and accordingly relevant MOs or multielectron states are not degenerate. There exists no excited state that is an eigenstate of $\hat{L}_z$ in a chiral aromatic molecule. However, it is possible to transiently create approximate eigenstates of $\hat{L}_z$ when ultrashort pulses prepare a linear combination of quasidegenerate excited states, which subsequently evolves in time. With the notations $|\mathrm{L}\rangle$ and $|\mathrm{H}\rangle$ for the lower and higher of the quasidegenerate real excited states, respectively, the approximate angular momentum eigenstates $|+\rangle$ and $|-\rangle$ are expressed as

$$|\pm\rangle = 2^{-1/2}(|\mathrm{L}\rangle \pm \mathrm{i}|\mathrm{H}\rangle) \tag{3.5}$$

and the coherent nonstationary states $|+(t)\rangle$ and $|-(t)\rangle$ are

$$|\pm(t)\rangle = 2^{-1/2}\left(\mathrm{e}^{-\mathrm{i}\omega_{\mathrm{L}}t}\,|\mathrm{L}\rangle \pm \mathrm{i}\mathrm{e}^{-\mathrm{i}\omega_{\mathrm{H}}t}\,|\mathrm{H}\rangle\right), \tag{3.6}$$

where $\omega_{\mathrm{L}}(\omega_{\mathrm{H}})$ is the angular frequency of $|\mathrm{L}\rangle(|\mathrm{H}\rangle)$. The approximate angular momentum eigenstates can be transiently created within a period of the electronic state change $T \equiv 2\pi/(\omega_{\mathrm{H}} - \omega_{\mathrm{L}})$. The matrix elements $\langle\pm|\hat{L}_z|\pm\rangle$ are close to the eigenvalues $\pm\hbar$. Selective generation of an approximate angular momentum eigenstate is expected to bring about transient rotation of π electrons along an aromatic ring. The strategy for generating predominantly either $|+\rangle$ or $|-\rangle$ by a linearly polarized laser pulse, which has no spin angular momentum, will be discussed in the next section.

3.3 Laser Control of π-Electron Rotation Within a Frozen-Nuclei Model

In this section, we show that the initial direction of π-electron rotation in a chiral aromatic molecule depends on the polarization direction of a linearly polarized laser pulse and then propose a pump-dump method for performing unidirectional rotation of π electrons [19, 20]. An ansa (planar-chiral) aromatic molecule with a six-membered ring, 2,5-dichloro[*n*](3,6)pyrazinophane (DCPH; Fig. 3.1), was chosen as a model system for demonstration. The term *ansa* or *planar-chiral* is used for a chiral molecule lacking a chiral center but possessing two noncoplanar rings. The positive integer *n* specifies the length of the ansa group, ethylene bridge $(CH_2)_n$, and is set as $n \simeq 10$ to avoid conversion between enantiomers through free rotation of the aromatic ring. The molecule is assumed to be preoriented, e.g., fixed to a surface by the ansa group, and all nuclei are treated as frozen.

3.3.1 Effective Hamiltonian Formalism

The first task is to describe how to evaluate the field-free π-electronic states of chiral aromatic molecules. To obtain the qualitative picture of optically induced

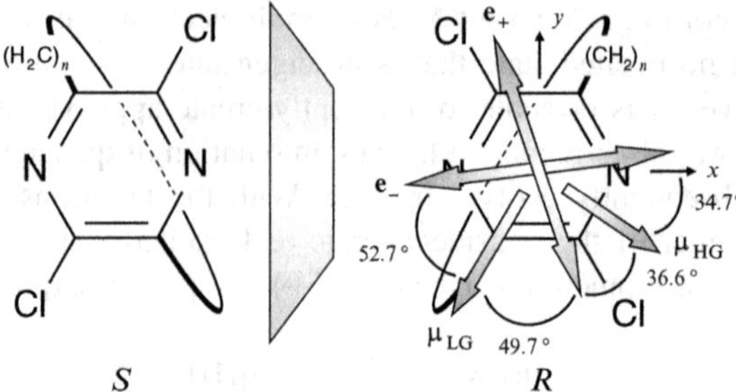

Fig. 3.1 S and R enantiomers of DCPH. The directions of transition moments $\boldsymbol{\mu}_{\mathrm{LG}}$ and $\boldsymbol{\mu}_{\mathrm{HG}}$ of an R enantiomer are shown as well as those of photon polarization vectors $\mathbf{e}_{\pm}$ defined as $\boldsymbol{\mu}_{\mathrm{LG}} \cdot \mathbf{e}_{\pm} = \pm \boldsymbol{\mu}_{\mathrm{HG}} \cdot \mathbf{e}_{\pm}$. The magnitudes of $\boldsymbol{\mu}_{\mathrm{LG}}$ and $\boldsymbol{\mu}_{\mathrm{HG}}$ are $2.02ea_0$ and $1.63ea_0$, respectively

π-electron rotation, we employ in this section the semiempirical Pariser–Parr–Pople (PPP) model to express the effective π-electronic Hamiltonian $\hat{H}_{\pi}$:

$$\hat{H}_{\pi} = \sum_{j,j',\sigma} h_{jj'} \hat{a}^{\dagger}_{j\sigma} \hat{a}_{j'\sigma} + U \sum_{j} \hat{n}_{j\uparrow} \hat{n}_{j\downarrow} + \sum_{j>j',\sigma,\sigma'} V_{jj'} \hat{n}_{j\sigma} \hat{n}_{j'\sigma'}, \tag{3.7}$$

where $\hat{a}^{\dagger}_{j\sigma}$ is a creation operator of a π electron in $|\mathrm{p}_{zj}\rangle$ with a spin σ, and $\hat{n}_{j\sigma} \equiv \hat{a}^{\dagger}_{j\sigma} \hat{a}_{j\sigma}$. The PPP model has been demonstrated to reproduce various optical properties of π-conjugated systems with an appropriate choice of parameters [27–30]. The effect of the nonaromatic ansa group is neglected in the PPP model. DCPH is therefore regarded as being of $C_{2\mathrm{h}}$ symmetry and having eight p_z orbitals (localized at four carbon, two nitrogen, and two chlorine atoms) and ten π electrons (one per carbon or nitrogen atom and two per chlorine atom) in this description of π-electron dynamics. The first term in (3.7) gives the one-electron energy and $h_{jj'}$ is the hopping integral. The second and third terms in (3.7) represent the on-site and long-range electron–electron repulsions, respectively. U is the so-called on-site Hubbard parameter. Several forms have been proposed for the long-range repulsion potential $V_{jj'}$, and we assume the Ohno form [31]

$$V_{jj'} = \frac{U}{\left[1 + \left(R_{jj'}/d\right)^2\right]^{1/2}}, \tag{3.8}$$

where $R_{jj'} \equiv |\mathbf{R}_j - \mathbf{R}_{j'}|$ with nuclear coordinates $\mathbf{R}_j$ of the jth atom and d is the unit bond length. The diagonal elements of the hopping integral $h_{jj'}$ are given by

$$h_{jj} = \alpha_j - \sum_{j'} V_{jj'} \tag{3.9}$$

and off-diagonal elements are assumed to have a nonzero value only between nearest neighbors: $h_{jj'} = \beta_{jj'}$ for $|j - j'| = 1$ and otherwise zero. Here, α_j and $\beta_{jj'}$ are determined as

$$\alpha_j = \alpha + \zeta_j \beta, \quad \beta_{jj'} = \eta_{jj'} \beta, \tag{3.10}$$

where α is the Coulomb integral at a carbon atom and β is the resonance integral between the nearest carbon atoms. Parameters ζ_j and $\eta_{jj'}$ are introduced for evaluation of the integrals for hetero atoms, halogen atoms, and carbon atoms adjacent to them.

For $U = 0$, the second and third terms in (3.7) vanish and one obtains $h_{jj} = \alpha_j$. The PPP Hamiltonian $\hat{H}_\pi$ is then reduced to the Hückel Hamiltonian $\hat{H}_{\mathrm{HMO}}$ [32]. To obtain eigenstates of $\hat{H}_\pi$, we first calculate those of $\hat{H}_{\mathrm{HMO}}$. Parameters for the one-electron energy are set as employed for linear polyenes [29, 30]: $\alpha = 0$ and $\beta = -2.40$ eV. The standard values presented by Streitwieser [33] are adopted for ζ_j and $\eta_{jj'}$: $\zeta_{\mathrm{N}} = 0.5, \zeta_{\mathrm{Cl}} = 2, \zeta_{\mathrm{C}} = 0.2$ (only for carbon atoms adjacent to a chlorine atom), $\eta_{\mathrm{CCl}} = 0.4$, and for the other atoms and bonds, $\zeta_j = 0$ and $\eta_{jj'} = 1$. Orbital energies and LCAO coefficients of eight π orbitals of DCPH are obtained by solving the secular equation with overlaps $\langle \mathrm{p}_{zj} | \mathrm{p}_{zj'} \rangle = \delta_{jj'}$ (Table 3.1). The eigenstate of $\hat{H}_{\mathrm{HMO}}$ is represented by a Slater determinant composed of LCAO-MOs.

Configuration interaction (CI) is taken into account by diagonalizing $\hat{H}_\pi$ including all singlet single and double excitations (CISD). The on-site Hubbard parameter U is set as in [29, 30]: $U = 11.1$ eV. The aromatic ring forms a planar structure, and we simply assume that the CC and CN bond lengths are 1.40 Å (∼ the CC bond length of benzene), the CCl bond length is 1.80 Å, and ∠NCCl = 120°. The unit bond length d is then 1.40 Å. Consequently, 136 singlet eigenstates of $\hat{H}_\pi$, each of which is a linear combination of those of $\hat{H}_{\mathrm{HMO}}$, are obtained by CISD using 15 single and 120 double excitations (Table 3.2).

DCPH has a pair of quasidegenerate π-electronic excited states, $|\mathrm{L}\rangle = |5^1\mathrm{B_u}\rangle$ and $|\mathrm{H}\rangle = |6^1\mathrm{B_u}\rangle$, with the energy gap $\hbar(\omega_{\mathrm{H}} - \omega_{\mathrm{L}}) = 0.11$ eV. In this semiempirical model, the angular momentum operator $\hat{L}_z$ is defined as

$$\hat{L}_z = \sum_{m,\sigma} m\hbar \hat{v}_{m\sigma}, \tag{3.11}$$

Table 3.1 Hückel MO energies of π orbitals of DCPH calculated by solving the secular equation

π Orbital	Orbital energy (eV)
$\lvert 4\mathrm{b_g}\rangle$	4.31
$\lvert 4\mathrm{a_u}\rangle$	2.25
$\lvert 3\mathrm{a_u}\rangle$ (LUMO)	1.56
$\lvert 3\mathrm{b_g}\rangle$ (HOMO)	−2.44
$\lvert 2\mathrm{b_g}\rangle$	−3.23
$\lvert 2\mathrm{a_u}\rangle$	−4.55
$\lvert 1\mathrm{b_g}\rangle$	−5.11
$\lvert 1\mathrm{a_u}\rangle$	−5.74

Table 3.2 Properties of optically allowed π-electronic excited states of DCPH whose excitation energies from the ground state $|G\rangle = |1^1A_g\rangle$ are less than 10.0 eV. The PPP Hamiltonian $\hat{H}_\pi$ was diagonalized at the level of CISD. Weight is a square of a CI coefficient of each electronic configuration

Excited state	Excitation energy (eV)	Oscillator strength	Dominant electronic configurations	Weight
$\|7^1B_u\rangle$	9.79	2.29×10^{-3}	$\|2a_u\rangle \to \|4b_g\rangle$	0.614
			$\|2a_u\rangle, \|3b_g\rangle \to \|3a_u\rangle, \|4a_u\rangle$[a]	0.140
$\|6^1B_u\rangle$	7.77	1.51	$\|1b_g\rangle \to \|4a_u\rangle$	0.460
			$\|2b_g\rangle \to \|3a_u\rangle$	0.144
			$\|2b_g\rangle \to \|4a_u\rangle$	0.103
$\|5^1B_u\rangle$	7.66	2.31	$\|2b_g\rangle \to \|4a_u\rangle$	0.334
			$\|1b_g\rangle \to \|3a_u\rangle$	0.201
			$\|3b_g\rangle \to \|4a_u\rangle$	0.167
			$\|2b_g\rangle \to \|3a_u\rangle$	0.124
$\|4^1B_u\rangle$	7.07	1.05	$\|1b_g\rangle \to \|4a_u\rangle$	0.340
			$\|2b_g\rangle \to \|3a_u\rangle$	0.241
			$\|3b_g\rangle \to \|4a_u\rangle$	0.138
$\|3^1B_u\rangle$	6.66	1.00	$\|1b_g\rangle \to \|3a_u\rangle$	0.605
			$\|2b_g\rangle \to \|4a_u\rangle$	0.169
$\|2^1B_u\rangle$	5.05	8.52×10^{-2}	$\|3b_g\rangle \to \|4a_u\rangle$	0.528
			$\|2b_g\rangle \to \|3a_u\rangle$	0.382
$\|1^1B_u\rangle$	4.31	3.46×10^{-1}	$\|3b_g\rangle \to \|3a_u\rangle$	0.662
			$\|2b_g\rangle \to \|4a_u\rangle$	0.211

[a] This double excitation consists of a singlet electron pair and a singlet hole pair [28].

where $\hat{\upsilon}_{m\sigma}$ is an occupation-number operator of π electrons in the orbital $|\pi_m\rangle$ defined by (3.1) with a spin σ and the summation is taken over $-N/2+1 \leq m \leq N/2-1$. The approximate eigenstates of $\hat{L}_z$, $|+\rangle$ and $|-\rangle$, in DCPH, consist of the quasidegenerate excited states $|L\rangle$ and $|H\rangle$ as in (3.5), where $\langle\pm|\hat{L}_z|\pm\rangle = \pm 0.86\hbar$. π electrons with positive (negative) angular momentum travel counterclockwise (clockwise) around the ring in Fig. 3.1.

3.3.2 *Time Evolution of π-Electron WPs*

The time-dependent Hamiltonian of an aromatic molecule interacting with a classical laser field $\boldsymbol{\varepsilon}(t)$ is expressed in the length gauge under the dipole approximation as

$$\hat{H}(t) = \hat{H}_\pi - \hat{\boldsymbol{\mu}} \cdot \boldsymbol{\varepsilon}(t), \tag{3.12}$$

where $\hat{\boldsymbol{\mu}}$ is the electric dipole moment operator. In this semiempirical model, $\hat{\boldsymbol{\mu}}$ is expanded in terms of $\{\hat{n}_{j\sigma}\}$ as

$$\hat{\boldsymbol{\mu}} = -e\sum_{j,\sigma} \mathbf{R}_j \hat{n}_{j\sigma}. \tag{3.13}$$

This implies that a π electron occupying $|p_{zj}\rangle$ is assumed to be localized just at the nuclear coordinate $\mathbf{R}_j$. The time-dependent Schrödinger equation (TDSE) for a π-electron WP is

$$i\hbar\frac{\partial}{\partial t}\left|\Psi(t)\right\rangle = \hat{H}(t)|\Psi(t)\rangle \tag{3.14}$$

with the initial condition $|\Psi(0)\rangle = |\mathrm{G}\rangle$, where $|\mathrm{G}\rangle = |1^1\mathrm{A_g}\rangle$ is the ground state. We solve (3.14) by expanding $|\Psi(t)\rangle$ in terms of 136 singlet eigenstates $\{|k\rangle\}$ of $\hat{H}_\pi$ obtained at the level of CISD:

$$|\Psi(t)\rangle = \sum_k c_k(t)\mathrm{e}^{-\mathrm{i}\omega_k t}|k\rangle \tag{3.15}$$

with $\omega_\mathrm{G} = 0$. By inserting (3.15) into (3.14), we derive the coupled equations of motion for the expansion coefficients $\{c_k(t)\}$:

$$i\hbar\frac{\mathrm{d}c_k(t)}{\mathrm{d}t} = -\sum_{k'} c_{k'}(t)\mathrm{e}^{\mathrm{i}(\omega_k-\omega_{k'})t}\boldsymbol{\mu}_{kk'}\cdot\boldsymbol{\varepsilon}(t), \tag{3.16}$$

where $\boldsymbol{\mu}_{kk'} \equiv \langle k|\hat{\boldsymbol{\mu}}|k'\rangle$. The coupled equations can be solved numerically with a conventional algorithm, e.g., Runge–Kutta method.

π-electron rotation can be quantified by the angular momentum expectation value $L_z(t) \equiv \langle\Psi(t)|\hat{L}_z|\Psi(t)\rangle$, which is calculated by additions and multiplications of LCAO coefficients, CI coefficients, phase factors $\{\mathrm{e}^{-\mathrm{i}\omega_k t}\}$, and expansion coefficients $\{c_k(t)\}$ in this semiempirical model. Here, in a circular motion of a particle, angular velocity of the particle is equivalent to its angular momentum divided by the mass of the particle and the square of the circulation radius. Thus, we also define the rotational angle of π electrons, $\phi(t)$, as

$$\phi(t) \equiv \frac{1}{m_\mathrm{e}b^2}\int_0^t \mathrm{d}t' L_z(t'), \tag{3.17}$$

where b is the radius of the ring. In the case of a six-membered ring, b is equal to the unit bond length d. Integration with respect to t' in (3.17) is implemented numerically using Simpson's rule. The expectation values $L_z(t)$ and $\phi(t)$ are utilized as measures of π-electron rotation.

3.3.3 Generation Scheme for Approximate Angular Momentum Eigenstates

We now design a linearly polarized laser pulse to transfer as much of the population as possible from $|\mathrm{G}\rangle$ to either $|+\rangle$ or $|-\rangle$. The linearly polarized laser pulse $\boldsymbol{\varepsilon}(t)$ is assumed to be of the form

$$\boldsymbol{\varepsilon}(t) = f\sin^2(\pi t/t_\mathrm{d})\cos(\omega t)\mathbf{e} \tag{3.18}$$

for $0 < t < t_d$ and otherwise zero. Here, f is the peak intensity, t_d is the pulse duration, ω is the central frequency, and $\mathbf{e}$ is the polarization unit vector.

First, the central frequency ω is resonant with the average energy of the quasidegenerate states: $\omega = \omega_L + \Delta\omega$, where $2\Delta\omega \equiv \omega_H - \omega_L$. Next, the polarization vector $\mathbf{e}$ is determined by its alignment with respect to the transition electric dipole moments $\boldsymbol{\mu}_{LG}$ and $\boldsymbol{\mu}_{HG}$. This is obtained from a three-level model analysis in the short-pulse limit: $|L\rangle$ and $|H\rangle$ are independently coupled to $|G\rangle$ by a linearly polarized laser pulse with a Dirac-delta-function-like envelope. The polarization vector $\mathbf{e}$ is chosen in two ways, $\mathbf{e}_+$ or $\mathbf{e}_-$ defined as $\boldsymbol{\mu}_{LG} \cdot \mathbf{e}_\pm = \pm \boldsymbol{\mu}_{HG} \cdot \mathbf{e}_\pm$ for each enantiomer. The directions of $\mathbf{e}_\pm$ for an R enantiomer as well as those of $\boldsymbol{\mu}_{LG}$ and $\boldsymbol{\mu}_{HG}$ are illustrated in Fig. 3.1. At the moment of irradiation ($t = t_i$), the pulse with $\mathbf{e}_+(\mathbf{e}_-)$ produces an in-phase superposition $|L\rangle + |H\rangle$ (out-of-phase superposition $|L\rangle - |H\rangle$) in $|\Psi(t_i)\rangle$. At $t > t_i$, the electron WP propagates freely. Hence, $|L\rangle \pm |H\rangle$ in $|\Psi(t_i)\rangle$ temporally evolves as $|L\rangle \pm e^{-i\theta(t)}|H\rangle$ except for the global phase factor, where $\theta(t) \equiv 2\Delta\omega(t - t_i)$. Relative phase factor $e^{-i\theta(t)}$ changes as $+1 \rightarrow -i \rightarrow -1 \rightarrow +i \rightarrow +1 \rightarrow \ldots$ with the progression of $t - t_i, 0 \rightarrow T/4 \rightarrow T/2 \rightarrow 3T/4 \rightarrow T \rightarrow \ldots$, where $T \equiv \pi/\Delta\omega$. This indicates that in the first quarter period of T after excitation $|L\rangle \mp i|H\rangle$, namely, $|\mp\rangle$ is created. The initial direction of π-electron rotation depends on the polarization direction. Afterward, the rotation direction switches between clockwise and counterclockwise with the period T. If a molecule is highly symmetric, e.g., benzene, $e^{-i\theta(t)}$ takes an infinite time to reach $-i$ since $\Delta\omega = 0$. That is, lowering the molecular symmetry is essential for the selective generation of either $|+\rangle$ or $|-\rangle$ by a linearly polarized laser pulse. Finally, the peak intensity f and the pulse duration t_d are determined following the idea of the so-called π pulse [34].

3.3.4 Single-Pulse Control

In this section, we briefly present theoretical results on a single-pulse control of π-electron rotations. To demonstrate the control method based on the three-level model analysis, a numerical simulation within 136-state expansion was carried out for an R enantiomer with a linearly polarized laser pulse $\boldsymbol{\varepsilon}(t)$ designed to initially create $|-\rangle$. The values of the laser parameters employed in the simulation were $f = 1.63\,\mathrm{GVm}^{-1}$, $t_d = 26.6\,\mathrm{fs}$, $\omega = 7.72\,\mathrm{eV}/\hbar$, and $\mathbf{e} = \mathbf{e}_+$.

Figure 3.2a shows the temporal behavior in $\boldsymbol{\varepsilon}(t)$. In Fig. 3.2b, the solid, dotted, and dash-dotted lines denote the temporal behavior in the populations of $|G\rangle, |+\rangle$, and $|-\rangle$, respectively. We use the notations $P_k(t) \equiv |\langle k|\Psi(t)\rangle|^2 (k = \mathrm{G}, +, \text{and} -)$ for those populations. The expectation values $L_z(t)$ and $\phi(t)$ are plotted in Fig. 3.2c, d, respectively. If the pulse duration t_d is less than the oscillation period T, the pulse peak $t_d/2$ can be regarded as the moment of irradiation t_i in the short-pulse limit, although $P_+(t_d/2)$ and $P_-(t_d/2)$ are not exactly equal. At $t > t_d/2 = 13.3\,\mathrm{fs}$, a significant amount of the population is transferred to $|L\rangle - i|H\rangle$, i.e., $|-\rangle$, and accordingly π electrons start to rotate clockwise. When the

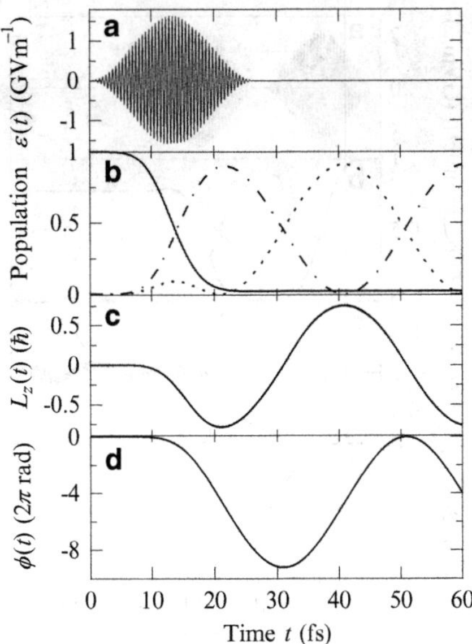

Fig. 3.2 (**a**) The linearly polarized laser pulse $\boldsymbol{\varepsilon}(t)$ to initially create $|-\rangle$ of an R enantiomer. The polarization vector of $\boldsymbol{\varepsilon}(t)$ is $\mathbf{e}_+$. (**b**) Temporal behavior in the populations of $|\mathrm{G}\rangle$ (*solid line*), $|+\rangle$ (*dotted line*), and $|-\rangle$ (*dash-dotted line*) denoted as $P_{\mathrm{G}}(t)$, $P_+(t)$, and $P_-(t)$, respectively. (**c**) Expectation value of angular momentum $L_z(t)$. (**d**) Expectation value of rotational angle $\phi(t)$

laser pulse ceases at $t = t_{\mathrm{d}} = 26.6\,\mathrm{fs}$, the total population of π electrons in $|+\rangle$ and $|-\rangle$, $P_+(t_{\mathrm{d}}) + P_-(t_{\mathrm{d}})$, reaches 0.91. From the energy-time uncertainty relation, $P_+(t_{\mathrm{d}}) + P_-(t_{\mathrm{d}})$ is maximum at the pulse duration $t_{\mathrm{d}} = 26.6\,\mathrm{fs}$: A smaller bandwidth of a longer pulse does not cover the energy gap $2\hbar\Delta\omega$ sufficiently, and, on the contrary, a broader bandwidth of a shorter pulse populates other excited states. At $t > 26.6$ fs, the population of 0.91 is exchanged between $|+\rangle$ and $|-\rangle$ since the system is isolated and the laser field is absent. $L_z(t)$ and $\phi(t)$ thus oscillate with the period of $T = 39.5\,\mathrm{fs}$, and π electrons are estimated to circulate around the ring more than nine times within this period.

3.3.5 Pump-Dump Control

In this section, we present an outline of a pump-dump method for performing consecutive unidirectional rotation of π electrons. The three-level model analysis in a short-pulse limit also provides a simple control scheme for determining unidirectional rotation of π electrons. As already stated, the pulse with $\mathbf{e}_+$ ($\mathbf{e}_-$) triggers the transition between $|\mathrm{G}\rangle$ and $|\mathrm{L}\rangle + |\mathrm{H}\rangle$($|\mathrm{L}\rangle - |\mathrm{H}\rangle$), and $|\mathrm{L}\rangle + |\mathrm{H}\rangle$ created by a pump pulse with $\mathbf{e}_+$ evolves as $|\mathrm{L}\rangle + |\mathrm{H}\rangle \rightarrow |\mathrm{L}\rangle - \mathrm{i}|\mathrm{H}\rangle$. Then, the population in $|-\rangle$ can be dumped to $|\mathrm{G}\rangle$ by applying a dump pulse with $\mathbf{e}_-$ just after the created state has

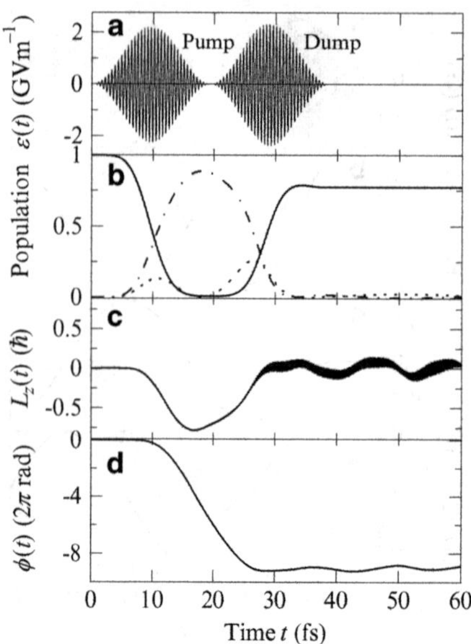

Fig. 3.3 (**a**) Pump and dump pulses for clockwise π-electron rotation in an R enantiomer. The polarization vectors of the pump and dump pulses are $\mathbf{e}_+$ and $\mathbf{e}_-$, respectively. (**b**) Temporal behavior in the populations of $|\mathrm{G}\rangle$ (*solid line*), $|+\rangle$ (*dotted line*), and $|-\rangle$ (*dash-dotted line*) denoted as $P_\mathrm{G}(t)$, $P_+(t)$, and $P_-(t)$, respectively. (**c**) Expectation value of angular momentum $L_z(t)$. (**d**) Expectation value of rotational angle $\phi(t)$

completely shifted as $|\mathrm{L}\rangle - \mathrm{i}|\mathrm{H}\rangle \rightarrow |\mathrm{L}\rangle - |\mathrm{H}\rangle$. Figure 3.3 shows the results of a pump-dump control simulation of an R enantiomer. The values of the parameters of the pump pulse were $f = 2.24\,\mathrm{GVm}^{-1}$, $t_\mathrm{d} = 19.4\,\mathrm{fs}$, $\omega = 7.72\,\mathrm{eV}/\hbar$, and $\mathbf{e} = \mathbf{e}_+$; those of the dump pulse were $f = 2.37\,\mathrm{GVm}^{-1}$, $t_\mathrm{d} = 19.4\,\mathrm{fs}$, $\omega = 7.72\,\mathrm{eV}/\hbar$ and $\mathbf{e} = \mathbf{e}_-$. The delay time between the pulses was 19.4 fs.

After $|-\rangle$ is generated, we have $P_+(t) \simeq P_-(t)$ around the peak of the dump pulse at $t = 29.1$ fs; in other words, an out-of-phase superposition $|\mathrm{L}\rangle - |\mathrm{H}\rangle$ is created. At $t > 29.1$ fs, most of the population is dumped to $|\mathrm{G}\rangle$; the rest is brought to higher excited states. Consequently, the value of $L_z(t)$ is almost zero and reverse rotation is successfully prevented. A pair of pump and dump pulses realizes unidirectional rotation of π electrons. Moreover, repetition of the unidirectional rotation can be achieved by a sequence of pulse pairs.

π-electron rotation in an S enantiomer can be controlled in the same way. By reflecting the polarization directions of the pump and dump pulses to a mirror plane as an R enantiomer is converted to an S enantiomer, π electrons in an S enantiomer are rotated counterclockwise in Fig. 3.1. Adjusting the alignments of S and R enantiomers with respect to the polarization directions makes it possible to produce photocurrents in opposite directions or a photocurrent in only one of the enantiomers.

3.4 Optimal Control of Unidirectional π-Electron Rotation

The pulse-design scheme developed in Sect. 3.3 is valid only for an ultrashort linearly polarized single-peaked laser pulse, e.g., a $\sin^2$ or Gaussian short pulse such as we have used in the calculations for Figs. 3.2 and 3.3. It is not clear yet what are the significant factors for determination of the rotation direction without restriction on the envelope of a laser pulse. In this section, we present the results of optimal control simulations of π-electron rotation in a chiral aromatic molecule [21]. Optimal control simulation, which reveals the best wave profile of a laser pulse for achieving a target state under a given constraint, is the most suitable approach to investigate how π-electron rotation can be controlled in an efficient way with an appropriate choice of laser parameters.

3.4.1 Optimal Control Theory

The optimal control theory (OCT) is a powerful mathematical tool for designing a laser pulse optimized to lead a quantum system to a desired target state [35–39]. In the OCT, a positive definite operator $\hat{O}$, which takes a maximum expectation value when the system reaches a target state, is introduced. The optimal control pulse $\boldsymbol{\varepsilon}(t)$ is designed so as to maximize the expectation value of $\hat{O}$ at the final time t_d subject to minimum laser energy. That is, $\boldsymbol{\varepsilon}(t)$ is defined as the laser pulse that gives a maximum value to the objective functional

$$J = \langle \Psi(t_\mathrm{d})|\hat{O}|\Psi(t_\mathrm{d})\rangle - \int_0^{t_\mathrm{d}} \mathrm{d}t \frac{[\boldsymbol{\varepsilon}(t)]^2}{\hbar A(t)} - 2\mathrm{Re} \int_0^{t_\mathrm{d}} \mathrm{d}t \left\langle \Lambda(t) \left| \left[\frac{\partial}{\partial t} - \frac{1}{\mathrm{i}\hbar}\hat{H}(t) \right] \right| \Psi(t) \right\rangle, \tag{3.19}$$

where $A(t)$ is a positive function to weigh the penalty for laser energy and $|\Lambda(t)\rangle$ is a Lagrange multiplier for $|\Psi(t)\rangle$ to satisfy (3.14). According to the variational procedure, the requirement of $\delta J = 0$ gives the equation that $|\Lambda(t)\rangle$ satisfies:

$$\mathrm{i}\hbar \frac{\partial}{\partial t}|\Lambda(t)\rangle = \hat{H}(t)|\Lambda(t)\rangle \tag{3.20}$$

with the final condition $|\Lambda(t_\mathrm{d})\rangle = \hat{O}|\Psi(t_\mathrm{d})\rangle$. The explicit form of the optimal control pulse $\boldsymbol{\varepsilon}(t)$ is eventually expressed as

$$\boldsymbol{\varepsilon}(t) = -A(t)\mathrm{Im}[\langle \Lambda(t)|\hat{\boldsymbol{\mu}}|\Psi(t)\rangle \cdot \mathbf{e}]\mathbf{e}. \tag{3.21}$$

The coupled equations (3.14), (3.20), and (3.21) need to be solved simultaneously by means of an iteration algorithm [36]. Starting with the initial condition, we perform time propagation of $|\Psi(t)\rangle$ forward by numerically integrating (3.16) with an initial trial pulse. Then, Lagrange multiplier at $t = t_\mathrm{d}$ is set using the π-electron WP at $t = t_\mathrm{d}$, and backward time propagation of $|\Lambda(t)\rangle$ is implemented; $|\Lambda(t)\rangle$ is expanded in terms of $\{|k\rangle\}$, and the counterpart of (3.16) for $|\Lambda(t)\rangle$ is solved. Forward time propagation of $|\Psi(t)\rangle$ is subsequently carried out. The laser pulse $\boldsymbol{\varepsilon}(t)$ is calculated from (3.21) in both the backward and forward time propagations. This iteration step is repeated until convergence is achieved. The laser pulse finally obtained is regarded as an optimal control pulse.

3.4.2 Optimal Control for Counterclockwise Rotation at $e = e_-$

The results in Figs. 3.2 and 3.3 indicate that the contribution of higher excited states to π-electron rotation in DCPH is very small. In this section, $|\Psi(t)\rangle$ is thus expanded in terms of 17 singlet eigenstates of $\hat{H}_\pi$ with $\hbar\omega_k < 10.0\,\text{eV}$, which is a typical value of the first ionization energy of aromatic molecules. We aim to obtain as much angular momentum as possible at the end of control and set $\hat{O}$ simply as the projection operator onto an approximate angular momentum eigenstate: $\hat{O} = |+\rangle\langle+|(|-\rangle\langle-|)$ for counterclockwise (clockwise) π-electron rotation. The control time (pulse duration) is chosen to be equal to the oscillation period corresponding to the energy gap between $|\text{L}\rangle$ and $|\text{H}\rangle$: $t_\text{d} = T = 39.5\,\text{fs}$. The penalty function $A(t)$ should be set to a smooth one starting and ending at zero so that the envelope of an optimal control pulse shows adiabatic ramp and decay. Here, it is taken to be of the form

$$A(t) = F\sin^2(\pi t/t_\text{d}). \tag{3.22}$$

We adopt $F = 1.5 \times 10^{-2} E_\text{h}/(ea_0)^2$. The initial trial pulse is a π pulse [34] of the form in (3.18). Convergence is assumed when ΔJ, the difference in J between the last two iteration steps, satisfies $\Delta J/J < 10^{-6}$.

The results of an optimal control simulation within 17-state expansion for counterclockwise π-electron rotation at $\mathbf{e} = \mathbf{e}_-$ in an R enantiomer are shown in Fig. 3.4. The wave profile of the optimal control pulse $\boldsymbol{\varepsilon}(t)$ in Fig. 3.4a is different from that of an ultrashort single-peaked laser pulse. The inset of Fig. 3.4a displays the power spectrum of $\boldsymbol{\varepsilon}(t), S(\omega)$, defined as the absolute square of the Fourier transform of $\boldsymbol{\varepsilon}(t)$:

$$S(\omega) \equiv \left| \int_0^{t_\text{d}} \text{d}t e^{-\text{i}\omega t} \boldsymbol{\varepsilon}(t) \right|^2. \tag{3.23}$$

$S(\omega)$ shows two main peaks with the same intensity at $\hbar\omega = 7.63$ and $7.80\,\text{eV}$ that are almost equal to $\hbar\omega_\text{L}$ and $\hbar\omega_\text{H}$, respectively. Therefore, $\boldsymbol{\varepsilon}(t)$ in Fig. 3.4a can be approximated by a sum of ultrashort single-peaked laser pulses with frequencies ω_L and ω_H:

$$\boldsymbol{\varepsilon}(t) \simeq f\sin^2(\pi t/t_\text{d})[\cos(\omega_\text{L}t + \varphi_\text{L}) + \cos(\omega_\text{H}t + \varphi_\text{H})]\mathbf{e}, \tag{3.24}$$

where φ_k (k = L and H) is the optical phase of the pulse with frequency ω_k. Equation (3.24) can be rewritten as

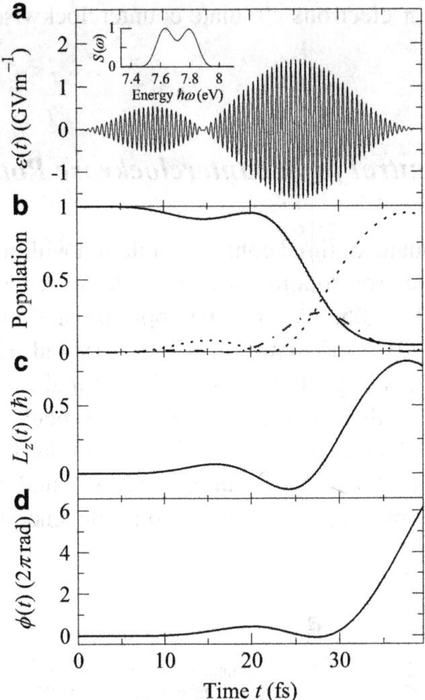

Fig. 3.4 (**a**) Optimal control pulse $\boldsymbol{\varepsilon}(t)$ for counterclockwise π-electron rotation at $\mathbf{e} = \mathbf{e}_-$ in an R enantiomer. *Inset*: Power spectrum of the optimal control pulse $\boldsymbol{\varepsilon}(t)$, $S(\omega)$, defined by (3.23). The values of $S(\omega)$ are scaled so that the maximum value is unity. (**b**) Temporal behavior in the populations of $|\mathrm{G}\rangle$ (*solid line*), $|+\rangle$ (*dotted line*), and $|-\rangle$ (*dash-dotted line*) denoted as $P_{\mathrm{G}}(t)$, $P_+(t)$, and $P_-(t)$, respectively. (**c**) Expectation value of angular momentum $L_z(t)$. (**d**) Expectation value of rotational angle $\phi(t)$

$$\boldsymbol{\varepsilon}(t) \simeq 2f \sin^2(\pi t/t_{\mathrm{d}}) \cos(\Delta\omega t + \Delta\varphi) \cos[(\omega_{\mathrm{L}} + \Delta\omega)t + \varphi_{\mathrm{L}} + \Delta\varphi]\mathbf{e}, \quad (3.25)$$

where $\Delta\varphi \equiv (\varphi_{\mathrm{H}} - \varphi_{\mathrm{L}})/2$. Equation (3.25) consists of two parts: a slowly varying part $2f \sin^2(\pi t/t_{\mathrm{d}}) \cos(\Delta\omega t + \Delta\varphi)$, which works as an envelope of $\boldsymbol{\varepsilon}(t)$, and a rapidly oscillating part $\cos[(\omega_{\mathrm{L}} + \Delta\omega)t + \varphi_{\mathrm{L}} + \Delta\varphi]$, resonant with the average energy of the quasidegenerate states. Now, f and $\Delta\varphi$ that specify the envelope of $\boldsymbol{\varepsilon}(t)$ are important. For $\boldsymbol{\varepsilon}(t)$ in Fig. 3.4a, we estimate that $f \simeq 1.34$ GVm^{-1} and $\Delta\varphi \simeq 0.13\pi$ rad.

Throughout the control, $P_{\mathrm{G}}(t) + P_+(t) + P_-(t) \simeq 1$, that is, the system behaves similar to a three-level one. At $t < 15$ fs, a small amount of the population is excited from $|\mathrm{G}\rangle$ and $P_+(t)$ exceeds $P_-(t)$. Accordingly, π electrons start to rotate counterclockwise around the ring in Fig. 3.1, following which the population created in $|+\rangle$ completely shifts to $|-\rangle$ and the rotation direction of π electrons is reversed. At $t > 23$ fs, the optimal control pulse $\boldsymbol{\varepsilon}(t)$ approaches and then passes its peak intensity. The population remaining in $|\mathrm{G}\rangle$ is drastically pumped up and that in $|-\rangle$ also moves to $|+\rangle$. In parallel, $P_+(t)$ exhibits a sharp rise and finally, we have $P_+(t_d) = 0.93$. π electrons circulate counterclockwise around the ring with $L_z(t_d) = 0.78\hbar$.

3.4.3 Optimal Control for Counterclockwise Rotation at $e = e_+$

We have also carried out an optimal control simulation within 17-state expansion for counterclockwise π-electron rotation at $\mathbf{e} = \mathbf{e}_+$ in an R enantiomer. The optimal control pulse $\boldsymbol{\varepsilon}(t)$ in Fig. 3.5a can also be approximated by (3.24) or (3.25). We estimate that $f \simeq 0.90\,\mathrm{GVm}^{-1}$ and $\Delta\varphi \simeq -0.24\pi$ rad. Here, the magnitude of $\Delta\varphi$ for $\boldsymbol{\varepsilon}(t)$ in Fig. 3.5a is larger than that for the optimal control pulse at $\mathbf{e} = \mathbf{e}_-$ in Fig. 3.4a, and thus the oscillation of $\boldsymbol{\varepsilon}(t)$ in Fig. 3.5a is negligibly small at $t > 29$ fs. Consequently, the envelope of $\boldsymbol{\varepsilon}(t)$ in Fig. 3.5a is similar to that in Fig. 3.2a and, in the power spectrum $S(\omega)$, only a single peak is found at $\hbar\omega = 7.71$ eV that is almost equal to the average energy of the quasidegenerate states $\hbar(\omega_L + \Delta\omega)$.

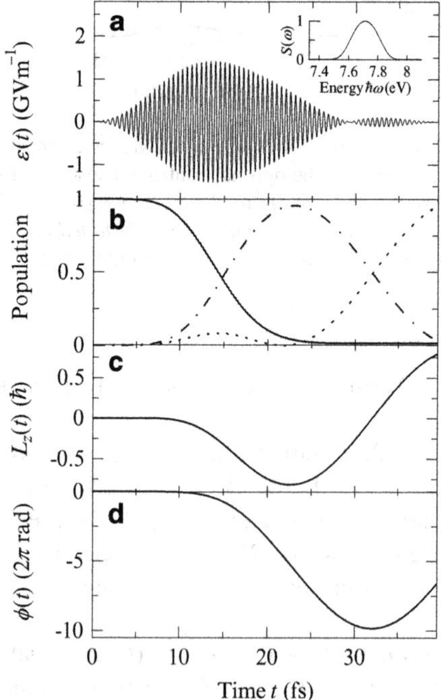

Fig. 3.5 (**a**) Optimal control pulse $\boldsymbol{\varepsilon}(t)$ for counterclockwise π-electron rotation at $\mathbf{e} = \mathbf{e}_+$ in an R enantiomer. *Inset*: Power spectrum of the optimal control pulse $\boldsymbol{\varepsilon}(t)$, $S(\omega)$, defined by (3.23). The values of $S(\omega)$ are scaled so that the maximum value is unity. (**b**) Temporal behavior in the populations of $|G\rangle$ (*solid line*), $|+\rangle$ (*dotted line*), and $|-\rangle$ (*dash-dotted line*) denoted as $P_G(t)$, $P_+(t)$, and $P_-(t)$, respectively. (**c**) Expectation value of angular momentum $L_z(t)$. (**d**) Expectation value of rotational angle $\phi(t)$

Temporal behaviors in $P_k(t)(k = \mathrm{G}, +, \text{and} -)$, $L_z(t)$, and $\phi(t)$ resemble those in Fig. 3.2b–d, respectively. Eventually, $P_+(t_\mathrm{d})$ and $L_z(t_\mathrm{d})$ are as large as those in Fig. 3.4b, c, respectively, although $\phi(t_\mathrm{d})$ still has a negative value.

3.4.4 *Optimal Control for Clockwise Rotation*

The optimal control pulse $\boldsymbol{\varepsilon}(t)$ for clockwise π-electron rotation at $\mathbf{e} = \mathbf{e}_+(\mathbf{e}_-)$ is very similar to that for counterclockwise π-electron rotation at $\mathbf{e} = \mathbf{e}_-(\mathbf{e}_+)$. Temporal behavior in $P_k(t)(k = G, +, \text{and} -)$ for clockwise π-electron rotation at $\mathbf{e} = \mathbf{e}_+(\mathbf{e}_-)$ is also analogous to that for counterclockwise π-electron rotation at $\mathbf{e} = \mathbf{e}_-(\mathbf{e}_+)$ except that the dotted and dash-dotted lines in Fig. 3.4b (Fig. 3.5b) are switched to signify $P_-(t)$ and $P_+(t)$, respectively. Hence, the values of $L_z(t)$ and $\phi(t)$ for clockwise π-electron rotation at $\mathbf{e} = \mathbf{e}_+(\mathbf{e}_-)$ are given just by inverting their signs in Fig. 3.4c, d (Fig. 3.5c, d), respectively.

3.4.5 *Characteristics of Control by a Single-Color or Two-Color Laser Pulse*

As clearly shown in Figs. 3.4 and 3.5, π-electron rotation can be controlled efficiently by applying ultrashort single-peaked laser pulses with central frequencies ω_L and ω_H simultaneously. The relative optical phase $\Delta\varphi$ in (3.25) is a crucial parameter for determination of the rotation direction of π electrons. By setting an appropriate value to $\Delta\varphi$, π electrons can be rotated in an intended direction at the final time t_d regardless of whether $\mathbf{e} = \mathbf{e}_+$ or $\mathbf{e}_-$. Moreover, as clearly seen in the insets of Figs. 3.4a and 3.5a, the value of $\Delta\varphi$ also determines the spectral distribution of the combined pulse: For the control time t_d close to $T \equiv \pi/\Delta\omega$, the optimal control pulse $\boldsymbol{\varepsilon}(t)$ for counterclockwise (clockwise) π-electron rotation at $\mathbf{e} = \mathbf{e}_+(\mathbf{e}_-)$ is a single-color laser pulse, while that at $\mathbf{e} = \mathbf{e}_-(\mathbf{e}_+)$ is a two-color laser pulse.

Each of the polarization vectors has its "preferred" rotation direction: counterclockwise for $\mathbf{e}_-$ and clockwise for $\mathbf{e}_+$. After a linearly polarized laser pulse is turned on, π electrons start to rotate in the preferred rotation direction defined by the polarization vector, regardless of whether the applied pulse is a single- or two-color laser pulse. Then population transfer occurs between $|+\rangle$ and $|-\rangle$ and, therefore $L_z(t)$ oscillates between positive and negative values. Finally, π electrons circulate in the rotation direction specified by the target state with a large amount of angular momentum at the end of the control. The difference between the two cases is that $P_+(t) + P_-(t)$ does not change throughout population oscillation between $|+\rangle$ and $|-\rangle$ for excitation with a single-color laser pulse but increases during control by a two-color laser pulse.

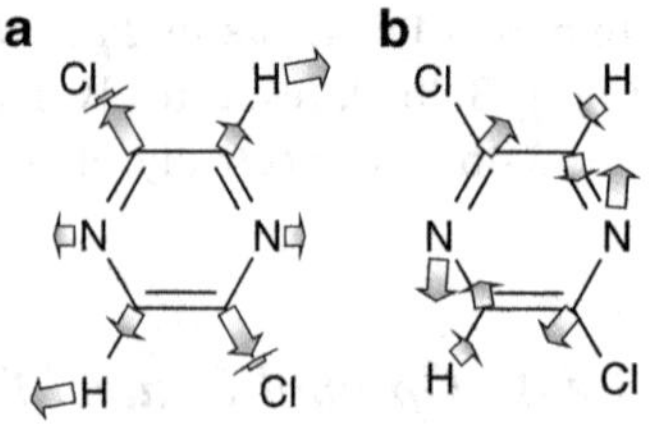

Fig. 3.6 Molecular formula of DCP. Vibrational vectors of the (**a**) breathing and (**b**) distortion modes of DCP are indicated by *arrows*

3.5 Nonadiabatic Effects

In previous sections, we treated π-electron rotation under a vibrationally frozen condition. When π-electron rotation lasts as long as the period of molecular vibrations (several tens of femtoseconds), the electronic and nuclear motions may be coupled to each other. Therefore, π-electron rotation should be subjected to undergo non-negligible nonadiabatic perturbations by nuclear motions. In this section, we present the results of WP simulations of nonadiabatic dynamics in a model chiral aromatic molecule irradiated by a linearly polarized laser pulse based on ab initio MO methods [22]. To reduce computational costs, we replaced the ansa group with hydrogen atoms. The simplified model molecule 2,5-dichloropyrazine (DCP; Fig. 3.6) is not chiral in a strict sense but valid because π electrons cannot be directly affected by the ansa group with σ electrons and the atomic configuration of the aromatic ring still differs between the two rotation directions.

3.5.1 Electronic Structure at the Optimized Geometry

All ab initio electronic structure calculations in this section were performed using the quantum chemistry program MOLPRO [40] with the 6–31 G* Gaussian basis set [41]. Geometry optimization for the ground state of DCP was carried out at the level of the second-order Møller–Plesset perturbation theory (MP2) [41] followed by a single-point ground- and excited-state calculation at the complete-active-space self-consistent field (CASSCF) [41] level with ten active electrons and eight active orbitals (Table 3.3). DCP is of $C_{2\mathrm{h}}$ symmetry at the optimized geometry of the ground state $|\mathrm{G}\rangle = |1^1\mathrm{A_g}\rangle$ and has a pair of optically allowed quasidegenerate excited states, $|\mathrm{L}\rangle = |3^1\mathrm{B_u}\rangle$ and $|\mathrm{H}\rangle = |4^1\mathrm{B_u}\rangle$, with the energy gap $2\hbar\Delta\omega = 0.44\,\mathrm{eV}$. In ab initio MO methods, the angular momentum operator $\hat{L}_z$ is expressed in terms of the partial differential operators with respect to electronic coordinates according to its strict definition. The approximate angular momentum eigenstates $|+\rangle$ and $|-\rangle$ in DCP are superpositions of the quasidegenerate excited states $|\mathrm{L}\rangle$ and $|\mathrm{H}\rangle$ as in (3.5), where $\langle\pm|\hat{L}_z|\pm\rangle = \pm 0.98\hbar$.π electrons with positive (negative) angular momentum travel counterclockwise (clockwise) around the ring in Fig. 3.6.

Table 3.3 Properties of optically allowed π-electronic excited states of DCP whose excitation energies from $|G\rangle = |1^1A_g\rangle$ are less than 10.0 eV. The ab initio geometry optimization for $|G\rangle$ and succeeding single-point calculation were done at the MP2/6–31G* and CASSCF(10,8)/6–31G* levels, respectively

Excited state	Excitation energy (eV)	Oscillator strength
$\|4^1B_u\rangle$	9.84	1.81
$\|3^1B_u\rangle$	9.40	1.90
$\|2^1B_u\rangle$	8.04	1.31
$\|1^1B_u\rangle$	4.78	1.73×10^{-1}

3.5.2 Effective Vibrational Degrees of Freedom

The effective vibrational degrees of freedom for the WP simulations were determined by performing geometry optimization for $|L\rangle$ and $|H\rangle$ at the CASSCF(10,8) level, and it was found that DCP is also of C_{2h} symmetry at the optimized geometry of $|L\rangle$ and that of $|H\rangle$. Hence, vibrational modes with displacements from the optimized geometry of $|G\rangle$ to that of $|L\rangle$ and $|H\rangle$ are totally symmetric modes. Furthermore, vibrational modes that couple two ^1B_u states are also totally symmetric A_g modes. For these reasons, we consider two types of A_g normal modes with large potential displacements and nonadiabatic coupling matrix element, namely, breathing and distortion modes (Fig. 3.6a, b), whose ground-state harmonic wave numbers are 1,160 and 1,570 cm^{-1}, respectively. The two-dimensional adiabatic PESs of $|L\rangle$ and $|H\rangle$ with respect to the breathing and distortion modes were computed at the CASSCF(10,8) level. There exists an avoided crossing between the PESs. We confirmed by a calculation at the level of the second-order CAS perturbation theory (CASPT2) [41] that the avoided crossing remains unchanged when dynamical electron correlation is taken into account, while the PESs are lowered by $\sim$ 3 eV.

3.5.3 Time Evolution of Nuclear WPs

The results in Figs. 3.2–3.5 indicate that the system can be treated as a three-level one consisting of $|G\rangle, |L\rangle$, and $|H\rangle$. The initial nuclear WP was set to be the vibrational ground-state wave function of $|G\rangle$ and the system is then electronically excited by a single-color linearly polarized laser pulse $\boldsymbol{\varepsilon}(t)$ of the form in (3.18). To include the effects of the nonadiabatic coupling on the WP propagation, we expanded the state vector of the system in terms of the three diabatic states $\{|k^D\rangle\}$, constructed as a linear combination of the adiabatic states $|G\rangle, |L\rangle$, and $|H\rangle$. The time evolution of the expansion coefficients for $|k^D\rangle$, $\psi_k^D(\mathbf{Q},t)$, where $\mathbf{Q}$ is the two-dimensional mass-weighted normal coordinate vector, can be obtained from the following coupled equations [38]:

$$i\hbar\frac{\partial}{\partial t}\psi_k^D(\mathbf{Q},t) = -\frac{\hbar^2}{2}\nabla^2\psi_k^D(\mathbf{Q},t) + \sum_{k'}[W_{kk'}^D(\mathbf{Q}) - \boldsymbol{\mu}_{kk'}^D(\mathbf{Q})\cdot\boldsymbol{\varepsilon}(t)]\psi_{k'}^D(\mathbf{Q},t), \tag{3.26}$$

where ∇^2 is the Laplacian with respect to $\mathbf{Q}$. $W^{\mathrm{D}}_{kk'}(\mathbf{Q})$ are the diabatic potentials and couplings and $\boldsymbol{\mu}^{\mathrm{D}}_{kk'}(\mathbf{Q})$ are the transition moments between the two diabatic states. The coupled equations were solved numerically with the split-operator method for a multisurface Hamiltonian [35]. The resultant diabatic WPs $\psi^{\mathrm{D}}_{k}(\mathbf{Q},t)$ are converted to adiabatic WPs $\psi_k(\mathbf{Q},t)$.

3.5.4 Electronic Angular Momentum and Vibrational Amplitude

Figure 3.7a, b show the temporal behavior in the expectation value of electronic angular momentum $L_z(t)$ and that of vibrational coordinate $\mathbf{Q}(t)$, respectively, by applying a laser pulse with $\mathbf{e} = \mathbf{e}_+$ and that with $\mathbf{e} = \mathbf{e}_-$ (hereafter termed $\mathbf{e}_+$ and $\mathbf{e}_-$ excitations). Here, the linear polarization vectors $\mathbf{e}_+$ and $\mathbf{e}_-$ are defined as $\boldsymbol{\mu}_{\mathrm{LG}}(0) \cdot \mathbf{e}_\pm = \pm\boldsymbol{\mu}_{\mathrm{HG}}(0) \cdot \mathbf{e}_\pm$, in which $\boldsymbol{\mu}_{\mathrm{LG}}(0)$ and $\boldsymbol{\mu}_{\mathrm{HG}}(0)$ are the transition moments evaluated at the optimized geometry of $|\mathrm{G}\rangle\,(\mathrm{Q} = 0)$. It should be noted that an ultrashort laser pulse $\boldsymbol{\varepsilon}(t)$ vanishes before the WPs excited on the two adiabatic PESs start to run and hence the coordinate dependence of the transition moments $\boldsymbol{\mu}_{\mathrm{LG}}(\mathbf{Q})$ and $\boldsymbol{\mu}_{\mathrm{HG}}(\mathbf{Q})$ is important only in the vicinity of the optimized geometry of $|\mathrm{G}\rangle$, in which they are almost constant. The values of the laser parameters were determined following the idea of π pulse [34]: For $\mathbf{e}_+$ excitation, $f = 5.53\,\mathrm{GVm}^{-1}$,

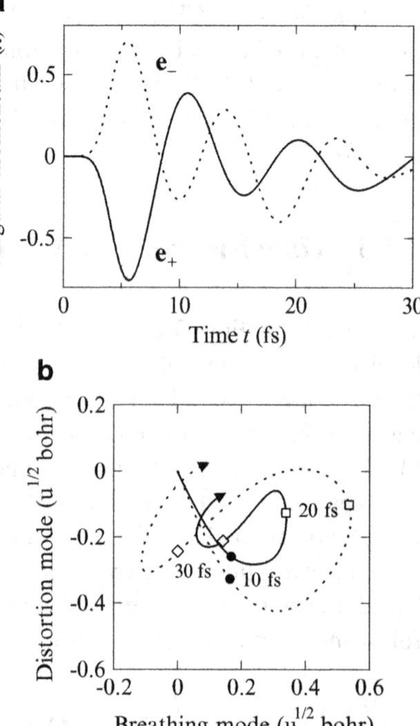

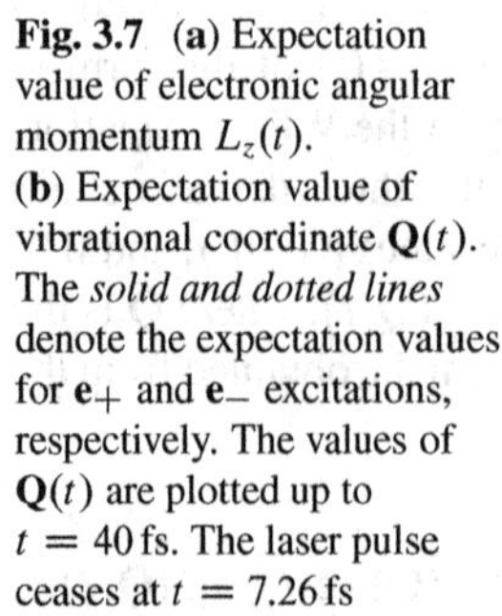

Fig. 3.7 (**a**) Expectation value of electronic angular momentum $L_z(t)$. (**b**) Expectation value of vibrational coordinate $\mathbf{Q}(t)$. The *solid and dotted lines* denote the expectation values for $\mathbf{e}_+$ and $\mathbf{e}_-$ excitations, respectively. The values of $\mathbf{Q}(t)$ are plotted up to $t = 40\,\mathrm{fs}$. The laser pulse ceases at $t = 7.26\,\mathrm{fs}$

$t_d = 7.26\,\text{fs}$, and $\omega = 9.62\,\text{eV}/\hbar$; for $\mathbf{e}_-$ excitation, $f = 9.02\,\text{GVm}^{-1}$, $t_d = 7.26\,\text{fs}$, and $\omega = 9.62\,\text{eV}/\hbar$.

In Fig. 3.7a, the initial rotation direction of π electrons depends on the photon polarization vector, i.e., clockwise (counterclockwise) direction for $\mathbf{e}_+(\mathbf{e}_-)$ excitation, which has been described in Sect. 3.3. The amplitudes of $L_z(t)$ gradually decay for both cases. The decay of the angular momentum originates from two factors: decrease of the overlap between the WPs moving on the relevant two adiabatic PESs, which occurs even within the Born–Oppenheimer approximation [42], and electronic relaxations due to nonadiabatic couplings, which is the major factor. This is one of the characteristic behaviors that are absent in a frozen-nuclei model. There are some differences between the oscillatory decays of the angular momentum for $\mathbf{e}_+$ and $\mathbf{e}_-$ excitations. The curve of $L_z(t)$ for $\mathbf{e}_+$ excitation can be approximately expressed in a sinusoidal exponential decay form with its oscillation period of $\sim T \equiv \pi/\Delta\omega = 9.4\,\text{fs}$ and lifetime of $\sim$7 fs. In contrast, the amplitude of $L_z(t)$ for $\mathbf{e}_-$ excitation does not undergo a monotonic decrease but makes a small transient recovery around $t \sim 14$–$20\,\text{fs}$. Its oscillation period is slightly shorter than that for $\mathbf{e}_+$ excitation in this time range. The difference in the oscillation period of the angular momentum for $\mathbf{e}_+$ and $\mathbf{e}_-$ excitations stems from that in the energy gap between the two adiabatic PESs for the regions in which the WPs run. Furthermore, it should be noted that the behaviors of $\mathbf{Q}(t)$ are strongly dependent on the polarization of the applied pulse. The amplitude of $\mathbf{Q}(t)$ for $\mathbf{e}_-$ excitation is more than two times larger than that for $\mathbf{e}_+$ excitation. This finding is remarkable in the sense that the initial rotation direction of π electrons controlled by the polarization direction of the laser pulse greatly affects the amplitude of subsequent molecular vibration through nonadiabatic couplings. This indicates that molecular chirality can be identified by analyzing vibrational spectra since the rotation direction basically differs between enantiomers according to their alignments with respect to the polarization direction as noted in Sect. 3.3.

3.5.5 *Interference Between Nuclear WPs in Nonadiabatic Transition*

Temporal behaviors in the population and WP dynamics on the relevant two adiabatic PESs are plotted in Fig. 3.8a, b. The populations on the PESs are defined as $P_k(t) \equiv \int d\mathbf{Q}|\psi_k(\mathbf{Q},t)|^2 (k = \text{L and H})$. For $\mathbf{e}_-$ excitation, the probability densities $|\psi_\text{L}(\mathbf{Q},t)|^2$ and $|\psi_\text{H}(\mathbf{Q},t)|^2$ at $t \sim 5\,\text{fs}$ have almost the same shape as that of the initial WP $|\psi_\text{G}(\mathbf{Q},0)|^2$, while the WPs created in the two excited states are out of phase from the definition of $\mathbf{e}_-$. As the WPs start to move along the gradient of each PES, significant population transfer occurs from $|\text{H}\rangle$ to $|\text{L}\rangle$ by nonadiabatic transition. Consequently, $P_\text{L}(t)$ is more than seven times larger than $P_\text{H}(t)$ at $t \sim 10\,\text{fs}$, although they are almost equal at $t \sim 5\,\text{fs}$. The loss of a superposition of $|\text{L}\rangle$ and $|\text{H}\rangle$ reduces the amplitude of $L_z(t)$ as in Fig. 3.7a. Afterward, the direction of the population transfer is reversed periodically despite the rather small

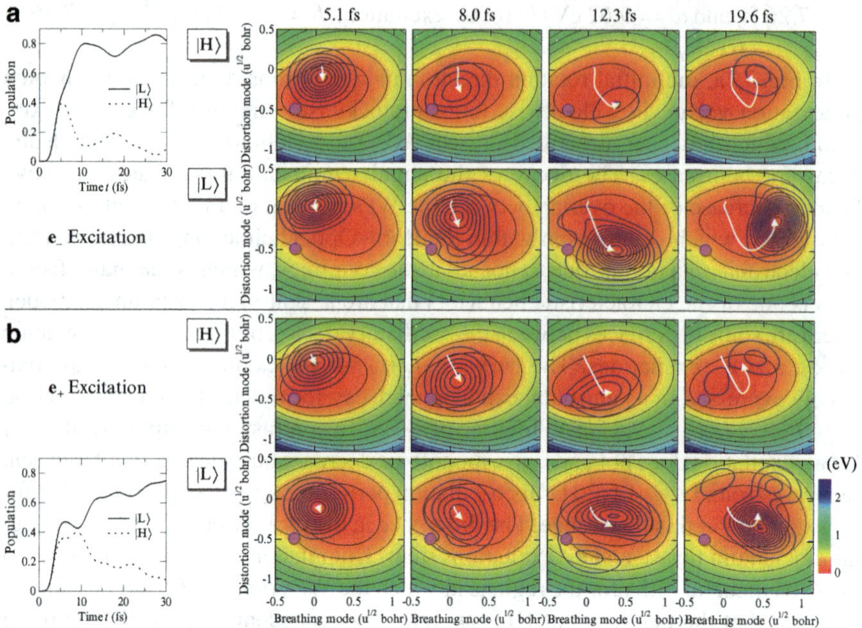

Fig. 3.8 *Left panel*: Temporal behavior in the populations of |L⟩ (*solid line*) and |H⟩ (*dotted line*) denoted as $P_L(t)$ and $P_H(t)$, respectively. *Right panels*: Propagation of the adiabatic WPs on the two-dimensional adiabatic PESs of |L⟩ and |H⟩. The origin of the PESs is the optimized geometry of |G⟩. The *bold* contours represent the probability densities $|\psi_L(\mathbf{Q}, t)|^2$ and $|\psi_H(\mathbf{Q}, t)|^2$ and the *arrows* indicate the motion of the center of the WPs. The avoided crossing is signified by a *circle*

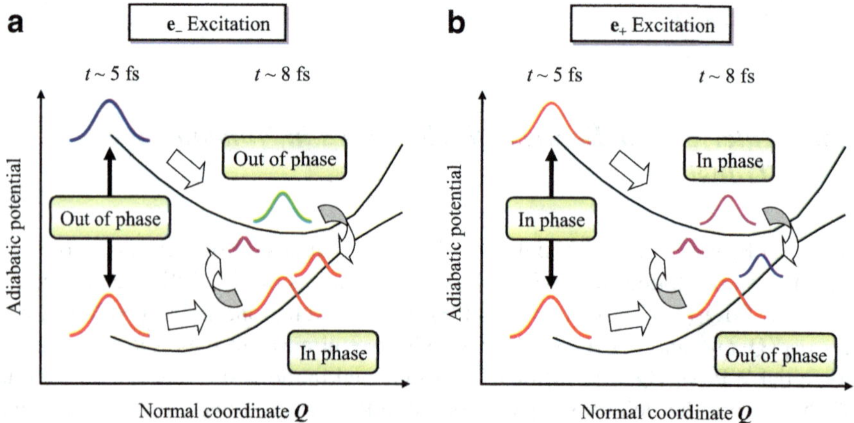

Fig. 3.9 One-dimensional conceptual diagrams illustrating the interference between the adiabatic WPs $\psi_L(\mathbf{Q}, t)$ and $\psi_H(\mathbf{Q}, t)$. The WPs and the adiabatic PESs of |L⟩ and |H⟩ are depicted by one-dimensional curves. The colors of the Gaussian-like curves represent the relative quantum phase between the WPs. In particular, the WPs drawn by *red* (*purple*) and *blue* (*green*) curves have the opposite phases

amount of the population transferred; regeneration of the superposition of $|\mathrm{L}\rangle$ and $|\mathrm{H}\rangle$ around $t \sim 14$–20 fs gives rise to the transient recovery of the angular momentum. $\psi_{\mathrm{L}}(\mathbf{Q}, t)$ moves in the high-potential region following the potential gradient of $|\mathrm{L}\rangle$, which leads to the large-amplitude vibration in Fig. 3.7b.

In contrast, $\psi_{\mathrm{L}}(\mathbf{Q}, t)$ and $\psi_{\mathrm{H}}(\mathbf{Q}, t)$ are in phase when DCP is excited by a pulse with $\mathbf{e}_{+}$. For this excitation, a small amount of the population shifts from $|\mathrm{L}\rangle$ to $|\mathrm{H}\rangle$ around $t \sim 5$–10 fs. Then, a considerable population transfer takes place in the reverse way around $t \sim 10$–14 fs when the WPs come closer to the avoided crossing. The contours of $|\psi_{\mathrm{L}}(\mathbf{Q}, t)|^2$ at $t = 12.3$ fs in Fig. 3.8b clearly exhibit the node arising from the interference between the WPs. At $t > 14$ fs, the upward population transfer is extremely small. The interference continues to increase (decrease) the probability density in the low-potential (high-potential) region, resulting in the small-amplitude vibration in Fig. 3.7b.

The photon polarization dependence of the populations and WPs in Fig. 3.8a, b can be explained in terms of interferences between the WP existing on the original PES and that created by nonadiabatic couplings. We briefly illustrate the interference effects in one-dimensional conceptual diagrams in Fig. 3.9. As mentioned above, a pulse with $\mathbf{e}_{-}$ produces $\psi_{\mathrm{L}}(\mathbf{Q}, t)$ and $\psi_{\mathrm{H}}(\mathbf{Q}, t)$ out of phase, and their relative quantum phase evolves as the WPs move on each PES. The WP created by nonadiabatic couplings gains an additional phase shift and interferes with that on the other PES. Around $t \sim 5$–10 fs, they are almost in phase (out of phase) and the interference is constructive (destructive) on the lower (higher) PES, which causes the downward population transfer in Fig. 3.8a. The constructive interference works particularly on high vibrational quantum states in $\psi_{\mathrm{L}}(\mathbf{Q}, t)$. The direction of the population transfer switches as the relative quantum phase evolves. For $\mathbf{e}_{+}$ excitation in which the two excited WPs are in phase, the interference effects are reversed: destructive (constructive) interference on the lower (higher) PES around $t \sim 5$–10 fs. The resultant upward population transfer is small because the interference effects on the two PESs cancel out each other. $\psi_{\mathrm{L}}(\mathbf{Q}, t)$ and $\psi_{\mathrm{H}}(\mathbf{Q}, t)$ thus reach the avoided crossing and the reverse population transfer occurs around $t \sim 10$–14 fs. The interference enhances low vibrational quantum states in $\psi_{\mathrm{L}}(\mathbf{Q}, t)$, exhibiting the clear node in Fig. 3.8b.

3.5.6 Vibrational Analysis for Observation of π-Electron Rotation

Finally, we verify that the initial rotation direction of π electrons can be determined by analyzing vibrational spectra. In general, the Fourier transform of the autocorrelation function of WPs gives its frequency spectrum [43]. The spectrum of $\psi_{\mathrm{L}}(\mathbf{Q}, t)$ after the nonadiabatic transition from $|\mathrm{H}\rangle$ to $|\mathrm{L}\rangle$ is defined as

$$s_{\mathrm{L}}(\omega) \equiv \mathrm{Re} \int_{t_{\mathrm{n}}}^{t_{\mathrm{f}}} \mathrm{d}t\, e^{(i\omega - 1/\tau)(t - t_{\mathrm{n}})} \int \mathrm{d}\mathbf{Q}\, \psi_{\mathrm{L}}^{*}(\mathbf{Q}, t_{\mathrm{n}}) \psi_{\mathrm{L}}(\mathbf{Q}, t). \tag{3.27}$$

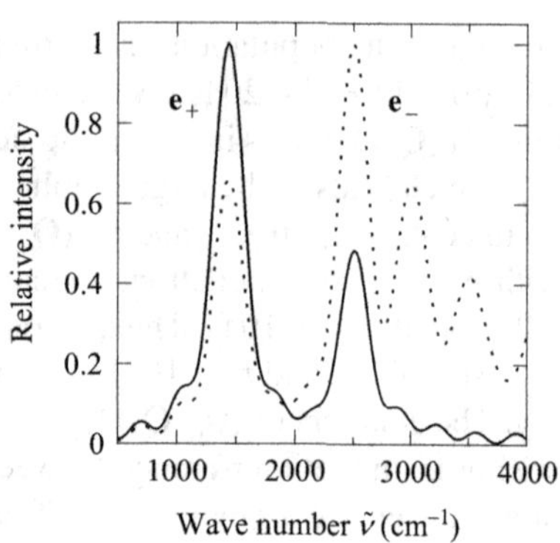

Fig. 3.10 The frequency spectra of $\psi_L(\mathbf{Q}, t)$, $s_L(\omega)$, defined by (3.27). The *solid and dotted lines* denote the spectra for $\mathbf{e}_+$ and $\mathbf{e}_-$ excitations, respectively. In each case, the values of $s_L(\omega)$ are scaled so that the maximum value is unity

The parameter τ was introduced to smooth the spectra and set at 39.6 fs, which is longer than the vibrational periods of the breathing and distortion modes (28.8 and 21.2 fs). The values of t_n for $\mathbf{e}_+$ and $\mathbf{e}_-$ excitations were 14.0 and 10.0 fs, respectively, and t_f–t_n = 99.1 fs for both cases. The spectra for $\mathbf{e}_+$ and $\mathbf{e}_-$ excitations are displayed in Fig. 3.10. In the former case, the maximum value of $s_L(\omega)$ appears at $\tilde{\nu} \sim$ 1,400 cm^{-1} and another peak is found at $\tilde{\nu} \sim$ 2,500 cm^{-1}; in the latter case, the spectrum reaches its maximum at $\tilde{\nu} \sim$ 2,500 cm^{-1} and also exhibits a couple of strong peaks at $\tilde{\nu} >$ 3,000 cm^{-1}. The wave numbers of 1,400, 2,500, and 3,000 cm^{-1} are very close to those of the lowest three vibrational states of $|G\rangle$ owing to the similarity between $|G\rangle$ and $|L\rangle$ in the PES around its minimum. The spectral features in Fig. 3.10 confirm that at $t > t_n \psi_L(\mathbf{Q}, t)$ mainly consists of low (high) vibrational quantum states for $\mathbf{e}_+$ ($\mathbf{e}_-$) excitation. The vibrational structure changes of DCP or DCPH can be measured experimentally with optical spectroscopic methods, e.g., transient impulsive Raman spectroscopy [44]. Thus, attosecond π-electron rotations can be observed by spectroscopic detection of femtosecond molecular vibrations that induce nonadiabatic couplings.

3.6 Summary and Perspectives

We have reviewed our recent studies on the theoretical foundations of switching on and off photocurrents in chiral systems. Quantum dynamical simulations have shown that π electrons can be rotated along the ring of a chiral aromatic molecule using a linearly polarized UV laser pulse. π-electron rotation originates from creation of an approximate angular momentum eigenstate consisting of optically allowed quasidegenerate π-electronic excited states. Lowering the molecular symmetry is essential for the selective generation of one of the approximate eigenstates. The rotation direction of π electrons depends on the spatial configuration of each enantiomer with respect to the polarization direction of the applied pulse. Unidirectional rotation of π electrons can be achieved by pump and dump pulses whose polarization directions are properly chosen.

Optimal control simulations assuming a linear polarization have revealed that for both counterclockwise and clockwise π-electron rotations the optimal control pulse at $\mathbf{e} = \mathbf{e}_+$ or $\mathbf{e}_-$ consists of ultrashort single-peaked laser pulses with frequencies ω_L and ω_H. Then, in addition to the photon polarization direction, the relative optical phase $\Delta\varphi$ is a crucial parameter for determination of the rotation direction of π electrons. For the control time close to $T \equiv \pi/\Delta\omega$, the combined pulse for counterclockwise (clockwise) π-electron rotation at $\mathbf{e} = \mathbf{e}_+(\mathbf{e}_-)$ is a single-color laser pulse, while that at $\mathbf{e} = \mathbf{e}_-(\mathbf{e}_+)$ is a two-color laser pulse.

Finally, the nonadiabatic coupling between π-electron rotation and molecular vibration has been analyzed with a simplified model of a chiral aromatic molecule excited by a linearly polarized laser pulse. The angular momentum of π electrons coupled to molecular vibration gradually decays, while the vibrational amplitude greatly depends on their rotation direction. The former is attributed mainly to the electronic relaxations caused by nonadiabatic transition; the latter results from the interference in nonadiabatic transition governed by the relative quantum phase between the WPs.

Our insight suggests that the rotation direction of π electrons traveling on an attosecond timescale can be identified by detecting femtosecond molecular vibrations with spectroscopy. This may open a way to observe ultrafast coherent electron motion in polyatomic molecules. In particular, this effect may be utilized for rapid identification of chiral compounds. Even in the presence of nonadiabatic couplings, π-electron rotations can produce angular momentum sufficient for ultrafast switching.

Acknowledgements We are grateful to Dr. K. Hoki and Professor S. H. Lin for their fruitful discussions. M. K. acknowledges financial support from a JSPS Research Fellowship for Young Scientists (No. 18 · 5252). This work was supported in part by a JSPS Research Grant (No. 17350004).

References

1. K. Yamanouchi, Science **295**, 1659 (2002)
2. M.F. Kling, Ch. Siedschlag, A.J. Verhoef, J.I. Khan, M. Schultze, Th. Uphues, Y. Ni, M. Uiberacker, M. Drescher, F. Krausz, M.J.J. Vrakking, Science **312**, 246 (2006)
3. P. Krause, T. Klamroth, P. Saalfrank, J. Chem. Phys. **123**, 074105 (2005)
4. P. Krause, T. Klamroth, P. Saalfrank, J. Chem. Phys. **127**, 034107 (2007)
5. K. Nobusada, K. Yabana, Phys. Rev. A **75**, 032518 (2007)
6. D. Geppert, P. von den Hoff, R. de Vivie-Riedle, J. Phys. B **41**, 074006 (2008)
7. M. Nest, F. Remacle, R.D. Levine, New J. Phys. **10**, 025019 (2008)
8. H. Kono, Y. Sato, M. Kanno, K. Nakai, T. Kato, Bull. Chem. Soc. Jpn. **79**, 196 (2006)
9. H. Kono, K. Nakai, M. Kanno, Y. Sato, S. Koseki, T. Kato, Y. Fujimura, in *Progress in Ultrafast Intense Laser Science*, ed. by K. Yamanouchi, A. Becker, R. Li, S.L. Chin, vol. 4, (Springer, Berlin, 2009), pp. 41–66
10. T. Kato, H. Kono, M. Kanno, Y. Fujimura, K. Yamanouchi, Laser Phys. **19**, 1712 (2009)
11. J. Itatani, J. Levesque, D. Zeidler, H. Niikura, H. Pépin, J.C. Kieffer, P.B. Corkum, D.M. Villeneuve, Nature (London) **432**, 867 (2004)

12. M. Lein, C.C. Chirila, in *Advances in Multi-Photon Processes and Spectroscopy*, ed. by S.H. Lin, A.A. Villaeys, Y. Fujimura, vol. 18 (World Scientific, Singapore, 2008), pp. 69–106 and references therein
13. E.P. Fowe, A.D. Bandrauk, Phys. Rev. A **81**, 023411 (2010)
14. A. Matos-Abiague, J. Berakdar, Phys. Rev. Lett, **94**, 166801 (2005)
15. E. Räsänen, A. Castro, J. Werschnik, A. Rubio, E.K.U. Gross, Phys. Rev. Lett. **98**, 157404 (2007)
16. I. Barth, J. Manz, Angew. Chem. **118**, 3028 (2006)
17. I. Barth, J. Manz, Angew. Chem. Int. Ed. **45**, 2962 (2006)
18. I. Barth, J. Manz, Y. Shigeta, K. Yagi, J. Am. Chem. Soc. **128**, 7043 (2006)
19. M. Kanno, H. Kono, Y. Fujimura, Angew. Chem. **118**, 8163 (2006)
20. M. Kanno, H. Kono, Y. Fujimura, Angew. Chem. Int. Ed. **45**, 7995 (2006)
21. M. Kanno, K. Hoki, H. Kono, Y. Fujimura, J. Chem. Phys. **127**, 204314 (2007)
22. M. Kanno, H. Kono, Y. Fujimura, S.H. Lin, Phys. Rev. Lett. **104**, 108302 (2010)
23. L. Salem, *The Molecular Orbital Theory of Conjugated Systems* (Benjamin, New York, 1966), pp. 112–116
24. A.A. Frost, B. Musulin, J. Chem. Phys. **21**, 572 (1953)
25. M. Rubio, B.O. Ross, L. Serrano-Andrés, M. Merchán, J. Chem. Phys. **110**, 7202 (1999)
26. D. Sundholm, Chem. Phys. Lett. **317**, 392 (2000)
27. S. Abe, J. Yu, W.P. Su, Phys. Rev. B **45**, 8264 (1992)
28. V.A. Shakin, S. Abe, Phys. Rev. B **50**, 4306 (1994)
29. M. Chandross, Y. Shimoi, S. Mazumdar, Phys. Rev. B **59**, 4822 (1999)
30. M. Suzuki, S. Mukamel, J. Chem. Phys. **119**, 4722 (2003)
31. K. Ohno, Theor. Chim. Acta **2**, 219 (1964)
32. E. Hückel, Z. Phys. **70**, 204 (1931), **72**, 310 (1931); **76**, 628 (1932)
33. A. Streitwieser Jr., *Molecular Orbital Theory for Organic Chemists* (Wiley, New York, 1961), p. 117
34. B.W. Shore, *The Theory of Coherent Atomic Excitation,* Vol. 1, (Wiley, New York, 1990), pp. 304–309
35. P. Gross, D. Neuhauser, H. Rabitz, J. Chem. Phys. **96**, 2834 (1992)
36. W. Zhu, H. Rabitz, J. Chem. Phys. **109**, 385 (1998)
37. J. Manz, K. Sundermann, R. de Vivie-Riedle, Chem. Phys. Lett. **290**, 415 (1998)
38. Y. Ohtsuki, K. Nakagami, Y. Fujimura, in *Advances in Multi-Photon Processes and Spectroscopy*, ed. by S.H. Lin, A.A. Villaeys, Y. Fujimura, Vol. 13, (World Scientific, Singapore, 2001), pp. 1–127
39. I. Serban, J. Werschnik, E.K.U. Gross, Phys. Rev. A **71**, 053810 (2005)
40. H.J. Werner, P.J. Knowles, R. Lindh, F.R. Manby, M. Schütz, P. Celani, T. Korona, G. Rauhut, R.D. Amos, A. Bernhardsson, A. Berning, D.L. Cooper, M.J.O. Deegan, A.J. Dobbyn, F. Eckert, C. Hampel, G. Hetzer, A.W. Lloyd, S.J. McNicholas, W. Meyer, M.E. Mura, A. Nicklass, P. Palmieri, R. Pitzer, U. Schumann, H. Stoll, A.J. Stone, R. Tarroni, T. Thorsteinsson, MOLPRO, version 2006.1 (Cardiff, UK, 2006)
41. I.N. Levine, *Quantum Chemistry*, 5th ed. (Prentice-Hall, Upper Saddle River, N.J., 2000), pp. 480–625
42. M. Born, J.R. Oppenheimer, Ann. Phys. **84**, 457 (1927)
43. D.J. Tannor, *Introduction to Quantum Mechanics: A Time-Dependent Perspective* (University Science Books, California, 2007), pp. 81–86
44. S. Takeuchi, S. Ruhman, T. Tsuneda, M. Chiba, T. Taketsugu, T. Tahara, Science **322**, 1073 (2008)

Chapter 4
Optically Probed Laser-Induced Field-Free Molecular Alignment

O. Faucher, B. Lavorel, E. Hertz, and F. Chaussard

Abstract Molecular alignment induced by laser fields has been investigated in research laboratories for over two decades. It led to a better understanding of the fundamental processes at play in the interaction of strong laser fields with molecules, and also provided significant contributions to the fields of high harmonic generation, laser spectroscopy, and laser filamentation. In this chapter, we discuss molecular alignment produced under field-free conditions, as resulting from the interaction of a laser pulse of duration shorter than the rotational period of the molecule. The experimental results presented will be confined to the optically probed alignment of linear as well as asymmetric top molecules. Special care will be taken to describe and compare various optical methods that can be employed to characterize laser-induced molecular alignment. Promising applications of optically probed molecular alignment will be also demonstrated.

4.1 Introduction

The use of external fields to manipulate the different degrees of freedom of molecules has been an efficient tool in the hands of physicists and chemists so far. In particular, during the last decade, these two communities have shown a growing interest in controlling alignment and orientation of molecules by laser light. Compared with traditional methods, which rely, for example, on the use of charged electrodes producing static or variable fields, the large electric field routinely available with pulsed lasers provides the opportunity to produce molecular sample with a high degree of alignment. For chemists, molecular alignment may provide a means to influence a chemical reaction by controlling, for example, its yield and the branching ratios. For physicists and physico-chemists, the interest is twofold.

O. Faucher (✉), B. Lavorel, E. Hertz, and F. Chaussard
Laboratoire Interdisciplinaire CARNOT de Bourgogne (ICB), UMR 5209 CNRS-Université de Bourgogne, BP 47870, 21078 Dijon Cedex, France
e-mail: olivier.faucher@u-bourgogne.fr, bruno.lavorel@u-bourgogne.fr, edouard.hertz@u-bourgogne.fr, frederic.chaussard@u-bourgogne.fr

K. Yamanouchi et al. (eds.), *Progress in Ultrafast Intense Laser Science VII*, Springer Series in Chemical Physics 100, DOI 10.1007/978-3-642-18327-0_4,

On the one hand, the fundamental interest for molecular alignment provides a better understanding of the processes occurring when molecules are exposed to intense laser fields. On the other hand, most of light–matter interactions, i.e., absorption, ionization, dissociation, and harmonic generation, depend on the angles formed between the principal axes of the molecule and the direction of the oscillating electric field. The alignment allows therefore not only to manipulate these processes but also to acquire new informations about the molecular system as well.

It is usual for laser-induced alignment to distinguish between the adiabatic and nonadiabatic regimes (for a review on this issue, see [1] and references therein). In the former case, the pulse duration is long compared to the picosecond timescale of the rotational motion of molecules, whereas in the latter it is the opposite. Since in this chapter we will focus on results obtained using femtosecond lasers, the adiabatic regime where the alignment is exclusively produced during the laser interaction will not be described. In most cases, the alignment proceeds from the nonresonant impulsive Raman excitation of the molecular polarizability. In this case, a rotational wave packet is produced in the ground vibronic state of the molecule. After the pulse turns off, the free evolution of this wave packet exhibits revivals at fractional times of the rotational period corresponding to a rephasing of the rotational components. Each revival produces an alignment as well as a planar delocalization [2] of the molecular axis with respect to the field polarization. Such laser-induced field-free molecular alignment lasts as long as the coherence of the medium is maintained. Its main advantage is that applications based on aligned molecules can be conducted in field-free conditions.

Direct measurements of laser-induced alignment can be obtained using a dissociation technique [3, 4]. According to this technique, the molecules after exposure to an aligning pulse are interacting with a second short laser pulse producing dissociation of the molecular bonds. A space-sensitive detection of the fragments allows then to reconstruct the angular distribution of the molecule. Alternative methods consist in probing the optical properties of the medium modified by the anisotropy induced by the aligned molecules [2, 5]. The aim of this chapter is to describe some optical methods used in the context of field-free molecular alignment and their applications.

4.2 Molecular Alignment Induced by Short Laser Pulses

4.2.1 Model

In quantum mechanics, the evolution of a molecular system exposed to a laser pulse is described by the time-dependent Hamiltonian

$$H = H_0 + H_{\text{int}}, \tag{4.1}$$

where H_0 represents the energy of the free molecule and H_{int} describes the interaction with the laser field. In the electric dipole approximation, the last term can be written in the following form [6]

$$H_{\text{int}} = -\int_0^E \vec{\mu} \cdot \mathrm{d}\vec{E} \tag{4.2}$$

with $\vec{\mu}$ the electric dipole coupled to the electric field

$$\vec{E}(t) = \vec{\mathcal{E}}(t) \cos \omega t, \tag{4.3}$$

where $\vec{\mathcal{E}}(t)$ is a slowly varying envelope and ω the angular frequency. Since here we are only concerned with nonresonant laser pulses, we can write the electric dipole term as a sum of a permanent component $\vec{\mu}_0$ and a field-induced component $\alpha\vec{E}(t)$

$$\vec{\mu} = \vec{\mu}_0 + \alpha \cdot \vec{E}(t). \tag{4.4}$$

In this expression, α stands for the first-order polarizability tensor written in the laboratory reference frame. Higher-order terms of the polarizability (hyperpolarizabilities) will be neglected in this chapter. According to (4.2) and (4.4), the interaction Hamiltonian is given by

$$H_{\text{int}} = -\vec{\mu}_0 \cdot \vec{E}(t) - \frac{1}{2}\vec{E}(t) \cdot \alpha \cdot \vec{E}(t). \tag{4.5}$$

α can be deduced from the molecular polarizability α' expressed in the molecule-fixed reference frame by using the following transformation

$$\alpha = \mathsf{R}\alpha'\mathsf{R}^{-1}, \tag{4.6}$$

where R is the rotation matrix [7] defined in terms of the three Euler angles ϕ, θ, and χ depicted in Fig. 4.1

$$\mathsf{R} = \begin{pmatrix} \mathrm{c}\phi\,\mathrm{c}\theta\,\mathrm{c}\chi - \mathrm{s}\phi\,\mathrm{s}\chi & \mathrm{s}\phi\,\mathrm{c}\theta\,\mathrm{c}\chi + \mathrm{c}\phi\,\mathrm{s}\chi & -\mathrm{s}\theta\,\mathrm{c}\chi \\ -\mathrm{c}\phi\,\mathrm{c}\theta\,\mathrm{s}\chi - \mathrm{s}\phi\,\mathrm{c}\chi & -\mathrm{s}\phi\,\mathrm{c}\theta\,\mathrm{s}\chi + \mathrm{c}\phi\,\mathrm{c}\chi & \mathrm{s}\theta\,\mathrm{s}\chi \\ \mathrm{c}\phi\,\mathrm{s}\theta & \mathrm{s}\phi\,\mathrm{s}\theta & \mathrm{c}\theta \end{pmatrix}, \tag{4.7}$$

where c and s represent the cos and sin functions, respectively.

To proceed further, we need to specify the molecular system as well as the polarization of the field. For simplicity, we can choose a linear molecule exposed to a laser pulse having a linear polarization. Nonlinear molecules with linearly, elliptically, or circularly polarized fields will be addressed in Sect. 4.3.1. If we assume that the laser field does not induce any transition between the different vibronic states of the molecule, we can reduce the expression of the free Hamiltonian to its rotational part H_{rot}

$$H_0 = H_{\text{rot}} = BJ^2 - DJ^4, \tag{4.8}$$

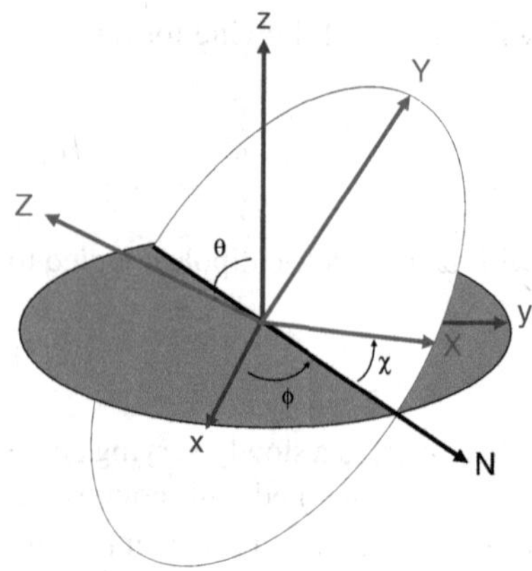

Fig. 4.1 Euler angles. x, y, z: laboratory frame. X, Y, Z: molecule-fixed frame. θ is the polar angle measured from the z- to Z-axis, whereas ϕ and χ describe a rotation around z- and Z, respectively

where B and D are the rotational constants [8], neglecting higher-order corrections, and J is the total angular momentum. The polarizability tensor for a linear rotor is given by

$$\alpha' = \begin{pmatrix} \alpha_\perp & 0 & 0 \\ 0 & \alpha_\perp & 0 \\ 0 & 0 & \alpha_\parallel \end{pmatrix}, \tag{4.9}$$

where $\alpha_\parallel$ and $\alpha_\perp$ are the polarizability components parallel and perpendicular to the molecular axis, respectively. So, for a field polarized along the z-axis and making use of (4.3)–(4.7) and (4.9), the interaction Hamiltonian can be written as

$$H_{\text{int}} = -\frac{1}{4}\alpha_{zz}\mathcal{E}^2(t) \tag{4.10a}$$

$$= -\frac{1}{4}\mathcal{E}^2(t)\left(\Delta\alpha\cos^2\theta + \alpha_\perp\right) \tag{4.10b}$$

with $\Delta\alpha = \alpha_\parallel - \alpha_\perp$ being the anisotropy of polarizability. It should be noted that (4.10) are valid in the high frequency limit approximation (with respect to the rotational frequency) for which the contribution of the electric permanent dipole introduced in (4.4) averages to zero. It is obvious from (4.10b) that the potential features a minimum at $\theta = 0$ or π corresponding to an angular confinement of the molecular axis along the field direction.

A convenient way to describe the dynamical alignment consists in inspecting the temporal evolution of the expectation value $\langle\cos^2\theta\rangle = \langle\psi|\cos^2\theta|\psi\rangle$, with ψ the state vector of the system. This quantity can be calculated by solving the time-dependent Schrödinger equation $i\hbar d\psi/dt = H\psi$. As an example, we present in Fig. 4.2 a simulation of N_2 molecule excited by a 100 fs pulse. To account for the finite temperature (10 K), $\langle\cos^2\theta\rangle$ has been averaged over the thermal distribution of the initial rotational states. The result is characterized by transients of field-free alignment corresponding to revivals of the rotational wave packet initially excited by the laser pulse. The revivals are spaced by $T_r/4$, with $T_r = \hbar\pi/B \approx 8.4$ ps

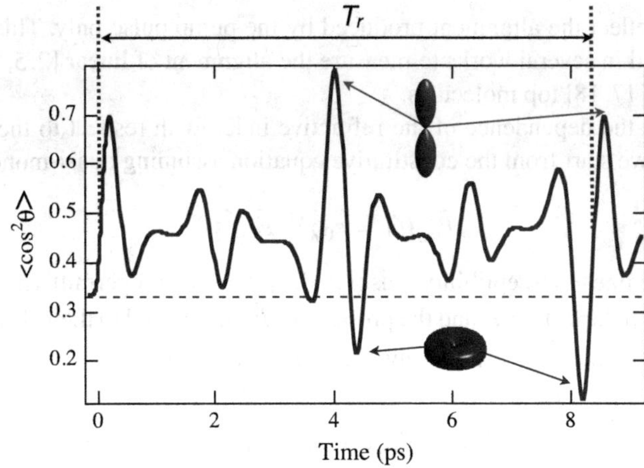

Fig. 4.2 Numerical simulation of $\langle \cos^2 \theta \rangle$ in N_2 at 10 K for a Gaussian pulse of 100 fs pulse duration and peak intensity 45 TW/cm^2. T_r is the rotational period of the molecule. The plots in the insets illustrate the angular distribution of the molecular axis when it is aligned (*red*) along the field or delocalized (*blue*) in the equatorial plane

being the rotational period, assuming that the rotational constant B has a value of 1.99 cm^{-1}. It should be noticed that quarter period revivals are only observed in molecules with inversion symmetry, such as N_2, CO_2 in its ground vibrational state, or O_2, for whom the population is affected by nuclear spin statistics [9, 10]. The dashed line corresponds to $\langle \cos^2 \theta \rangle = 1/3$. As it has been reported in several studies [2,11,12], the rotational dynamics of a linear molecule leads to a regular alternation of the molecular axis between alignment ($\langle \cos^2 \theta \rangle > 1/3$) and planar delocalization ($\langle \cos^2 \theta \rangle < 1/3$). Between the revivals, the angular distribution remains slightly elongated along the field direction leading to a small permanent alignment with $\langle \cos^2 \theta \rangle > 1/3$ indicated by the elevated baseline.

4.2.2 Principle of the Optical Detections

Optical detection of molecular alignment is based on the optical Kerr effect resulting from the orientation of the field-induced molecular dipoles. The results that we present in this chapter have been obtained by measuring the alignment of molecules using a weak laser pulse that interacts with the molecules subsequently to their exposure to a strong aligning pulse. In the next, for convenience, the first and the second pulses will be called the pump and the probe pulses, respectively. The anisotropic angular distribution of the molecular axes results in a modification of the refractive index experienced by the probe. This change can be observed in time by changing the temporal delay between the two laser pulses. Since the probe pulse is weak, we can safely assume that the alignment is not affected and therefore the probe signal will reflect the alignment produced by the pump pulse only. This technique has been used in several works to measure the alignment of linear [2, 5, 13–16] and asymmetric [17, 18] top molecules.

To obtain the dependence of the refractive index with respect to the molecular orientation, we start from the constitutive equation assuming quasi-monochromatic waves [19]

$$\vec{P}_{\omega_p}(t) = \varepsilon_0 \chi^{(1)} \vec{E}_{\omega_p}(t) \tag{4.11}$$

with $\chi^{(1)}$ the linear susceptibility tensor and ε_0 the vacuum permittivity. The linear macroscopic polarization $\vec{P}$ and the probe field $\vec{E}_p$ are related to their slowly varying envelopes $\vec{E}_{\omega_p}$ and $\vec{P}_{\omega_p}$, respectively, through the relations

$$\vec{E}_p = \frac{1}{2}\vec{E}_{\omega_p}(t)\exp(\mathrm{i}\omega_p t) + \text{c.c.} \tag{4.12a}$$

$$\vec{P} = \frac{1}{2}\vec{P}_{\omega_p}(t)\exp(\mathrm{i}\omega_p t) + \text{c.c.} \tag{4.12b}$$

Projected along the i-axis in the laboratory frame, (4.11) becomes

$$P^{(1)}_{\mathrm{i}\omega_p}(t) = \varepsilon_0 \chi^{(1)}_{ij}\left(-\omega_p; \omega_p\right) E_{j\omega_p}(t). \tag{4.13}$$

If we analyze the polarization induced by the field along the k-axis, we can deduce from the latter relation the refractive index produced along this direction

$$n^2_{kk} - 1 = \chi^{(1)}_{kk}\left(-\omega_p; \omega_p\right) = \frac{\varrho}{\varepsilon_0}\alpha_k, \tag{4.14}$$

where we have introduced the relation between the linear susceptibility and the first-order polarizability [20], with ϱ the number density of molecules. The susceptibility can be written in the molecule-fixed frame using the rotation matrix given by (4.7). If we consider a pump field, i.e., the alignment field, polarized along the z-axis, so that the Euler angles ϕ and χ are conserved during the field interaction ($\langle\cos^2\phi\rangle = \langle\cos^2\chi\rangle = 1/2$), we can show that the last equation finally leads to

$$\chi^{(1)}_{xx} = \chi^{(1)}_{yy} = \frac{\varrho}{\varepsilon_0}\left\{\bar{\alpha} - \frac{\Delta\alpha}{2}\left(\langle\cos^2\theta\rangle - 1/3\right)\right\} \tag{4.15a}$$

$$\chi^{(1)}_{zz} = \frac{\varrho}{\varepsilon_0}\left\{\bar{\alpha} + \Delta\alpha\left(\langle\cos^2\theta\rangle - 1/3\right)\right\} \tag{4.15b}$$

with $\bar{\alpha} = (\alpha_\parallel + 2\alpha_\perp)/3$ the average polarizability.

The second term in the right-hand side of both equations represents the orientational contribution to the optical Kerr effect (also known as the Langevin contribution [20]). For an ensemble of randomly oriented molecules ($\langle\cos^2\theta\rangle = 1/3$), this contribution averages to zero. The same applies for molecules that do not exhibit an anisotropy of their polarizability ($\Delta\alpha = 0$), e.g., CH_4 in its ground state. In general, any optical measurement whose result will depend on the refractive index along a given space direction will provide information about the alignment of the molecular axis with respect to this direction. This is the main idea that stands behind the optical techniques employed for monitoring molecular alignment.

Finally, we can deduced from (4.15) the difference of refractive indices between two directions, which is directly proportional to the produced alignment $\langle\cos^2\theta\rangle - 1/3$

$$\Delta n = n_z - n_y \simeq \frac{n_z^2 - n_y^2}{2\overline{n}} = \frac{3\varrho\Delta\alpha}{4\overline{n}\varepsilon_0}\left(\langle\cos^2\theta\rangle - 1/3\right) \qquad (4.16)$$

with $\overline{n} = 1/2(n_z + n_y)$. This relation will be exploited in the following section where the birefringence measurements will be presented.

4.3 Observation of Molecular Alignment by Time-Resolved Birefringence

One of the most sensitive optical schemes used for monitoring alignment is based on transient birefringence measurements. The principle is to observe the change of polarization experienced by a weak probe pulse, time-delayed with respect to the alignment pulse, as it propagates through a sample of aligned molecules. Two different configurations have been implemented, providing space-averaged and space-resolved single-shot detection.

4.3.1 Space-Averaged Detection

The required experimental arrangement for space-averaged time-resolved birefringence measurements is based on a standard pump-probe setup employed in time-resolved polarization spectroscopy. In the setup shown in Fig. 4.3, the pump and probe pulses are delivered by a chirped pulsed amplified Ti:sapphire femtosecond laser. The laser beam is split in two parts with 98% of the total energy used in the pump beam, the remaining 2% being used in the probe beam. A time delay line made of a corner cube mounted on a motorized stage is used to adjust the temporal delay between the pump and the probe pulses. Both pulses, linearly polarized at 45° to each other, are focused with the same lens and are crossed with a small angle (~3°) in the gas sample. At the exit of the interaction chamber, the pump is blocked by a beam stop and the depolarization of the probe is detected through an analyzer set at 90° with respect to the initial polarization of the probe. The depolarized probe field is then collected with a photomultiplier, sampled by a boxcar integrator, and sent to a computer, which is also used for the control of the delay line. The different polarization directions are shown in the inset of Fig. 4.3. The experiment can be conducted under static cell conditions or in a supersonic expansion of a molecular jet.

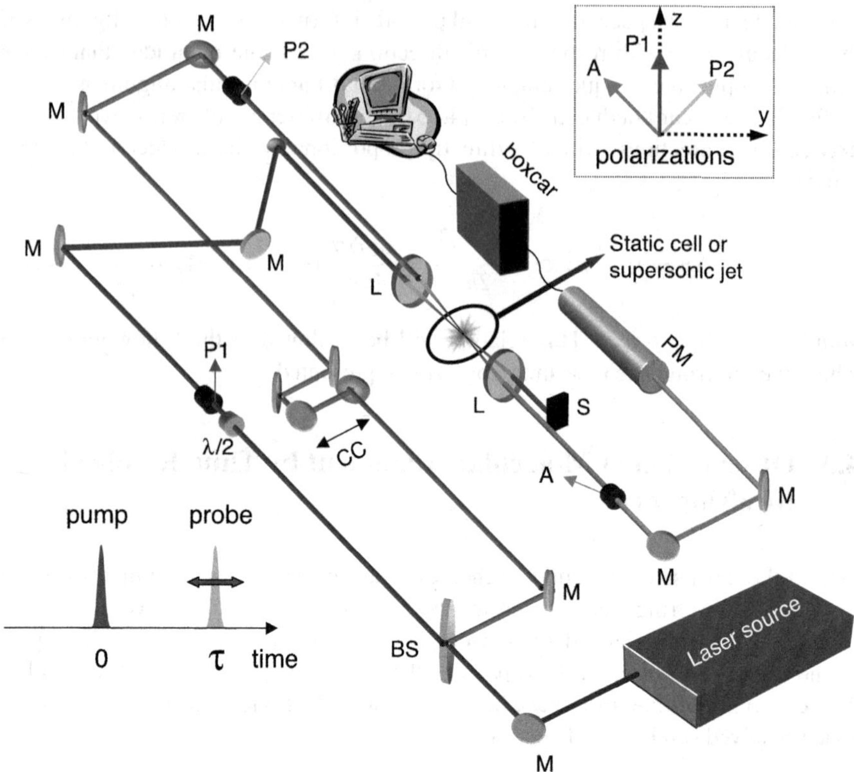

Fig. 4.3 Experimental setup for time-resolved birefringence measurements. *M*: Mirror, *BS*: Beam Splitter, *L*: Lens, *CC*: Corner Cube, *P*: Polarizer, *A*: Analyzer, *S*: Beam stop, *PM*: Photomultiplier. The relative polarizations of the pump (P1), probe (P2), and signal-field (A) are shown in the inset

The delayed probe field $\vec{E}_{\mathrm{p}}$ before entering the interaction chamber is polarized along $\vec{\hat{P}}_2 = \frac{1}{\sqrt{2}}(\vec{\hat{y}}+\vec{\hat{z}})$ (see Fig. 4.3). As it propagates through the aligned sample, its components accumulate a relative phase retardation φ. At the exit of the cell sample, after having traveled a sample length l, the complex probe field can be described by

$$\vec{E}_{\mathrm{p}}(t-\tau, x=l) = \frac{E_{\omega_{\mathrm{p}}}(t-\tau)}{2\sqrt{2}} \exp\left\{\mathrm{i}\omega_{\mathrm{p}}(t-\tau)\right\} \left\{\vec{\hat{z}} + \exp\left(-\mathrm{i}\varphi\right)\vec{\hat{y}}\right\} \tag{4.17}$$

with $\varphi = \frac{\omega_{\mathrm{p}} l}{c}\Delta n$ and τ the time delay between the pump and probe pulses. After passing through the analyzer $\vec{\hat{A}} = \frac{1}{\sqrt{2}}(\vec{\hat{z}} - \vec{\hat{y}})$, the probe field intensity can be written as

$$\left|\boldsymbol{E}_{\mathrm{p}} \cdot \hat{\boldsymbol{A}}\right|^2 (t-\tau, x=l) = \frac{\left|E_{\omega_{\mathrm{p}}}(t-\tau)\right|^2}{4} \left(1-\cos\varphi\right), \tag{4.18}$$

which can be approximated by the following relation:

$$\left|\boldsymbol{E}_\mathrm{p} \cdot \hat{\boldsymbol{A}}\right|^2 (t-\tau, x=l) \approx \frac{\left|E_{\omega_\mathrm{p}}(t-\tau)\right|^2}{8}\left(\frac{\omega_\mathrm{p} l}{c}\Delta n\right)^2 \tag{4.19}$$

if we assume $\varphi \ll 1$. So we see that the intensity before detection is directly proportional to the birefringence $(\Delta n)^2$ and as a result to $(\langle\cos^2\theta\rangle - \frac{1}{3})^2$, considering the pump pulse being linearly polarized along the $\vec{\hat{P}}_1 = \vec{\hat{z}}$ direction. Δn resulting from the pump applied at time t, the signal delivered by the photomultiplier is given by the convolution product

$$\mathscr{S}(\tau) \propto \int_{-\infty}^{+\infty} \mathrm{d}t\, I_\mathrm{p}(t-\tau)\,(\Delta n(t))^2 \tag{4.20}$$

with I_p the probe intensity. This expression is valid for homodyne detection. For heterodyne detection, a local oscillator is introduced between the gas sample and the analyzer [21, 22]. In this case, the signal is described by

$$\mathscr{S}(\tau) \propto \int_{-\infty}^{+\infty} \mathrm{d}t\, I_\mathrm{p}(t-\tau)\,(\Delta n(t) + \mathscr{P})^2 \tag{4.21}$$

with $\mathscr{P}$ a parameter that accounts for an additional static birefringence acting as a local oscillator [22].

Figure 4.4 presents the polarization signal versus the pump-probe delay as measured at room temperature in CO_2 with an homodyne detection. The experiment was performed under static cell conditions. The temporal trace is characterized by

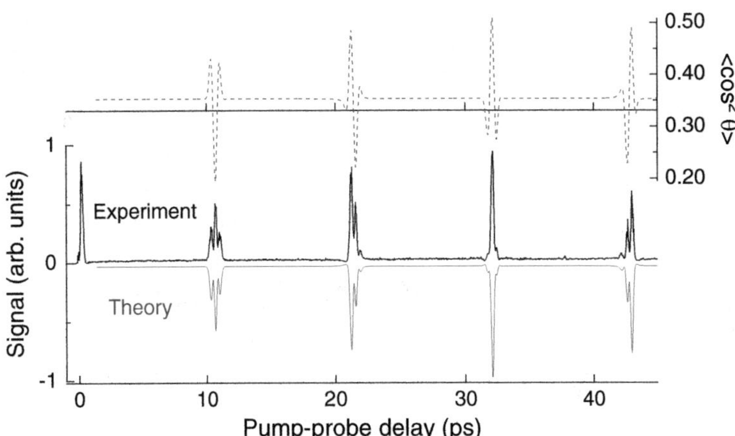

Fig. 4.4 *Lower graph*: homodyne signal (Experiment) versus pump-probe delay recorded in CO_2 at room temperature and 0.15 bar. Numerical simulation (Theory) for a peak intensity of 40 TW/cm^2 (plotted up-side down). *Upper graph*: corresponding values of $\langle\cos^2\theta\rangle$ (*dotted line*)

transients equally spaced by $T_r/4$ (see Sect. 4.2.1). The signal recorded close to the zero delay results from the quasi-instantaneous electronic response (compared to the pulse duration) combined with the rotational retarded response of the molecules. The convolution of these two contributions results in a transient signal occurring just after the peak intensity of the pump. The nonzero background signal observed between the revivals stems from the permanent alignment of the molecules [2]. The observed signal is compared with the numerical simulation of (4.20) based on the model described in Sect. 4.2. The calculation uses an effective pump peak intensity of 40 TW/cm^2, for which the structural shape of the transients reproduces the observation. This intensity is consistent with the measured value. We emphasize that the degree of alignment is determined through the shape of the pump-probe signal and not through the amplitude (although a calibration procedure is possible [13]). The temporal shape of the quantity $\left(\langle\cos^2\theta\rangle - 1/3\right)^2$ is indeed very sensitive to the degree of alignment [i.e., to small changes in $\langle\cos^2\theta\rangle(t)$] due to the occurrence of permanent alignment that modifies the asymmetry of the revivals [2]. The upper graph of Fig. 4.4 displays the corresponding value of $\langle\cos^2\theta\rangle$. Its horizontal scale axis is shifted by the vertical offset 1/3 corresponding to the isotropic value of $\langle\cos^2\theta\rangle$. The figure shows that each revival results from alignment ($\langle\cos^2\theta\rangle > 1/3$) and planar delocalization ($\langle\cos^2\theta\rangle < 1/3$) of the molecular axis. In the first case, the angular distribution is squeezed along the electric field direction, whereas it is flattened around the plane perpendicular to the field axis, in the second case.

The effect of the temperature on the molecular alignment is shown in Fig. 4.5, where the experiment was conducted in a molecular jet [13]. The signal is heterodyned due to combined effects of low signal-to-noise ratio and static birefringence introduced by the cell windows. The obtained values of $\langle\cos^2\theta\rangle$ can be compared

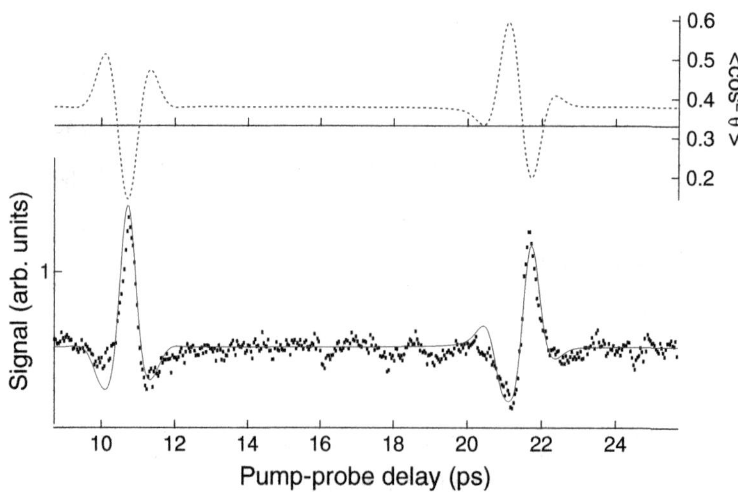

Fig. 4.5 *Lower graph*: heterodyne signal (*dots*) versus pump-probe delay recorded in a supersonic jet of CO_2. Numerical simulation (*full line*) of (4.21) performed at 60 K for a peak intensity of 30 TW/cm^2. *Upper graph*: corresponding values of $\langle\cos^2\theta\rangle$ (*dotted line*)

to those presented in Fig. 4.4 obtained at 300 K. As shown, lowering the temperature, i.e., narrowing the thermal distribution of the rotational states, leads to a higher degree of alignment [23].

The polarization technique has also been used to investigate asymmetric top molecules. For nonsymmetric top molecules, field-free alignment is generally more difficult to be achieved compared to symmetric top molecules as, e.g., the linear ones. One reason is that the rotational components of the wave packet do not fully rephase after the excitation because of the nonperiodic rotational motion of the asymmetric top. Another reason is that the laser field exerts a torque simultaneously upon two molecular axes as it will be shown next. For a laser field polarized along the z-axis, it can be shown from Sect. 4.2.1 and the polarizability tensor of an asymmetric top molecule

$$\alpha' = \begin{pmatrix} \alpha_{XX} & 0 & 0 \\ 0 & \alpha_{YY} & 0 \\ 0 & 0 & \alpha_{ZZ} \end{pmatrix} \tag{4.22}$$

that the interaction Hamiltonian is

$$H_{\text{int}} = -\frac{1}{4}\mathscr{E}^2(t)\alpha_{zz} = -\frac{1}{4}\mathscr{E}^2(t)\left(\Delta\alpha_{XY}\sin^2\theta\cos^2\chi + \Delta\alpha_{ZY}\cos^2\theta\right) \tag{4.23}$$

with $\Delta\alpha_{II'} = \alpha_{II} - \alpha_{I'I'}$ the polarizability anisotropy of the molecular axes $I = X, Y, Z$ depicted in Fig. 4.6 for the ethylene molecule. We see therefore that the field interaction produces a potential, where both Euler angles θ and χ are confined, with nevertheless a more pronounced effect for θ in case of a near symmetric top molecule with $|\Delta\alpha_{ZY}| > |\Delta\alpha_{XY}|$, implying that two molecular axes are constrained. Figure 4.7 shows the intensity-dependence signal of the ethylene molecule after it has been exposed to a linearly polarized laser pulse. For an asymmetric top molecule, the birefringence measured by the probe pulse depends on the angles θ and χ

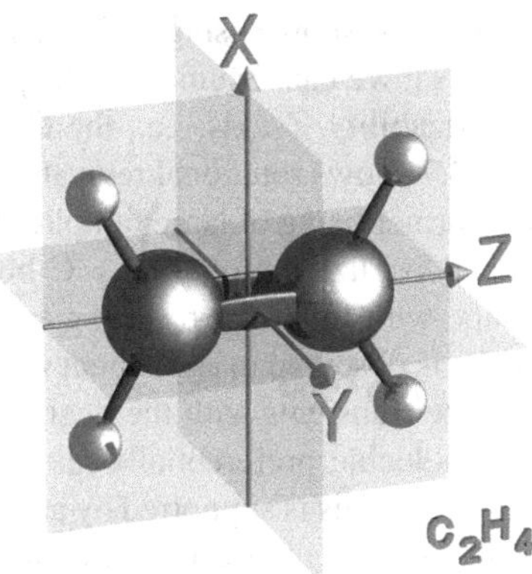

Fig. 4.6 Ethylene molecule depicted with its principal axes of inertia

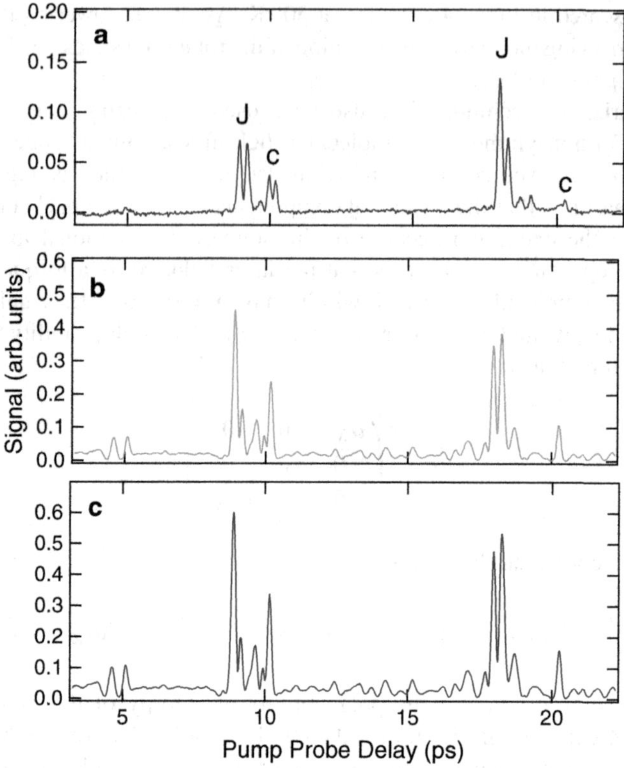

Fig. 4.7 Probe signals recorded in ethylene at room temperature for different intensities of the alignment laser: (**a**) 2×10^{12} W/cm^2, (**b**) 25×10^{12} W/cm^2, (**c**) 50×10^{12} W/cm^2

$$\Delta n(t) = \frac{3\varrho}{4\bar{n}\varepsilon_0}\left\{\Delta\alpha_{ZX}\left(\langle\cos^2\theta\rangle - \frac{1}{3}\right) + \Delta\alpha_{YX}\left(\langle\sin^2\theta\sin^2\chi\rangle - \frac{1}{3}\right)\right\}, \tag{4.24}$$

as it can be shown using (4.7), (4.14–4.16), and (4.22). For ethylene with $|\Delta\alpha_{ZX}| > |\Delta\alpha_{YX}|$, we can assume that the probe mainly measures the alignment of the major polarizability Z-axis, i.e., the C=C bond axis (see Fig. 4.8). The probe signal in Fig. 4.7 shows rotational revivals with identified J-type and C-type transients [24], the former being of larger amplitude. Each transient results from alignment and planar delocalization of the C=C bond axis. A clear modification of the transients' shape with intensity, resulting essentially from the creation of a permanent alignment, is observed in agreement with the theory [17]. The increasing amplitude of the C-type transients with the degree of alignment suggests that the laser field rotates the molecule preferentially about the C-axis, i.e., within the molecular plane of ethylene. This is supported by an analysis of the populations transferred by the laser field during the pulse and the repartition of the quantum levels of the molecule on the rotational energy surface. More details about this work can be found elsewhere [17].

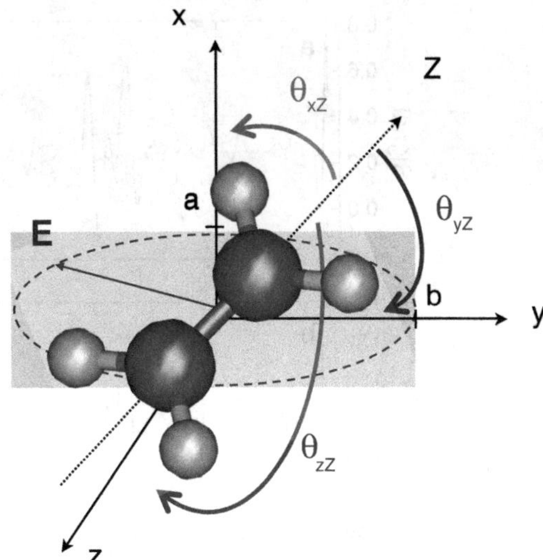

Fig. 4.8 Ethylene molecule depicted in the laboratory reference frame with the direction cosines and the elliptic field (see text)

So far we have been concerned with linearly polarized excitation, which restricts the angular confinement to only one molecular axis. A more appealing issue is the possibility of embedding the molecule in a potential where two molecular axes would be confined so as to lead to three-dimensional alignment [25–27]. For this purpose, we have investigated the ethylene molecule exposed to an elliptically polarized laser pulse [28] with the polarization detection scheme of Fig. 4.3. For an elliptic field propagating along the z-axis

$$\vec{E}(t) = \vec{\mathscr{E}}(t)\left(a\vec{\hat{x}}\cos\omega t + b\vec{\hat{y}}\sin\omega t\right) \tag{4.25}$$

with $a^2 + b^2 = 1$ and $b^2 > a^2$ (see Fig. 4.8), we can show from Sect. 4.2 and using the polarizability tensor (4.22) that the interaction Hamiltonian takes the form

$$\begin{aligned} H_{\text{int}} = \frac{1}{4}\mathscr{E}^2(t)\left\{\Delta\alpha_{ZX}\left[a^2\cos^2\theta_{zZ} - (b^2 - a^2)\cos^2\theta_{yZ}\right]\right. \\ \left. - \Delta\alpha_{XY}\left[a^2\cos^2\theta_{zY} - (b^2 - a^2)\cos^2\theta_{yY}\right]\right\}. \end{aligned} \tag{4.26}$$

For convenience, we use here the direction cosines $\cos\theta_{\gamma\Gamma}$ ($\gamma = x, y, z$, $\Gamma = X, Y, Z$) that are the elements appearing in the 3×3 matrix of (4.7). They describe the orientation of the molecular axes Γ with respect to the laboratory frame axes γ (see Fig. 4.8). By looking at the minima of the potential (4.26), we see that the interaction results in a simultaneous alignment of the molecular axes Z and Y along the largest y (term $\cos^2\theta_{yZ}$) and the smallest x component $\cos^2\theta_{zY}$ of the field respectively. The three-dimensional alignment can be therefore maximized if the energy between the two direction cosines is balanced. This requires that the field ellipticity satisfies the following condition:

$$a^2 = \Delta\alpha_{ZX}/\left(\Delta\alpha_{XY} + 2\Delta\alpha_{ZX}\right) \tag{4.27}$$

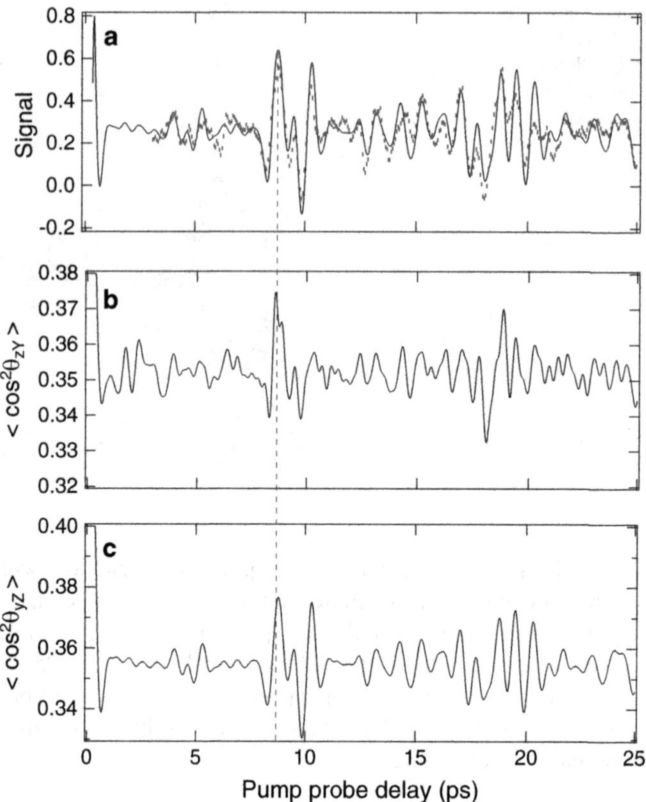

Fig. 4.9 (**a**) Alignment signal produced in a molecular jet of ethylene with a field ellipticity $a^2 = 0.44$ (*full line*) and numerical simulation (*dotted line*). Averaged direction cosines $\langle\cos^2\theta_{yZ}\rangle$ (**b**) and $\langle\cos^2\theta_{zY}\rangle$ (**c**) calculated for $T = 40$ K and 30 TW/cm^2 using the model presented in [28]

with $a^2 \approx 0.44$ in the case of ethylene. The birefringence signal recorded for this ellipticity is shown in Fig. 4.9. The maximum of alignment is observed at 8.7 ps (see the vertical dashed line) with $\langle\cos^2\theta_{yZ}\rangle = 0.37$, $\langle\cos^2\theta_{zY}\rangle = 0.37$, and $\langle\cos^2\theta_{xX}\rangle = 0.336$. The relatively weak alignment of the X molecular axis along the x-axis is due to the relatively high temperature at which the experiment was conducted. As shown in (4.24), the birefringence signal produced by molecules of small polarizability asymmetry like ethylene mainly provides information about the alignment of the Z-axis. In this respect, the birefringence technique does not allow a direct measurement of the field-free 3D alignment [27].

4.3.2 Space-Resolved Single-Shot Detection

The arrangement for space-resolved transient birefringence measurements is similar to the one presented in the previous section, except for the two laser beams that must be crossed at right angle to meet the requirement for space-resolved detection

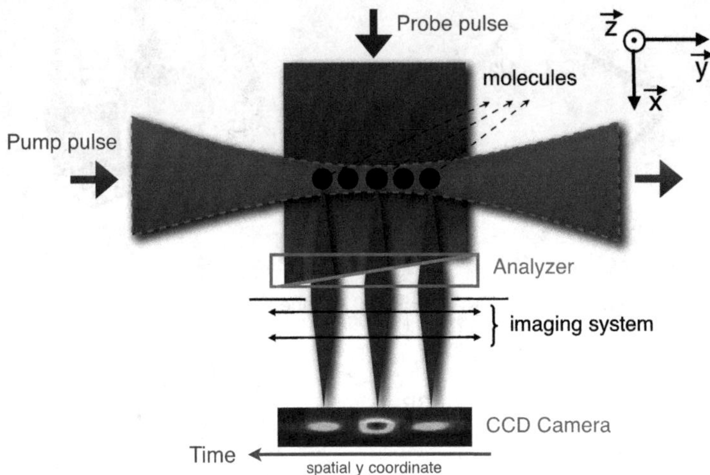

Fig. 4.10 Schematic diagram of the space-resolved transient birefringence technique

on a single-shot basis. The principle of the technique is presented in Fig. 4.10. A linearly polarized focused beam (i.e., the pump) propagates along the gas cell direction coinciding with the y-axis, while a second collimated beam (i.e., the probe) intersects it at a right angle and then passes through an analyzer set at 45° with respect to the polarization of the pump beam. The depolarized probe field generated within the overlapping region of the two intersecting beams is imaged onto a CCD camera. As the probe beam travels through the gas sample, it interacts with the molecules that have been previously aligned by the pump beam at specific times that correspond to their respective y-coordinates. The dynamical alignment resulting from the pump beam interaction is thus converted by the probe pulse into the horizontal axis of the CCD image. In addition, the vertical axis of the image provides information about the intensity distribution along the transverse profile of the focused pump beam. A schematic of the experimental setup is shown in Fig. 4.11 [29]. The laser source was a chirped pulsed amplified Ti:sapphire laser delivering pulses of 100 fs duration at 1 kHz repetition rate. The delay line allowed for the adjustment of the temporal position of the detection window by setting the retardation of the probe pulse with respect to the pump. The telescope (L_1, L_2) was used to image the overlapping region of the two laser beams on the CCD camera. It should be mentioned that the method, called Femtosecond Time-resolved Optical Polarigraphy (FTOP), had been originally developed to image the propagation of femtosecond pulses in a transparent medium [30].

Snapshot imaging of postpulse transient molecular alignment is shown in Fig. 4.12b for a detection window centered on the first revival of CO_2. As already mentioned, the advantage of the FTOP technique is that both time and intensity dependencies can be simultaneously obtained from a single image. The signals measured along the different horizontal lines of the CCD camera reflect the alignment produced at different intensities within the pump beam transverse profile.

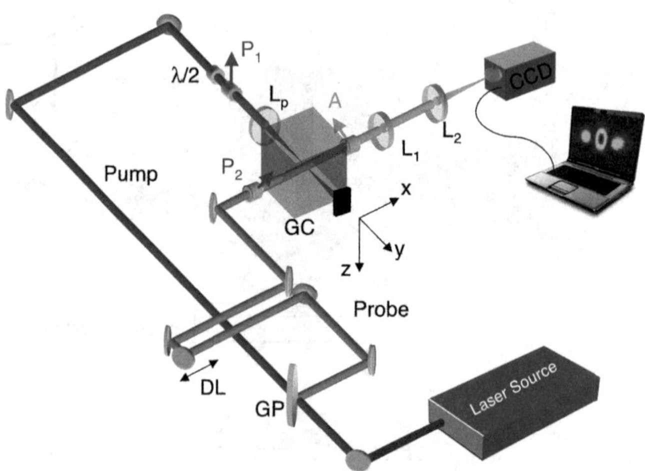

Fig. 4.11 Experimental setup for space-resolved transient birefringence measurements. L_i: Lenses, P_j: Polarizers, A: Analyzer, GP: Glass Plate, DL: Delay Line, GC: Gas Cell, CCD: Charged-Coupled Device camera

To highlight this aspect, the signals recorded along the central (line 0), the third (line 3), and the fifth horizontal lines (line 5) of the CCD camera are shown in Fig. 4.12a. The comparison between the three graphs shows the alteration of the signal produced by molecules located at different radial distances $|z| = 5$ pixel (line 5) and 3 pixel (line 3) from the beam center $z = 0$ pixel (line 0). To obtain the values of $\langle \cos^2 \theta \rangle$, we have calculated the signal versus the delay τ

$$\mathcal{S}(y, z, \tau) \propto \int_{-\infty}^{\infty} \mathrm{d}t \left\{ \int_{-\infty}^{\infty} \mathrm{d}x \, E_{\mathrm{pr}}(x, y, z, t - \tau - x/c) \Delta n \, (x, y, z, t) \right\}^2 \tag{4.28}$$

at the image coordinate (y, z). In this expression, the integration along the x-axis accounts for the depth of the imaging system being larger than the spot of the pump beam. Each experimental data point presented in Fig. 4.12a results therefore from a one-dimensional spatial averaging of the molecular alignment produced along the propagation axis of the probe beam. Δn has been obtained according to the model described in Sect. 4.2 and assuming a Gaussian temporal intensity profile for the pump beam, while the pulse duration has been determined from the corresponding autocorrelation measurements. The beam waist was inferred from analyzing the signal along the vertical lines of the recorded images. The simulated signals are shown inverted in Fig. 4.12a together with the corresponding expectation values $\langle \cos^2 \theta \rangle$ shown in the upper parts of the figure. The latter has been calculated at a pump intensity $I = 62$ (line 0), 34 (line 3), and 7 TW/cm^2 (line 5), respectively, which corresponds to the intensity experienced by the molecules located on the pump beam axis ($x = 0$). The largest signal is observed at line 0, where the molecules are

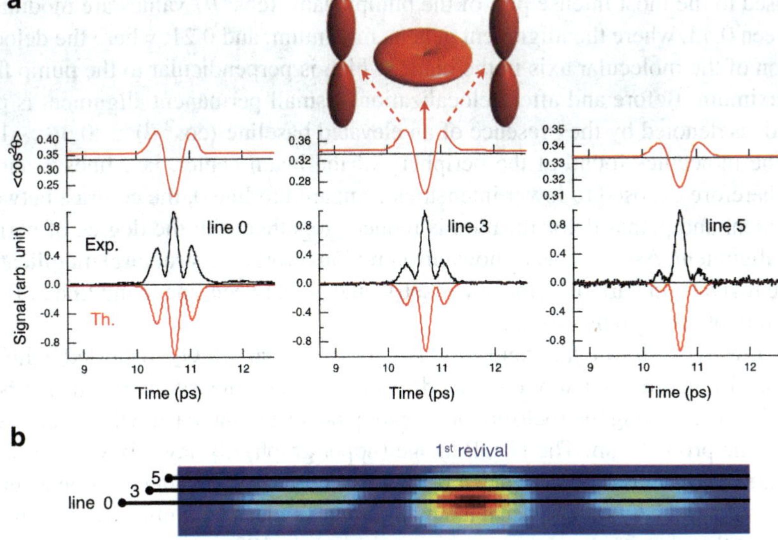

Fig. 4.12 Transient molecular alignment revivals of CO_2 at room temperature and 1 bar. The delay τ between the two pulses is 10.7 ps. **(a)** Signals extracted from three different lines of the CCD camera (see text) and comparison with numerical simulations. **(b)** Image of the CCD camera averaged over 100 laser shots

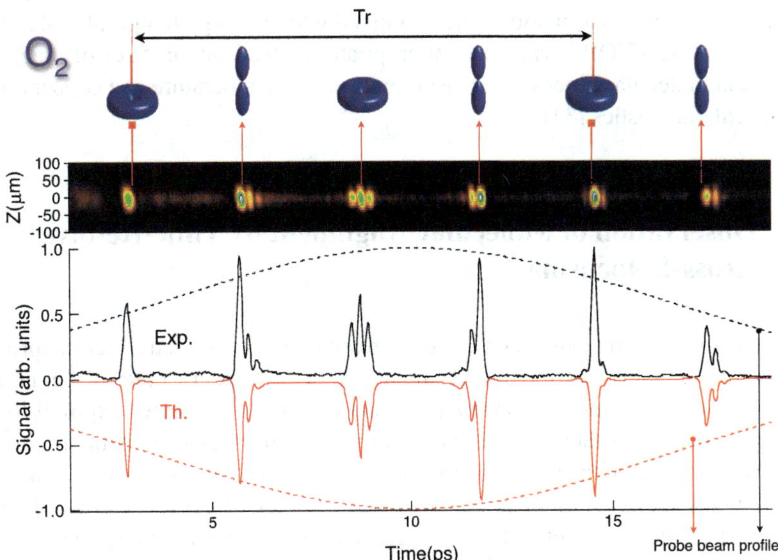

Fig. 4.13 *Upper graph*: Averaged image (100 shots) recorded in O_2 at room temperature and atmospheric pressure. The pump-probe delay was set at $\tau = 10$ ps. *Lower graph*: signal (*full line*) extracted from the horizontal central line of the CCD image ($z = 0$). The theoretical signal (shown reversed) has been calculated at a peak intensity of 49 TW/cm^2. Transverse profile of the probe beam (*dashed lines*)

exposed to the most intense part of the pump beam. $\langle\cos^2\theta\rangle$ values are modulated between 0.43, where the alignment gets its maximum, and 0.21, where the delocalization of the molecular axis in the plane, which is perpendicular to the pump field is maximum. Before and after delocalization, a small permanent alignment is produced, as denoted by the presence of an elevated baseline $\langle\cos^2\theta\rangle \simeq 0.36 > 1/3$. For the molecules found at the periphery of the beam center, i.e., lines 3 and 5, and therefore exposed to lower intensities compared to line 0, the contrast between alignment and planar delocalization is reduced together with the degree of permanent alignment. As it has been shown in a previous work, the structural modification of the revivals, in particular the asymmetry change between line 0 and line 5, is the signature of strong field alignment.

Figure 4.13 shows the field-free alignment signal of O_2 recorded with an expanded temporal window compared to the previous measurement. It has been recorded by reducing the focusing of the pump beam and increasing the beam diameter of the probe beam. The FTOP image (upper graph) displays six well-resolved alignment transients recorded over a temporal window of 18 ps (the rotational period of O_2 is $T_r \approx 11.6$ ps). The permanent alignment is clearly identifiable. We should point out that despite the short acquisition time (only 100 ms), the excellent resolution and signal-to-noise ratio make the measurement fully exploitable for a detailed analysis of the alignment dynamics over a long timescale. The curves shown with dashed lines (lower graph) indicate the spatial profile of the probe beam along the y coordinates. The variation of the probe intensity is mainly responsible for the signal attenuation observed at the limits of the region recorded by the camera. The relatively short acquisition time combined with the large temporal scale of the technique make FTOP particularly appropriate for feedback control of alignment dependent molecular processes or fast monitoring of temperature and concentration for optical diagnostics [31].

4.4 Observation of Molecular Alignment by Time-Resolved Cross-Defocusing

The birefringence-related techniques addressed in the previous section provide information about the difference of alignments with respect to two different axes of the laboratory frame. We have shown that for linear molecules exposed to linearly polarized fields they are adequate to provide a direct measurement of $\langle\cos^2\theta\rangle$. However, in case of asymmetric top molecules or linear molecules exposed to elliptic fields, a more complete characterization of the alignment can be obtained through measurements related to a single axis of the laboratory frame. In [32], we have shown that this can be achieved by employing the cross-defocussing technique. The basic idea is to exploit the modification experienced by a laser beam as it propagates through the refractive index gradient produced by the alignment of the molecular gas sample. The principle of the method is schematically

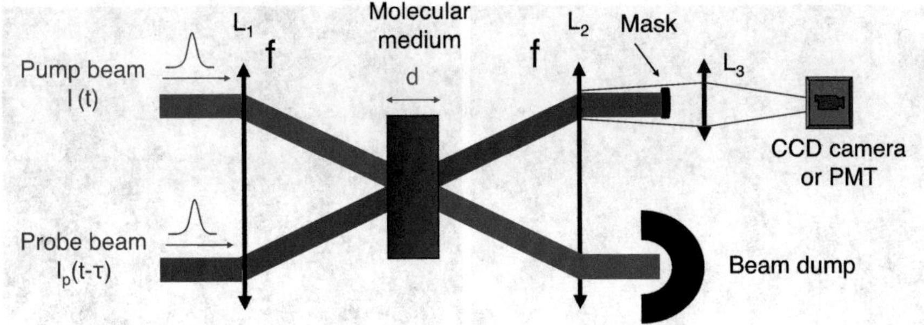

Fig. 4.14 Schematic diagram of the coronograph used to observe cross-defocusing. d represents the interaction length defined by the overlapping of the two beams. Lenses L_1 and L_2 have a focal length $f = 300$ mm. The lens L_3 is used to focus the light on the CCD camera or a photomultiplier tube (PMT)

depicted in Fig. 4.14. A linearly polarized pump pulse is focused in a molecular gas and aligns the molecules. The subsequent field-free alignment leads to a spatial gradient of the time-dependent refractive index. The effect can be seen as a formation of a time-dependent nonlinear lens in the medium, also known as Kerr lens effect. A time-delayed probe pulse crosses the pump pulse at the focus and experiences the gradient of the refractive index. As a consequence, it undergoes a modification of its spatial properties. This modification is directly related to the degree of alignment, which can be inferred from the measurement. As the change on the beam profile is small, the effect cannot be accurately measured directly on the beam size. To improve the sensitivity of the method, we use a mask to block the central part of the probe beam, like in a coronographic technique, and focus the light passing around the mask on a detector. The change of the beam size is then detected as a variation of the amount of light reaching the detector. In practice, the diameter of the mask is chosen large enough to prevent most of the light corresponding to a signal close to zero without defocusing. The method is thus nearly background free. The detector can be either a CCD camera or a photomultiplier tube. Typical images detected by a CCD camera are shown in Fig. 4.15. The nonuniform distribution of light observed in Fig. 4.15b results from the small crossing angle of about 2° between the two beams contained in an horizontal plane perpendicular to the image. The results are shown in Fig. 4.16 for the case of CO_2 at room temperature and at a pressure of 0.3 bar. The signal is compared to a numerical simulation based on the model described below.

The refractive index produced along the polarization axis of the pump field, i.e., the z-axis, is given by (4.14) and (4.15b)

$$n = 1 + \frac{\varrho}{2\bar{n}\varepsilon_0} \left\{ \bar{\alpha} + \Delta\alpha \left(\langle \cos^2\theta \rangle(t) - 1/3 \right) \right\}, \tag{4.29}$$

where we recall that

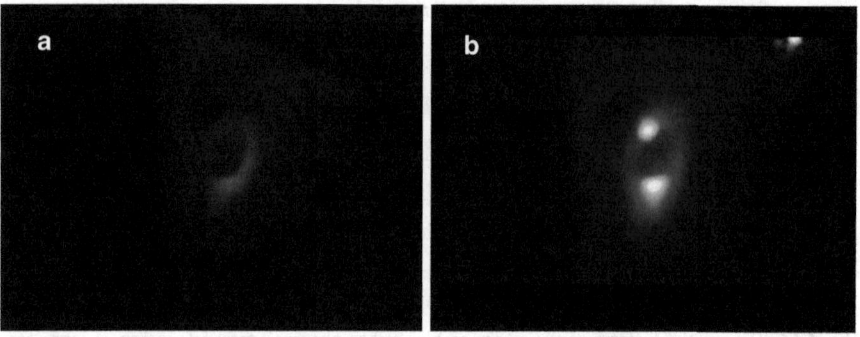

Fig. 4.15 Images of the probe field recorded after the coronograph (**a**) without and (**b**) with the aligning laser, at a probe delay corresponding to a revival of alignment

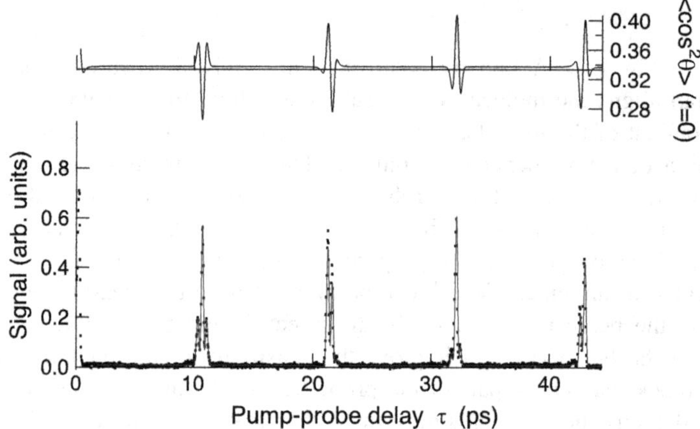

Fig. 4.16 Cross-defocusing signal (*dotted curve*) of CO_2 for a peak intensity of 15 TW/cm^2 and pulse duration of 100 fs. Numerical simulation (*full curve*) and the corresponding value of $\langle\cos^2\theta\rangle$ depicted on the right line scale

$$\frac{\varrho\Delta\alpha}{2\overline{n}\varepsilon_0}\left(\langle\cos^2\theta\rangle(t) - 1/3\right) = \delta n \tag{4.30}$$

is the orientational contribution to the Kerr index introduced in Sect. 4.2.2. We assume that the pump beam, linearly polarized along the z-axis, propagates along the x-axis with a Gaussian intensity profile given by

$$I = I(r = 0, x, t)\exp\left(\frac{-2r^2}{w^2(x)}\right) \tag{4.31}$$

with $r = \sqrt{y^2 + z^2}$ the radial distance, $I(r = 0, x, t)$ the on-axis intensity, and w the beam radius. The intensity distribution over the transverse profile of the beam results in an index gradient produced along the beam radius. If we assume that the alignment is produced far below saturation, which is generally fulfilled, then $\langle\cos^2\theta\rangle$ in (4.30) depends linearly on the intensity. Under such conditions, the variation of the index in the transverse direction can be approximated from (4.30) and (4.31) by

$$\delta_{\perp}n \approx \frac{\varrho\Delta\alpha}{2\overline{n}\varepsilon_0}\exp\left(\frac{-2r^2}{w^2(x)}\right)\left(\langle\cos^2\theta\rangle(r=0,x,t)-1/3\right) \tag{4.32a}$$

$$\approx \delta_{\perp}n(r=0,x,t)\exp\left(\frac{-2r^2}{w^2(x)}\right) \tag{4.32b}$$

with $\delta_{\perp}n(r = 0,x,t)$ the on-axis Kerr index resulting from the orientation of the molecule. The time-delayed probe beam traveling through this index gradient undergoes a focusing or defocusing, depending on whether the molecular sample is aligned ($\langle\cos^2\theta\rangle > 1/3$) or delocalized ($(\langle\cos^2\theta\rangle < 1/3)$), respectively. In the former case, the gradient is positive, whereas in the latter case it is negative. An analytic solution of the probe beam profile can be obtained by approximating the index gradient (4.32b) by a parabolic function in the limit $r \ll w(x)$

$$\delta_{\perp}n = \delta_{\perp}n(r=0,x,t)\left(1-\frac{2r^2}{w_0^2}\right). \tag{4.33}$$

Here, it has been further assumed that the Rayleigh length z_R of the beam is much larger than the interaction length d (see Fig. 4.14), so that the beam profile remains nearly constant over the interaction length $w(x) \approx w_0$. Using a Gaussian optics calculation, it can be shown that (4.33) results in an effective Kerr lens of focal length [33]

$$f_{\mathrm{Kerr}} = \frac{n_L w_0^2}{4d}\frac{1}{\delta_{\perp}n(r=0,t)}, \tag{4.34}$$

which modifies the beam radius w_p of the probe according to the relation

$$w_p'(x,t) = w_p(x,t)\left(1+\frac{16d^2z_R^2}{n_L^2w_0^4}\delta_{\perp}^2n(r=0,t)\right) \tag{4.35}$$

with n_L the linear refractive index and x the distance after the lens. This expression is valid for $d < f_{\mathrm{Kerr}} \ll f$, which is generally verified [32]. We notice that since w_p' does not depend on the sign of the $\delta_{\perp}n(r = 0,t)$, alignment, and planar delocalization both lead to an increase of the beam size. The signal recorded by the detector located after the mask of radius a is

$$\mathscr{S}(\tau) \equiv\propto \int_{-\infty}^{+\infty} \mathrm{d}t\, I_p(t-\tau)\int_a^{+\infty}\mathrm{d}r\frac{r\exp\left(-2\left(\frac{r}{w_p'(x,t)}\right)^2\right)}{\left(w_p'(x,t)\right)^2} \tag{4.36}$$

and leads according to (4.33) and (4.35), after spatial integration, to

$$\mathscr{S}(\tau) \propto \int_{-\infty}^{+\infty} \mathrm{d}t I_{\mathrm{p}}(t-\tau)\delta_{\perp}^{2} n(r=0,t) \tag{4.37a}$$

$$\propto \int_{-\infty}^{+\infty} \mathrm{d}t I_{\mathrm{p}}(t-\tau)\left(\langle\cos^{2}\theta\rangle(r=0,t)-1/3\right)^{2}. \tag{4.37b}$$

As it is shown, the signal provided by the coronograph allows to obtain the alignment produced by the pump beam at a peak intensity ($r = 0$). The calculated signal according to (4.37) is depicted in Fig. 4.16 together with $\langle\cos^2\theta\rangle(r = 0)$. The cross-defocusing has been used to perform three-dimensional characterization of field-free alignment induced by an elliptically polarized field [34]. In particular, measurements performed along two different axes of the laboratory frame have permitted the observation of two-direction alignment alternation of a linear molecule [35].

4.5 Applications

The molecular alignment probed by optical methods can be used to determine molecular parameters and/or to obtain information about the surrounding environment of the aligned molecules. In this section, we will present some applications of molecular alignment that illustrate these aspects.

4.5.1 Determination of Ionization Probabilities

The cross-defocusing technique presented in Sect. 4.4 is sensitive to the presence of free electrons resulting from laser-induced ionization of the gas sample. Field-free molecular alignment conducted at large intensity can lead to plasma formation that affects the alignment signal measured by cross-defocusing. In [36], we have shown that one can take advantage of this feature to obtain absolute probabilities of single-ionization based on field-free alignment. Figure 4.17a displays the cross-defocusing signal of N_2 aligned with a short pulse of peak intensity sufficiently high to lead to significant ionization of the medium. The dotted line of Fig. 4.17b corresponds to the mirror image of the simulation performed under plasma-free conditions. A comparison of the two graphs reveals a discrepancy between observed and calculated signals. As can be seen, this discrepancy is particularly pronounced for the baseline, as well as for the small transients, where double peak structures are observed in the simulated signals compared to the single peak structures experimentally observed and vice versa. This disagreement is due to the contribution of the free electrons, which act as a diverging lens. The refractive index induced by ionization can be written [37]

$$\delta_{\perp} n_{\mathrm{p}} = -\frac{n_{\mathrm{e}}}{2n_{\mathrm{cr}}} = -\frac{\varrho}{2n_{\mathrm{cr}}} P_{\mathrm{ion}} \tag{4.38}$$

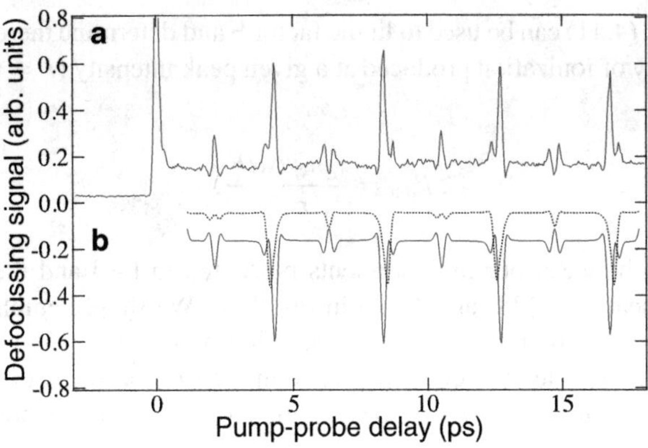

Fig. 4.17 (**a**) Cross-defocusing signal (*full line*) recorded in N_2 at 300 K, at a static pressure of 0.3 bar, a peak intensity of 58 TW/cm^2, and a pulse duration of 100 fs. (**b**) Numerical simulations shown reversed with (*full line*) and without (*dotted line*) plasma contribution

with n_e the electron density and $n_{cr} = 1.75 \times 10^{21}$ cm^{-3} the critical electron density at 800 nm. Assuming that the ionization occurs with a nonlinear factor N_L [38, 39], the ionization probability

$$P_{ion} \propto I^{N_L} \tag{4.39}$$

leads to a negative index gradient that can be written, using the same approximations as in Sect. 4.4,

$$\delta_\perp n_p = \delta_\perp n_p(r = 0)\left(1 - N_L \frac{2r^2}{w_0^2}\right) \tag{4.40}$$

with $\delta_p n(r = 0) < 0$ the refractive index of the plasma produced at the beam center. Following the demonstration made to derive (4.37), we can show that the cross-defocusing signal modified by the free electrons writes

$$\mathscr{S}(\tau) \propto \int_{-\infty}^{+\infty} \mathrm{d}t\, I_p(t-\tau)\left(\delta_\perp n(r=0,t) + \delta_\perp n_p(r=0)\right)^2 \tag{4.41a}$$

$$\propto \int_{-\infty}^{+\infty} \mathrm{d}t\, I_p(t-\tau)\left(\langle\cos^2\theta\rangle(r=0,t) - 1/3 + F\right)^2 \tag{4.41b}$$

with $F = 2\varepsilon_0 N_L \delta_\perp n_p(r = 0)/\varrho\Delta\alpha$ a negative factor. Compared to expression (4.21), we see that the plasma contribution F acts as a local oscillator with respect to the alignment contribution $\langle\cos^2\theta\rangle(r = 0, t) - 1/3$. The alignment signal heterodyned by the free electrons explains the discrepancy observed in Fig. 4.17. The small revivals are more affected by the plasma, since for them the alignment and plasma contributions are comparable. The simulation of (4.41) is presented by the full line in Fig. 4.17b. We see that the observation is well reproduced by the model, not only the shape of the transients but also the baseline resulting from the combined effects of permanent alignment and plasma contribution. For molecules whose polarizability is well known, (4.41) can be used to fit the factor F and determine therefore the absolute probability of ionization produced at a given peak intensity ($r = 0$) through the relation

$$P_{\text{ion}} = -\frac{n_{\text{cr}} \Delta\alpha F}{\varepsilon_0 N_L}. \tag{4.42}$$

A comparison between our measurements presented in [40] and the results from the model appeared in [38] are shown in Fig. 4.18. We should emphasize that the present method is experimentally simple and does not require any precise calibration of the overall detection efficiency encountered in ion mass or photoelectron spectroscopy. Furthermore, it can also be applied for calibration of atomic ionization probabilities [40].

4.5.2 Determination of Nonlinear Refractive Indices

In the previous section, we have discussed how field-free alignment can be used for absolute calibration of ionization probabilities. Similarly, electronic Kerr indices can also be determined by a thorough analysis of the alignment signal measured with an optical method. So far we have focused the discussions on the alignment occurring after turnoff of the laser pulse. During interaction with a femtosecond pulse, the slow rotational response of the molecule is excited together with the fast electronic response. The former leads to an alignment of the molecules that survives the pulse turnoff, whereas the last results in a prompt induced dipole that vanishes together with the pulse. When the pump pulse is present, the nonlinear refractive index felt by the probe pulse proceeds from the excitation of both electronic and

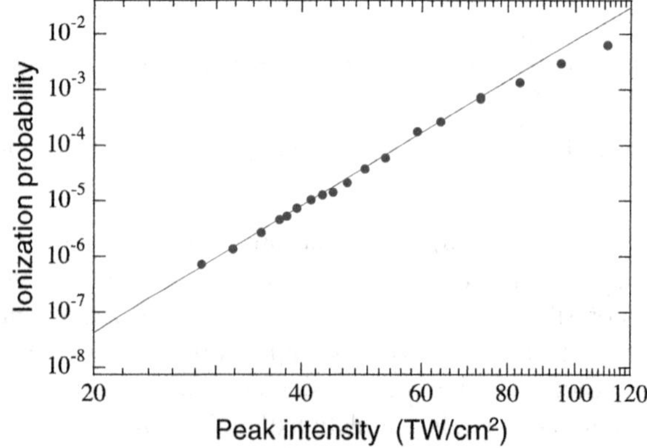

Fig. 4.18 Ionization probability of N_2 (*dots*). Results are compared with the semiempirical model (*full line*) from [38] for which the intensity has been multiplied by a factor 1.2

rotational responses. If the rotational response is known, then the electronic one can be deconvolved from the probe signal. We have taken advantage of this approach to measure high-order electronic Kerr index of major air components [41,42]. For this purpose, we have used the birefringence technique, which is insensitive to the plasma and allows to obtain the phase of the signal through heterodyne detection as well. The pure heterodyne detection is performed by recording twice the signal described in (4.21), but with opposite phase of the local oscillator $\mathscr{P}$, and then subtracting the two records so as to provide the signal

$$\mathscr{S}(\tau) \propto \int_{-\infty}^{+\infty} \mathrm{d}t \; I_\mathrm{p}(t-\tau)\Delta n(t) \tag{4.43}$$

with $\Delta n = \Delta n_\mathrm{rot} + \Delta n_\mathrm{el}$ for overlapped pulses. Figure 4.19 shows the normalized signal recorded in N_2 for three different intensities of the pump pulse. The positive signals observed close to the zero delay come from the alignment of the molecules involving the retarded rotational response. The intensity effect is evidenced through the permanent alignment indicated by an elevated baseline in the signals of Fig. 4.19b,c. Together with this effect, we observe in the same figures, at

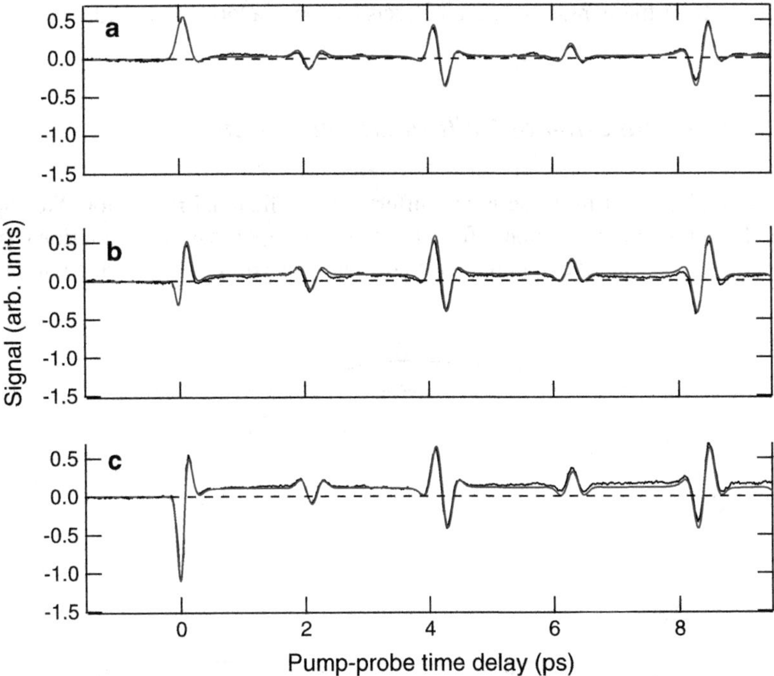

Fig. 4.19 Normalized pure heterodyne signal (*black lines*) recorded in N_2 at a static pressure of 100 mbar. The effective intensity is (**a**) 22 TW/cm^2, (**b**) 42 TW/cm^2, and (**c**) 49 TW/cm^2, respectively. All traces are normalized to the first positive transient. Simulations of (4.43) (*red lines*)

zero delay, a negative signal of increasing amplitude. This feature can be ascribed to an electronic Kerr response of opposite phase with respect to the rotational response. At low intensity, the two contributions are in phase, leading to a broader transient in Fig. 4.19a as compared to the other intensities. The inversion of the signal observed around 26 TW/cm^2 is attributed to a saturation of the electronic Kerr response.

The calculations shown in Fig. 4.19 have been performed using an electronic Kerr index written as a power series expansion $n_{\text{el}} = n_2 I + n_4 I^2 + \cdots n_8 I^4$ leading to the birefringence [41]

$$\Delta n_{\text{el}} = \frac{2}{3} n_2 I + \frac{4}{5} n_4 I^2 + \frac{6}{7} n_6 I^3 + \frac{8}{9} n_8 I^4. \tag{4.44}$$

The simulations are adjusted to the signals by fitting the Kerr terms n_i following the procedure described in [41]. The intensity dependence of the Kerr index measured in different gases is shown in Fig. 4.20. The sign inversion occurring above an intensity of few tens of TW/cm^2 is due to the higher order Kerr terms n_4 and n_8 of negative signs, never observed so far in gases. The same effect has been observed in Argon [41]. It should be mentioned that the values reported in Fig. 4.20 refer to the cross-Kerr effect. The corresponding values for the Kerr effect, as it would be observed in a single laser beam experiment, are given in [42]. The deep impact of the present results on laser filamentation related effects has been addressed in [43, 44].

4.5.3 Determination of Collisional Relaxation

Field-free alignment in dense gases suffers from collisional relaxation. We should remind that alignment originates from the coherent superposition of rotational states and is therefore sensitive to both amplitude and phase of the rotational components.

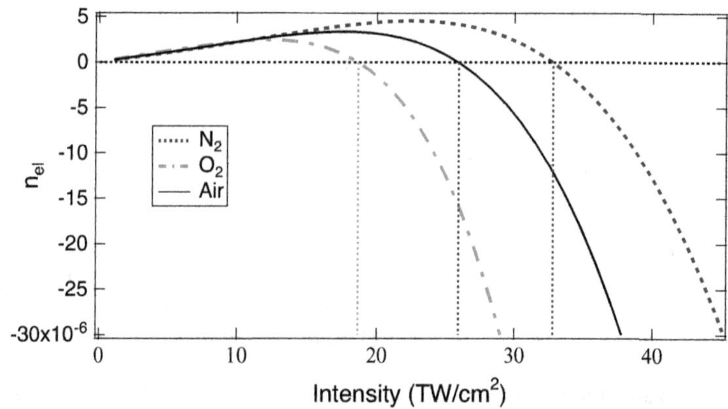

Fig. 4.20 Electronic cross-Kerr index at 800 nm versus intensity for different gases at Normal Temperature and Pressure (NTP)

Overall, the contact of the molecules with a dissipative environment results in a temporal decay of the alignment. However, amplitudes and phases are differently affected by this environment. In this respect, using the Liouville equation, it has been shown in [45] that the observable $\langle \cos^2\theta \rangle$ can be written as a sum of two terms

$$\langle \cos^2\theta \rangle = \langle \cos^2\theta \rangle_p + \langle \cos^2\theta \rangle_c, \tag{4.45}$$

where $\langle \cos^2\theta \rangle_p$ and $\langle \cos^2\theta \rangle_c$ are the permanent and transient alignment contributions, respectively. These terms are related to the diagonal (population) and off-diagonal (coherence) elements of the density matrix. Assuming that the former is sensitive to inelastic collisions (with time constant T_1), whereas the latter is sensitive to both elastic and inelastic collisions (with time constant T_2), it is expected that permanent and transient alignments exhibit a different temporal decay [45] depending on the nature of the molecules involved, the pressure, and the temperature. Figure 4.21 depicts such situation where the alignment signal of CO_2 is measured in a gas mixture having a 90:10 (CO_2:Ar) concentration ratio. The decay of permanent and transient alignments is clearly observable. The experiment is qualitatively well described by the simulation, which succeeds to reproduce correctly the shapes of the transients. However, due to the relatively low pressure it is not possible to deconvolve the effect of pure dephasing collisions (elastic collisions) from population transfers (inelastic collisions). CO_2-Ar mixture is a convenient case as all parameters encountered are relatively well known, but with a counterpart that T_1 and T_2 do not significantly differ close to NTP. So, experiments performed at larger pressure or/and temperature would be necessary to distinguish between the contribution of each effect.

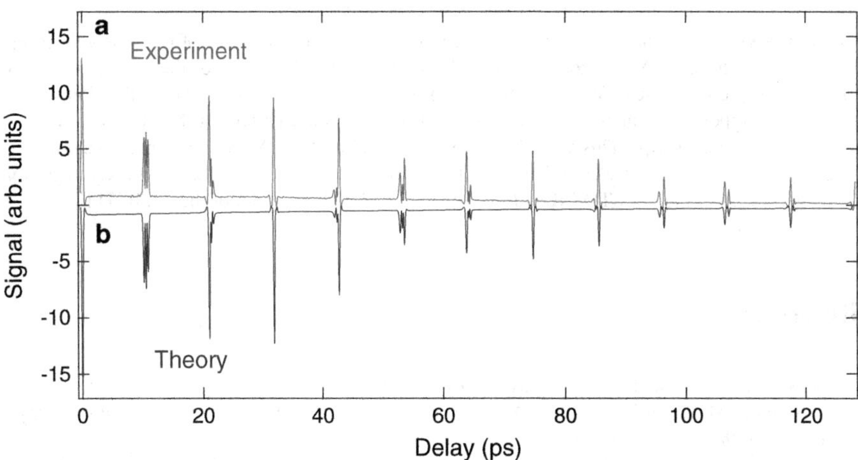

Fig. 4.21 (**a**) Alignment signal of CO_2-Ar at room temperature and 1.1 bar measured with the birefringence technique. (**b**) Simulation according to [46] shown inverted

4.6 Conclusion

In this chapter, we have reviewed the work performed by our group on field-free molecular alignment observed with noninvasive optical methods. After a brief description of the theory of laser-induced alignment restricted to linear molecules, we have introduced the principle of the optical measurements applied to the observation of field-free alignment. We have also shown that by measuring the alteration of weak probe pulse propagating through an aligned sample of molecules, information about the alignment of the molecular axes with respect to a space-fixed frame can be deduced. This modification can be observed through optical detection of either the birefringence or the characteristic Kerr lens of the aligned sample. We notice that alignment measured by time-resolved transient grating technique is not addressed here due to space limitations [47,48]. The corresponding techniques made to implement the optical detections have been then addressed by reporting a number of studies. Linear and asymmetric top molecules aligned with linear, circular, or elliptic fields have also been discussed. Multi-shot as well as single-shot measurements have been also presented. These investigations have shed light on the processes involved in molecular alignment, highlighting new features. The last part of the chapter has been devoted to present some applications of the field-free alignment based on optical measurements.

The main advantage of using such optical methods lies in their sensitivities and simplicities of implementation. The main disadvantage is due to the need of sufficiently good knowledge of the system parameters as they are mandatory for the theoretical modeling part. A second limitation is related to the gas density that usually prevents to perform experiments at low rotational temperatures. However, this last can be circumvented by employing high-density supersonic expansions. This should pave the way for new applications.

Acknowledgements Thanks are due to present and former PhD students of our research group, M. Renard, V. Renard, A. Rouzée, V. Loriot, R. Tehini, and T. Vieillard for their contributions to the research presented here. We thank S. Guérin, D. Sugny, H. R. Jauslin, and V. Boudon for the theoretical support that our work has benefited. The authors are grateful to S. Couris for critical reading of the manuscript. This work has been supported by the CNRS, the Conseil Regional de Bourgogne, the ACI *PHOTONIQUE*, the ANR *COMOC*, the Marie Curie European Reintegration Grant *MODISH* of the 6th RTD FP, and the *FASTQUAST* ITN Program of the 7th RTD FP.

References

1. H. Stapelfeldt, T. Seideman, Rev. Mod. Phys. **75**(2), 543 (2003)
2. V. Renard, M. Renard, S. Guérin, Y.T. Pashayan, B. Lavorel, O. Faucher, H.R. Jauslin, Phys. Rev. Lett. **90**(15), 153601 (2003)
3. J.J. Larsen, H. Sakai, C.P. Safvan, I. Wendt-Larsen, H. Stapelfeldt, J. Chem. Phys. **111**(17), 7774 (1999)
4. P.W. Dooley, I.V. Litvinyuk, K.F. Lee, D.M. Rayner, M. Spanner, D.M. Villeneuve, P.B. Corkum, Phys. Rev. A **68**(2), 23406 (2003)

5. J.G. Underwood, M. Spanner, M.Y. Ivanov, J. Mottershead, B.J. Sussman, A. Stolow, Phys. Rev. Lett. **90**(22), 223001 (2003)
6. J.O. Hirschfelder, C.F. Curtiss, R.B. Bird, *Molecular Theory of Gases and Liquids* (Wiley, New York, 1954)
7. R.N. Zare, *Angular Momentum: Understanding Spatial Aspects in Chemistry and Physics* (Wiley-Interscience, New York, 1988)
8. G. Herzberg, *Spectra of Diatomic Molecules* (Van Nostrand Reinhold, New York, 1950)
9. M. Morgen, W. Price, L. Hunziker, P. Ludowise, M. Blackwell, Y. Chen, Chem. Phys. Lett. **209**(1,2), 1 (1993)
10. M. Renard, E. Hertz, S. Guérin, H.R. Jauslin, B. Lavorel, O. Faucher, Phys. Rev. A **72**(2), 025401 (2005)
11. F. Rosca-Pruna, M.J.J. Vrakking, J. Chem. Phys. **116**(15), 6579 (2002)
12. P.W. Dooley, I.V. Litvinyuk, K.F. Lee, D.M. Rayner, M. Spanner, D.M. Villeneuve, P.B. Corkum, Phys. Rev. A **68**(2), 23406 (2003)
13. V. Renard, M. Renard, A. Rouzée, S. Guérin, H.R. Jauslin, B. Lavorel, O. Faucher, Phys. Rev. A **70**, 033420 (2004)
14. S. Zamith, Z. Ansari, F. Lépine, M.J.J. Vrakking, Opt. Lett. **30**(17), 2326 (2005)
15. D.W. Broege, R.N. Coffee, P.H. Bucksbaum, Phys. Rev. A **78**(3), 035401 (2008)
16. J.P. Cryan, P.H. Bucksbaum, R.N. Coffee, Phys. Rev. A **80**(6), 063412 (2009)
17. A. Rouzée, S. Guérin, V. Boudon, B. Lavorel, O. Faucher, Phys. Rev. A **73**(3), 033418 (2006)
18. B.J. Sussman, J.G. Underwood, R. Lausten, M. Ivanov, A. Stolow, Phys. Rev. A **73**(5), 53403 (2006)
19. P.N. Butcher, D. Cotter, *The Elements of Nonlinear Optics* (Cambridge University Press, New York, 1990)
20. J.R. Lalanne, A. Ducasse, S. Kielich, *Laser-Molecule Interaction: Laser Physics and Molecular Nonlinear Optics* (Wiley, New York, 1996)
21. C. Minhaeng, D. Mei, N.F. Scherer, G.R. Fleming, S. Mukamel, J. Chem. Phys. **99**(4), 2410 (1993)
22. B. Lavorel, O. Faucher, M. Morgen, R. Chaux, J. Raman Spectrosc. **31**(1-2), 77 (2000)
23. T. Seideman, J. Chem. Phys. **115**(13), 5965 (2001)
24. P.M. Felker, A.H. Zewail, in *Femtosecond Chemistry*, vol. 1, ed. by J. Manz, L. Waste (VCH, Weinheim, New York, 1995), p. 193
25. E. Péronne, M.D. Poulsen, C.Z. Bisgaard, H. Stapelfeldt, T. Seideman, Phys. Rev. Lett. **91**(4), 043003 (2003)
26. M.D. Poulsen, E. Péronne, H. Stapelfeldt, C.Z. Bisgaard, S.S. Viftrup, E. Hamilton, T. Seideman, J. Chem. Phys. **121**(2), 783 (2004)
27. K.F. Lee, D.M. Villeneuve, P.B. Corkum, S. Albert, G.U. Jonathan, Phys. Rev. Lett. **97**(17), 173001 (2006)
28. A. Rouzée, S. Guérin, O. Faucher, B. Lavorel, Phys. Rev. A **77**(4), 043412 (2008)
29. V. Loriot, R. Tehini, E. Hertz, B. Lavorel, O. Faucher, Phys. Rev. A **78**(1), 013412 (2008)
30. M. Fujimoto, S. Aoshima, M. Hosoda, Y. Tsuchiya, Phys. Rev. A **64**(3), 033813 (2001)
31. V. Loriot, E. Hertz, B. Lavorel, O. Faucher, J. Chem. Phys. **132**(18), 184303 (2010)
32. V. Renard, O. Faucher, B. Lavorel, Opt. Lett. **30**(1), 70 (2005)
33. J.C. Diels, W. Rudolph, *Ultrashort Laser Pulse Phenomena* (Academic, San Diego, 1996)
34. E. Hertz, D. Daems, S. Guérin, H.R. Jauslin, B. Lavorel, O. Faucher, Phys. Rev. A **76**(4), 043423 (2007)
35. D. Daems, S. Guérin, E. Hertz, H.R. Jauslin, B. Lavorel, O. Faucher, Phys. Rev. Lett. **95**(6), 063005 (2005)
36. V. Loriot, E. Hertz, A. Rouzée, B. Sinardet, B. Lavorel, O. Faucher, Opt. Lett. **31**(19), 2897 (2006)
37. C.W. Siders, G. Rodriguez, J.L.W. Siders, F.G. Omenetto, A.J. Taylor, Phys. Rev. Lett. **87**(26), 263002 (2001)
38. A. Talebpour, J. Yang, S.L. Chin, Optic. Comm. **163**(1-3), 29 (1999)
39. J. Kasparian, R. Sauerbrey, S.L. Chin, Appl. Phys. B **71**(6), 877 (2000)

40. V. Loriot, E. Hertz, B. Lavorel, O. Faucher, J. Phys. B **41**, 015604 (2008)
41. V. Loriot, E. Hertz, O. Faucher, B. Lavorel, Opt. Express **17**(16), 13429 (2009)
42. V. Loriot, E. Hertz, O. Faucher, B. Lavorel, Opt. Express **18**(3), 3011 (2010)
43. P. Béjot, J. Kasparian, S. Henin, V. Loriot, T. Vieillard, E. Hertz, O. Faucher, B. Lavorel, J.P. Wolf, Phys. Rev. Lett. **104**(10), 103903 (2010)
44. W. Ettoumi, P. Béjot, Y. Petit, V. Loriot, E. Hertz, O. Faucher, B. Lavorel, J. Kasparian, J.P. Wolf, Phys. Rev. A **82**(3), 039905 (2010)
45. S. Ramakrishna, T. Seideman, J. Chem. Phys. **124**(3), 034101 (2006)
46. T. Vieillard, F. Chaussard, D. Sugny, B. Lavorel, O. Faucher, J. Raman Spectrosc. **39**(6), 694 (2008)
47. A. Rouzée, V. Renard, S. Guérin, O. Faucher, B. Lavorel, Phys. Rev. A **75**(1), 013419 (2007)
48. A. Rouzée, V. Renard, S. Guérin, O. Faucher, B. Lavorel, J. Raman Spectrosc. **38**(8), 969 (2007)

Chapter 5
Directionally Asymmetric Tunneling Ionization and Control of Molecular Orientation by Phase-Controlled Laser Fields

Hideki Ohmura

Abstract Intense ($10^{12} - 10^{13}\text{W/cm}^2$) phase-controlled laser fields consisting of a fundamental light and a second-harmonic light induce directionally asymmetric tunneling ionization and the resultant selective ionization of oriented molecules. It is demonstrated that selective ionization of oriented molecules induced by phase-controlled $\omega+2\omega$ laser fields reflects the geometric structure of the highest occupied molecular orbital. This method is robust, being free of both laser wavelength and pulse-duration constraints, and thus can be applied to a wide range of molecules.

5.1 Introduction

Molecular manipulation by an intense laser field has made rapid progress owing to the recent advent of techniques to generate intense, ultrashort laser pulses. When laser intensity is around 10^{12-13}W/cm^2, molecules are either aligned or oriented (i.e., without or with discrimination of the head–tail order of the molecule, respectively) along the laser polarization direction by the torque generated by the interaction between the laser field and the dipole moment of the molecules (for a review of this subject, see [1]). Molecular alignment/orientation has the potential to be used for applications such as precision spectroscopy and chemical reaction control because one can eliminate the orientational averaging that leads to loss of information or disturbs the homogeneous molecular manipulation of a given species.

Recent investigations have revealed that an intense, ultrashort laser field can handle not only nuclear motions but also the motion of ionized electrons in the tunneling ionization (TI) process. Irradiation of an atom or molecule by an intense laser field causes a modification of the Coulomb potential to suppress the potential barrier of electrons, leading to electron removal via tunneling [2]. The strong nonlinear

H. Ohmura
National Institute of Advanced Industrial Science and Technology (AIST),
1-1-1, Higashi, Tsukuba, Ibaraki 305-8565, Japan
e-mail: hideki-ohmura@aist.go.jp

K. Yamanouchi et al. (eds.), *Progress in Ultrafast Intense Laser Science VII*, Springer Series in Chemical Physics 100, DOI 10.1007/978-3-642-18327-0_5,

tunneling rate leads to the generation of ionized electrons for short periods of time (attosecond timescale) at each laser field oscillation maximum [2]. Keldysh first suggested that an increase in laser intensity causes a transition from multiphoton ionization (MPI) to TI [3]. The Keldysh adiabatic parameter $\gamma = (2\omega^2 I_p/I)^{1/2}$ is used to judge whether a given observed phenomenon is of the MPI or TI type, where ω and I are the field frequency and intensity, and I_p is the ionization potential of the matter. If $\gamma > 1$, MPI is dominant; if $\gamma < 1$, then TI is dominant. The theory of atomic ionization in intense laser fields has been expanded, as in the Keldysh–Faisal–Reiss (KFR) [3–5], Perelomov–Popov–Terent'ev (PPT) [6], and Ammosov–Delone–Krainov (ADK) [7] models, and has been experimentally tested by many researchers [2]. Recently, a successful real-time observation of an optical TI process was made by Uiberacker et al. using attosecond extreme ultraviolet (XUV) pulse pumping and a near infrared (NIR), few-cycle pulse probing technique [8]. The molecular ionization process in intense laser fields has also been investigated experimentally, and theories developed for atomic ionization in such fields have been extended to molecules. The molecular ADK model [9, 10], which takes into account the TI of a single active electron from the highest occupied molecular orbital (HOMO), or the molecular Keldysh–Faisal–Reiss (KFR) model [11], which interprets results in terms of MPI modified by interference from a multicenter of molecules, have successfully described the observed results for diatomic or simple molecules. According to the molecular ADK model, the removal of electrons from the molecule via TI by an intense laser field depends on the relative angle between the polarization direction of the laser field and the geometry of the HOMO [9, 10, 12, 13]. As a consequence of the angular dependence of the TI rate, aligned/oriented molecules are selectively ionized and the fragment-emission pattern caused by dissociative ionization reflects the structure of the molecular orbital [9, 10, 12, 13].

Moreover, when laser intensity is around $10^{14}\mathrm{W/cm^2}$, ionized electrons in atoms and molecules show a remarkable behavior known as the *recollision process*: as a consequence of accelerated motion synchronized with an oscillating electric field, ionized electrons are pulled away from, pulled back to, and then recollided with the parent ions within one optical cycle [2, 14]. This recollision process plays an essential role not only in the high nonsequential double ionization probability in atoms [15, 16] and molecules [17] but also in the highly efficient generation of XUV light, soft X-rays, and attosecond pulses [2, 14, 18].

Such coherent nuclear motions triggered by an ultrashort laser pulse and coherent motions of electrons synchronized with an oscillating laser field are strongly affected by the laser's phase. Therefore, so-called coherent control or quantum control, which is the direct manipulation of the wavefunction and its quantum dynamics through the coherent nature of a laser field, is expected to be a powerful tool (for reviews, see [19] and [20]). Among various methods, we have investigated the coherent control of molecular ionization processes using phase-controlled, two-color laser pulses consisting of a fundamental pulse and its second-harmonic pulse (the $\omega + 2\omega$ pulse) [21–27, 27–35, 35–49].

In this review, we report the use of phase-controlled laser fields to achieve quantum control of molecular TI in the space domain and the resultant selective ionization of oriented molecules in the gas phase [44–49]. First, the characteristics of $\omega + 2\omega$ laser fields are described in Chap. 2. Then, descriptions of the experimental apparatus and method for detection of oriented molecules are presented in Chap. 3. The investigation of phase-sensitive molecular ionization processes related to the molecular orientation induced by the $\omega + 2\omega$ laser fields through various molecules is discussed in Chap. 4. Finally, a brief summary of this topic is provided.

5.2 Characteristics of the Two-Color $\omega + 2\omega$ Laser Fields

For weak laser fields ($<10^{12}\,\mathrm{W/cm^2}$), a phototransition scheme can be used to explain the phase-dependent phenomena induced by the phase-controlled $\omega + 2\omega$ laser fields. The quantum interference effect occurs between a one-photon transition of the second-harmonic photon and a two-photon transition of the fundamental photon (Fig. 5.1a). In this scheme, because the final states differ in parity owing to the different selection rules for the one- and two-photon transitions, the interference between the two transitions induces the breaking of spatial symmetry [19, 21]. This type of interference has been observed elsewhere, including in photoelectrons associated with the ionization of atoms [22, 23] and in photocurrents in semiconductors [24, 25]. Charron et al. have suggested the application of the $\omega + 2\omega$ scheme to the photodissociation of HD^+ molecules, and the scheme can also be used to explain directional separation of photofragments [26, 27]. Experimental attempts to control molecular photodissociation by means of an $\omega + 2\omega$ scheme have been reported for HD^+ [28], ${H_2}^+$ [29], and D_2 [30] molecules.

In contrast, for strong laser fields ($>10^{12}\,\mathrm{W/cm^2}$), a scheme involving electric fields that induce motion of charges or dipoles has been presented. The total electric field of the linearly polarized optical fields of the two frequencies is given by $E(t) = E_1 \cos(\omega t) + E_2 \cos(2\omega t + \phi)$, where E_1 and E_2 are the amplitudes of the electric fields and ϕ is the relative phase difference between the fundamental and the second-harmonic light. The amplitude of the electric field in the positive (negative) direction is about twice that in the negative (positive) direction when $\phi = 0(\pi)$ (Fig. 5.1b, c). The phase-controlled $\omega + 2\omega$ laser fields have a characteristic, phase-dependent asymmetric waveform, in contrast to single-frequency laser fields, which have symmetric waveforms. Figure 5.1d shows the degree of asymmetry $|E(t)_{\max}/E(t)_{\min}|$ plotted as a function of the ratio E_2/E_1 at $\phi = 0$. The $|E(t)_{\max}/E(t)_{\min}|$ reaches a maximum at $E_2/E_1 = 0.5$. The solid line in Fig. 5.1e shows $|E(t)_{\max}/E(t)_{\min}|$ plotted as a function of ϕ at $E_2/E_1 = 0.5$. Maximum asymmetry is achieved at ϕ values of 0 or π, with corresponding $|E(t)_{\max}/E(t)_{\min}|$ values of 2.0 and 0.5, respectively. The dotted line in Fig. 5.1e shows the absolute value $|E(t)_{\max}|$ plotted as a function of ϕ at $E_2/E_1 = 0.5$. The $|E(t)_{\max}|$ value reaches maxima at $\phi = 0$ and π. The asymmetric electric fields induce phase-sensitive phenomena, especially those related to the breaking of spatial symmetry

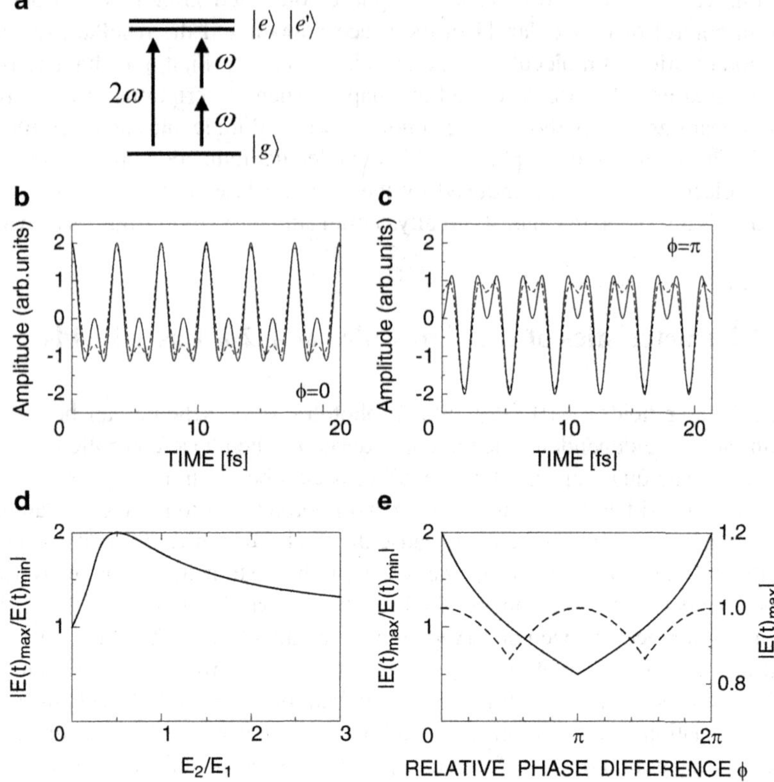

Fig. 5.1 (**a**) Schematic of the energy diagram corresponding to the interference between a one-photon transition induced by second-harmonic light and a two-photon transition induced by fundamental light (the $\omega + 2\omega$ scheme). Because the final states differ in parity owing to the different selection rules for the one- and two-photon transitions, the coherent superposition of different quantum states can be achieved. (**b, c**) Waveforms of phase-controlled, two-color $\omega + 2\omega$ laser fields at a relative phase difference of (**b**) $\phi = 0$ and (**c**) $\phi = \pi$ (*solid line*, $E_2/E_1 = 1.0$; *dotted line*, $E_2/E_1 = 0.5$). (**d**) $|E(t)_{\max}/E(t)_{\min}|$ at $\phi = 0$ as a function of E_2/E_1. (**e**) $|E(t)_{\max}/E(t)_{\min}|$ (*solid line*) and $|E(t)_{\max}|$ (*dotted line*) at $E_2/E_1 = 0.5$ as a function of ϕ

in the direction of electron motion or dipoles. Control of electron motion in TI by the intense phase-controlled $\omega + 2\omega$ pulse has been discussed [32–35]. The application of the $\omega + 2\omega$ scheme to molecules leads to further effective control of molecular orientation with discrimination of the molecules' head–tail order; such control is impossible to achieve with a monochromatic laser field with a symmetric waveform [36–43]. We have investigated the interaction between gas-phase molecules with asymmetric structures and intense (10^{12}–10^{13}W/cm^2) phase-controlled $\omega + 2\omega$ pulses with asymmetric waveforms. We observed phase-sensitive ionization related to molecular orientation induced by the intense phase-controlled $\omega + 2\omega$ pulses [44–49].

5.3 Experimental Apparatus and Detection of the Oriented Molecules

The experimental apparatus used in the studies described herein is shown in Fig. 5.2a. The experiments were performed with a Ti:sapphire laser system (Spectra-Physics, Hurricane) operating at 20 Hz or with a Q-switched Nd:YAG laser (Spectra-Physics, LAB-150) operating at 10 Hz. The Ti:sapphire laser system provided pulses of energy at 1 mJ/pulse and at a duration of 130 femtoseconds (fs) at an 800 nm central wavelength. The Q-switched Nd:YAG Laser provided pulses of energy at 100 mJ/pulse and at a duration 10 ns at a 1,064 nm wavelength. We inserted a frequency-doubling crystal (β-barium borate, type-I phase-matching) into the path of the laser beam. The second-harmonic light was separated from the fundamental light by a dielectric mirror in a Mach–Zehnder interferometer after second-harmonic generation. The configuration of the Mach–Zehnder interferometer is shown in Fig. 5.2a. We inserted a half-wave plate that rotated the polarization

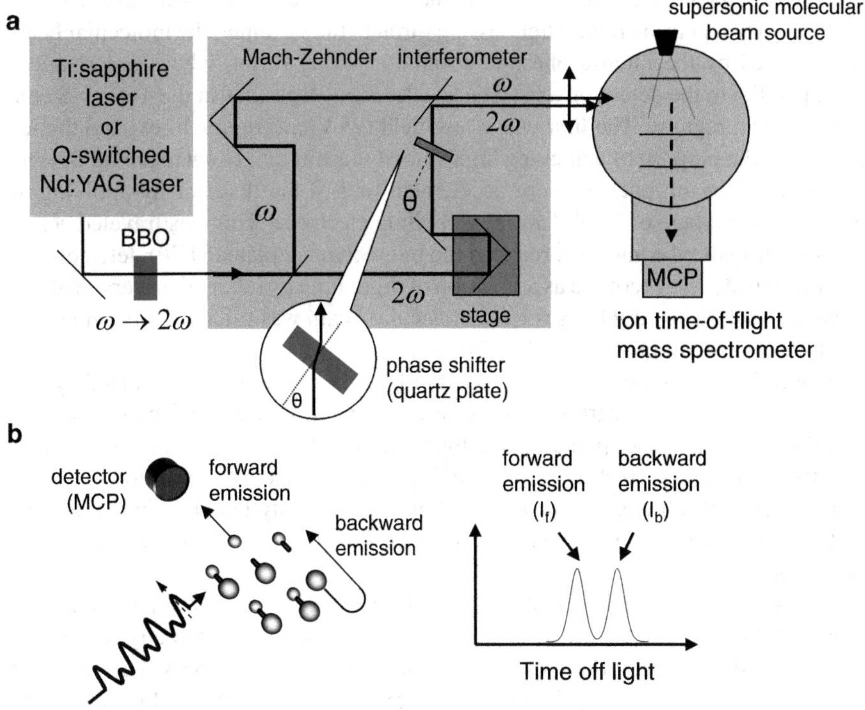

Fig. 5.2 (**a**) Schematic diagram of the experimental apparatus, which consisted of a laser source, a Mach–Zehnder interferometer, and an ion time-of-flight mass spectrometer (TOF-MS). *Arrows* indicate the paths of the laser beams and the *double-headed arrow* indicates their polarization. The *dotted arrow* indicates the path of ions induced by laser irradiation. (**b**) Schematic of the detection of the oriented molecular ions

of the fundamental light by 90° so that the polarizations of the two fields were parallel. The delay time of the two pulses was controlled by a translation stage located in the fundamental light path with a resolution of about 4 fs. The second-harmonic beams passed through an antireflection coated quartz plate (3 mm thickness) that could be rotated. This quartz plate was used to change ϕ of the two fields with a resolution of about 20 attoseconds (0.02π). The ratio of the light intensities (I_2/I_1) was adjusted to be around 0.25 ($E_2/E_1 = 0.5$) by rotating the phase-matching angle of the BBO crystal while keeping the total intensity $I = I_1 + I_2$ constant, where I_1 and I_2 are the intensities of the ω and 2ω pulses, respectively. After being recombined, the phase-controlled $\omega + 2\omega$ beams were directed toward the reaction chamber and were focused on the molecular beam by a concave mirror of 120 mm focal length.

The reaction chamber consisted of a supersonic molecular beam source and a homemade ion time-of-flight mass spectrometer (TOF-MS). The supersonic molecular beam was introduced into the chamber through a pulsed valve (General Valve; 0.5 mm diameter), which was differentially pumped by a diffusion pump and a turbo pump. The molecular beam passed through a skimmer (diameter 0.5 mm) located 20 mm from the nozzle. The pressure in the chamber was kept below 3.0×10^{-5} Pa with a 20-Hz repetition rate. After passing through the skimmer, the molecular beam was ionized by the intense phase-controlled $\omega + 2\omega$ beam, whose polarizations were parallel to the detection axis. The acceleration electrodes in the TOF-MS consisted of two regions. The first was a low-field (25 V/cm) region to expand the ion packet for the purpose of achieving high velocity resolution. Upon reaching the second region, the ion packets were accelerated toward the detector by applying an acceleration voltage of 300 V/cm to a repelling electrode. The ions traveled across a field-free drift tube and then reached the microchannel plate (MCP) detector. The signal intensity was recorded as a function of flight time by taking the average of 512 measurements. The target gas for the molecular beam was diluted 5% with helium gas to obtain a total pressure of 10^5–10^6 Pa.

Figure 5.3 shows the TOF mass spectra of singly charged ions produced by dissociative ionization of carbon monoxide (CO) when it was irradiated with $\omega + 2\omega(800 + 400\,\text{nm})$ laser pulses with a total intensity $I = I_1 + I_2$ of approximately 5×10^{13} ($I_1 = 4 \times 10^{13}, I_2 = 1 \times 10^{13}$ W/cm^2) and a pulse duration of 130 fs. The experiments all were performed with a laser intensity under or in the vicinity of the regime where doubly charged fragment ions caused by Coulomb explosion were observed.

We observed singly and doubly charged photofragment ions. Each photofragment exhibited a pair of peaks, one resulting from ions flying directly toward the detector, and the other from those ions, which first flew in the backward direction relative to the detector before being reversed by the extraction fields (Fig. 5.2b). The spacing of the forward and backward peaks reflects the kinetic energy release. The assignment of each dissociation channel has been reported as a Coulomb explosion process $CO^{+(p+q)} \rightarrow C^{+p} + O^{+q}$ (where p and q are integers) [50]. As seen in Fig. 5.3, forward–backward asymmetry was clearly observed in the TOF spectrum. O^+ (O^{2+}) ions were preferentially emitted toward the detector and C^+ (C^{2+})

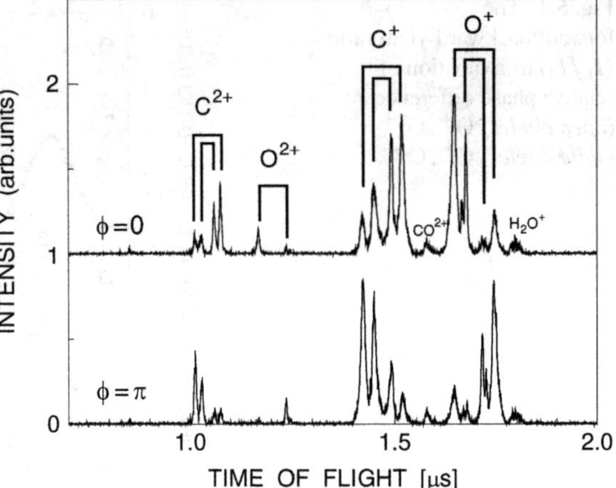

Fig. 5.3 TOF mass spectra of ions produced by dissociative ionization of CO when they were irradiated with the femtosecond $\omega + 2\omega$ laser pulses at relative phase differences $\phi = 0$ and $\phi = \pi$. The *solid lines* indicate the pair of forward and backward peaks

ions were emitted preferentially away from the detector at $\phi = 0$, when the optical electric field maximum pointed toward the detector. Conversely, O^+ (O^{2+}) ions were emitted preferentially in the backward direction and C^+ (C^{2+}) ions were emitted preferentially in the forward direction at $\phi = \pi$, when the optical electric field maximum pointed in the backward direction relative to the detector.

Figure 5.4 shows the ratio of forward and backward yields, I_f/I_b, obtained when we changed ϕ between the fundamental and the second-harmonic light by rotating the quartz plate. A clear periodicity of 2π was observed in the I_f/I_b ratio for all photofragments displayed. The oscillation of the signal was carefully examined to confirm that it was caused not by an artifact, such as fluctuation of the laser intensity, but by interference between the simultaneous excitations. The phase dependence between C^+ (C^{2+}) and O^+ (O^{2+}) indicates that they are completely out of phase with each other. This result shows that phase-controlled $\omega + 2\omega$ fields can discriminate molecular orientation from head–tail order and that the direction of orientation is flipped by the relative phase difference ϕ. Such flipping is impossible to achieve with a monochromatic laser field with a symmetric waveform. Furthermore, the phase dependence between C^+(O^+) and C^{2+}(O^{2+}) shows completely in-phase behavior. This indicates that the directions of molecular orientation for singly charged and doubly charged CO are the same.

5.4 Mechanism of Detection of Oriented Molecules

There are two conceivable mechanisms related to the detection of oriented molecules: (1) *dynamic molecular orientation (DMO)* and (2) *selective ionization of oriented molecules (SIOM)*.

Several theoretical investigations have reported that molecules can be dynamically oriented along the laser polarization direction by the torque generated by the

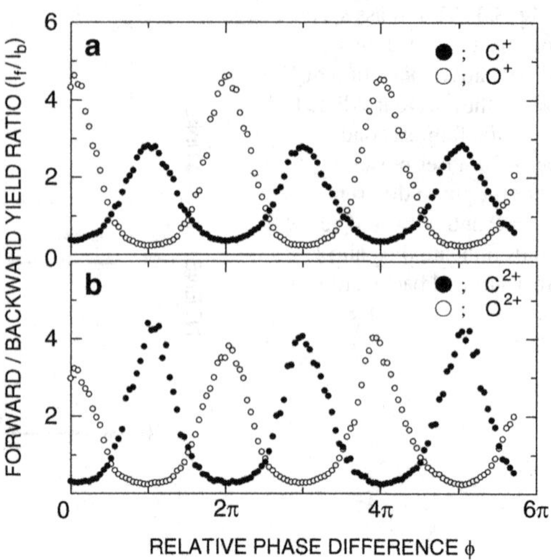

Fig. 5.4 The forward/backward yield ratio (I_f/I_b) as a function of relative phase difference ϕ: (*open circles*) O^+, O^{2+}, (*solid circles*) C^+, C^{2+}

nonlinear interaction between a nonresonant $\omega + 2\omega$ laser field and the hyperpolarizability of molecules (the *linear* interaction between an $\omega + 2\omega$ laser field and the permanent dipole of molecules averages to zero over an optical.) [37, 42]. If the laser pulse is longer than the rotational period of the molecules, then molecules orient adiabatically during laser irradiation (adiabatic molecular orientation) [37]. If the laser pulse is shorter than the rotational period of the molecules, then rotational wave packets are formed and dynamical alignment is reconstructed at revival times even after the laser irradiation (nonadiabatic molecular orientation) [42]. Recently, the DMO based on the hyperpolarizability of the molecules have been succeeded adiabatically [39] and nonadiabatically [43]. However, the degrees of orientation have observed to be very small. Therefore, the contribution of DMO based on the hyperpolarizability of the molecules during *the pulse duration* in our experiments can be negligible.

In contrast to DMO, which is a consequence of controlling nuclear motion, SIOM is a result of controlling electron motion. Molecules initially oriented in a certain direction with respect to the laser polarization have a high ionization rate, resulting in SIOM in randomly oriented molecular ensembles. The ionization process in an intense laser field can be well described by the molecular ADK model [9, 10]. According to this model, ionized electrons are much more strongly extracted via the tunneling process from electronic clouds of HOMO along the direction of an electric field. The removal of electrons from the molecule via TI by an intense laser field depends on the relative angle between the polarization direction of the laser field and the geometry of the HOMO [9, 10]. As a consequence of the angular dependence of the TI rate, aligned molecules are selectively ionized and the fragment-emission pattern caused by dissociative ionization reflects the structure of the molecular orbital [9, 10]. For example, a butterfly shaped pattern reflecting the structure of π orbital in O_2 molecules and dumbbell-shaped pattern reflecting σ orbital in N_2 molecules in the 2-demensional photofragment–emission pattern imaging have been observed

[12, 13]. When TI of molecules with an asymmetric HOMO structure is induced by an asymmetric $\omega + 2\omega$ field, electrons are much more likely to be removed from the large-amplitude part of the HOMO [9, 10, 12, 13]. Therefore, it has been logically deduced that molecules initially oriented in a certain direction with respect to the asymmetric $\omega + 2\omega$ field are selectively ionized among randomly oriented molecules [46]. The HOMO of CO shows an asymmetric σ structure. The angle dependence of the TI rate for CO at a laser intensity of $6 \times 10^{13}\mathrm{W/cm^2}$ reflecting the geometry of the HOMO has been calculated using molecular ADK theory, where electrons are much more likely to be removed from the large-amplitude part (carbon) than the small-amplitude part (oxygen) [10]. The result of CO (Figs. 5.3 and 5.4) supports that the direction of the detected molecules was consistent with that expected by the molecular ADK model; that is, electrons are much more strongly removed by the tunneling process from the large-amplitude part (carbon) of the HOMO opposite to the direction of the electric field vector at its maxima for $\phi = 0$ and $\phi = \pi$.

To confirm that the main mechanism of the detection of oriented molecules is SIOM, we investigated the dissociative ionization of molecules induced by $\omega + 2\omega$ laser fields with pulse durations of 130 fs and 10 ns by changing the parameters of molecules systematically, as shown in Fig. 5.5a–d.

Before we introduce various experimental examples, we note two controversial points. First, we are aware of the controversy concerning the boundary between MPI and TI. As indicated earlier, the Keldysh theory states that if $\gamma > 1$, MPI is dominant [3]. Since there is no absolute boundary between MPI and TI, in the intermediate region $\gamma \sim 1$, phenomena can often be successfully explained by both MPI and TI. Our experimental condition corresponds to $\gamma \sim 2$, where MPI is considered to be dominant. Nonetheless, in terms of directionally asymmetric TI, the expression can be useful for intuitive understanding of our experimental results. Some relevant investigations concerning the boundary between MPI and TI are listed in the following: (a) Uiberacker et al. has reported a real-time observation of the optical TI process by using attosecond XUV pulse pumping and NIR few-cycle pulse probing technique, and has shown that TI remains the dominant ionization mechanism even at $\gamma \sim 3$ [8]; (b) Dewitt and Levis have observed that a transition from MPI to TI occurs in polyatomic molecules by changing the electron delocalization through the molecular structure, and have shown that a large electronic orbital size reduces γ effectively, or in other words, TI can be dominant even for $\gamma > 1$ [51]; (c) Reiss has pointed out that there are disqualifying features in categorization using γ where ionization with $\gamma \gg 1$ can occur only by TI, and ionization with $\gamma \ll 1$ must be a more intense laser regime such as the over-the-barrier process [52]. Further theoretical studies are required to understand the ionization process of atoms and molecules in intense laser fields. Recently, using the molecular KFR model, total ionization yields of a molecule with an intense linearly polarized laser field was calculated as a function of the relative orientation between the molecular and the laser polarization and was found to be qualitatively similar to that calculated using the molecular ADK model [53, 54]. This result may mean that the molecular KFR model includes the expression of TI in terms of phototransition.

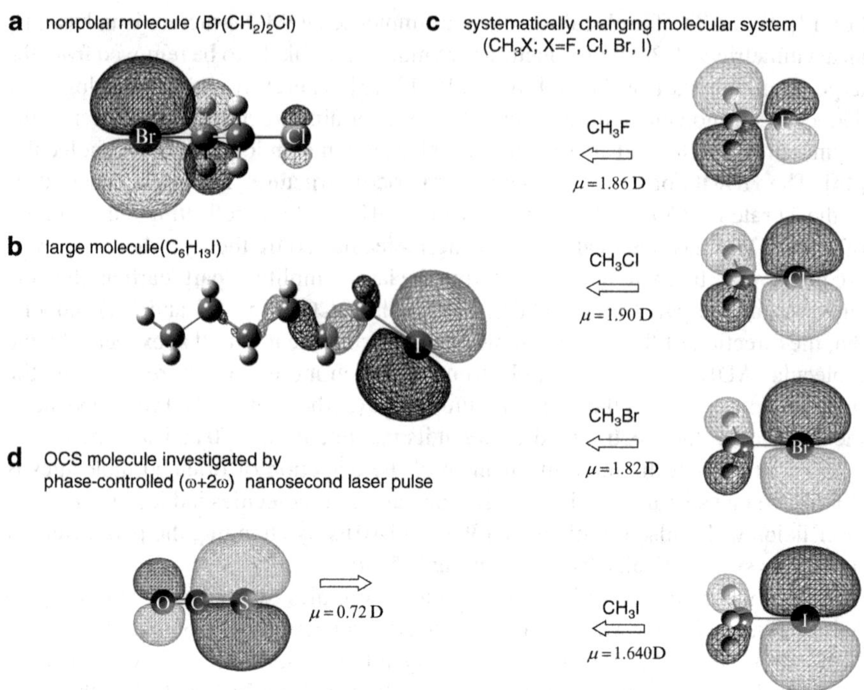

Fig. 5.5 Molecular structures and isocontours of the highest occupied molecular orbitals (HOMO) of the investigated molecules as determined by ab initio calculations using the Gaussian 03W software package. The *shadings* indicate the signs of the wavefunctions. The directions of permanent dipoles are shown by *thick arrows*

Second, some kind of complexity and uncertainty about monitoring the ionization process through dissociative ionization channels should be noted. The dissociative ionization processes include several entangled nonsequential processes such as (1) the generation of the parent ion in the electronic ground state followed by photoexcitation toward repulsive electronic states; (2) recollision-induced electron excitation; and (3) direct generation of the parent ion through an electron ejection from HOMO-1. These processes might induce the deviation from the molecular ADK model. Considering the laser intensity in the experiment shown in Figs. 5.3 and 5.4, these nonsequential processes should have certain contributions in the TOF spectra. Despite the situation, the experimental result of CO supports that the direction of the detected molecules was consistent with that expected by the molecular ADK model; that is, electrons are much more strongly removed by the tunneling process from the large-amplitude part of the HOMO opposite to the direction of the electric field vector at its maxima for $\phi = 0$ and $\phi = \pi$. From an experiment viewpoint, therefore, TI based on the molecular ADK model seems to be the main process, and some effects that induce deviations from the molecular ADK model seem to be smaller.

5.4.1 Nonpolar Molecule with Asymmetric Structure: $Br(CH_2)_2Cl$

As an example of a nonpolar molecule with an asymmetric structure, we have chosen 1-bromo-2-chloroethane (BCE) (ionization potential: 10.55 eV) [46]. Figure 5.5a shows the molecular structure and HOMO of BCE as determined by ab initio calculations using the Gaussian 03W software package (method: MP2; basis set: 6–311 + G(d,p)) [55]. Among the three possible rotational isomers, the trans isomer shown in Fig. 5.5a is the most stable in the gas phase. The BCE molecule has a very small permanent dipole moment (calculated value: 0.0057 Debye, pointing from Cl to Br) due to cancelation of two halogen atoms with large electronegativities (Cl: 3.0; Br: 2.8) located on opposite sides of the molecule. However, the HOMO shows a π structure with large asymmetry along the molecular frame (Fig. 5.5a).

When BCE molecules were irradiated with $\omega + 2\omega$ laser pulses, various singly charged photofragment ions and parent ions were detected in the TOF mass spectrum. Directional asymmetries in the forward–backward emissions were observed in various photofragment ions, and a clear periodicity of 2π was observed in the I_f/I_b ratio for all photofragments. The phase dependencies between the Cl^+ (Br^+) ions and counter cations were completely out of phase with each other. This result shows that the phase-controlled $\omega + 2\omega$ fields can discriminate molecular orientation from head–tail order. Moreover, the phase dependencies between Cl^+ and Br^+ were also out of phase with each other. It is evident from all phase dependencies that phase-controlled $\omega + 2\omega$ fields discriminate the molecular orientation of the head–tail order [46]. We define the relative phase difference as $\phi = 0$ when the electric field maxima pointed toward the detector and as $\phi = \pi$ when the optical electric field maxima pointed away from the detector. Br^+ ions were preferentially emitted toward the detector and Cl^+ ions were preferentially emitted away from the detector at $\phi = \pi$, when the electric field maxima pointed in the backward direction relative to the detector. These observed results show that the direction of the detected molecules is consistent with that expected by the molecular ADK model. Therefore, we concluded that the phase-controlled $\omega + 2\omega$ field achieves selective ionization of oriented molecules, reflecting the asymmetry of the HOMO structure [46]. *Even for nonpolar molecules, SIOM can achieved through discrimination of the wavefunction in the space domain by the enhancement of nonlinear interaction between the asymmetric laser fields and the asymmetric HOMO structure.*

In addition, we mention the relative angle between the oriented molecules and the polarization direction of the laser fields, and the contribution of the induced dipole moment. Alnaser et al. observed a butterfly shaped pattern reflecting the structure of π orbitals in O_2 molecules by using 8-fs optical pulses and 2D photofragment-emission pattern imaging, where the direction of selectively ionized molecules was 40° relative to the polarization direction [12, 13]. When 35-fs pulses were used instead of 8-fs pulses, the butterfly shaped pattern changed to a dumbbell-shaped pattern, indicating that the direction of the selectively ionized molecules was along the direction of the laser polarization due to dynamic alignment (not orientation) by the induced dipole during the laser pulse [12, 13]. Almost all molecules experience some contribution of dynamic molecular alignment due to an induced dipole. Thus, it is possible that our 130-fs $\omega + 2\omega$ pulse induces dynamic alignment, even for relatively heavy BCE molecules, and that our measurement is a result of the selective ionization of oriented molecules in aligned molecules, rather than in randomly oriented molecules, during the laser pulse.

5.4.2 *Large Molecule: $C_6H_{13}I$*

We studied a large polyatomic molecule, 1-iodohexane ($C_6H_{13}I$), to determine whether SIOM could be achieved [49]. The molecular structure and isocontour of the HOMO of $C_6H_{13}I$ (Fig. 5.5b) were determined by ab initio calculations using the Gaussian 03W software package (method: MP2; basis sets: LanL2DZ augmented by polarization functions and diffuse functions) [55]. The HOMO of $C_6H_{13}I(I_p : 9.20\,\text{eV})$ was remarkably asymmetric due to "squeezing" by the iodine atom (Fig. 5.5b). Although iodination is a simple chemical treatment, it can induce a dramatic change at the wavefunction level and can therefore be used to "quantum mark" molecules when designing wavefunctions.

When $C_6H_{13}I$ molecules were irradiated with femtosecond $\omega + 2\omega$ pulses, various singly charged photofragment ions such as hydrocarbon cations and iodine-containing cations, as well as parent ions, were detected in the TOF mass spectrum. Directional asymmetries in the forward–backward emissions were observed for various photofragment ions, and a clear periodicity of 2π was observed in the I_f/I_b ratio for all photofragments except $C_6H_{13}{}^+$. The phase dependencies of iodine and iodine-containing cations were completely out of phase with those for the carbon and hydrocarbon cations. This result shows that a phase-controlled $\omega + 2\omega$ optical field discriminates the head–tail order of molecules [49].

We draw two direct conclusions from these experimental results. First, the phase dependencies of all photofragments except $C_6H_{13}{}^+$ were consistent with the molecular structure of 1-iodohexane. Therefore, we can reasonably conclude that the prompt axial recoil approximation is valid even for large polyatomic molecules, and that the photofragment emission pattern reflects the molecular structure. Regarding $C_6H_{13}{}^+$, there are two conceivable explanations for the fact that the $C_6H_{13}{}^+$ fragments did not show a phase-dependent behavior: (1) a slow dissociation process, in which $C_6H_{13}{}^+$ was produced on a timescale longer than the rotational period, allowing orientation averaging; and (2) a kinetic energy that was too low to show a photofragment emission pattern. The excess energy that molecular cations obtain during the ionization process is divided between the translational and internal (vibrational and rotational) energy of the photofragments. The translational energy of the photofragments decreases with increasing number of constituent atoms, because the number of internal degrees of freedom increases. Although the observed phase difference decreased with increasing number of constituent atoms, favoring explanation (2), at present we cannot entirely exclude either explanation.

Second, our results indicate that the direction of the detected molecules was consistent with that expected by the molecular ADK model. Therefore, it is reasonable to conclude that the molecular ADK model is valid for SIOM even for large polyatomic molecules [49]. *SIOM is free of the constraints of size and weight of molecules, and this is an advantage compared to DMO, with which it is difficult to orient large heavy molecules that require large torques at practical laser intensities.*

5.4.3 *Systematically Changing Molecular System:* $CH_3X(X = F, Cl, Br, I)$

We have investigated the phase-sensitive ionization related to molecular orientation induced by intense phase-controlled $\omega + 2\omega$ pulses in the case of systematically changing orbital asymmetry. Figure 5.5c shows the molecular structures and isocontours of the HOMO of four methyl halide molecules we tested (CH_3I, CH_3Br, CH_3Cl, and CH_3F) as determined by ab initio calculations using the Gaussian 03W software package [55] (method: MP2; basis sets: 6–31++G(2df,p) for CH_3F and CH_3Cl, and LanL2DZ augmented by polarization functions and diffuse functions for CH_3Br and CH_3I). The HOMO of all the methyl halide molecules shows an asymmetric π structure, and the degree of asymmetry changes systematically with respect to the halogen atom. The wavefunctions for the halogen-atom side are larger than that of the methyl parts in CH_3I, CH_3Br, and CH_3Cl. The degree of asymmetry decreases gradually from CH_3I to CH_3Cl, and then reverses for CH_3F. Thus, if SIOM based on the molecular ADK model is the main orientation process, the orientation direction of selectively ionized CH_3F is opposite of that for CH_3Cl, CH_3Br, and CH_3I.

When methyl halide molecules were irradiated with femtosecond $\omega + 2\omega$ pulses, various singly charged photofragment ions and parent ions were detected in the TOF mass spectrum. The directional asymmetries in the forward–backward emissions were observed in various photofragment ions, and a clear periodicity of 2π was observed in the I_f/I_b ratio for all photofragments. The phase dependencies between the halogen ions and the CH_3^+ cations were completely out of phase with each other for CH_3I, and approximately out of phase with each other for CH_3Br, CH_3Cl, and CH_3F. This result shows that a phase-controlled $\omega + 2\omega$ laser field discriminates the head–tail order of oriented molecules [47].

To classify the direction of the oriented molecules, we performed a simultaneous measurement using gas mixtures of CH_3I/CH_3Br, CH_3I/CH_3Cl, and CH_3I/CH_3F [47]. The I^+, Br^+, and Cl^+ ions exhibited complete in-phase behavior, while the F^+ ion was approximately out of phase with the other three halogen atoms. This result indicates that the direction of oriented molecules is the same in CH_3I, CH_3Br, and CH_3Cl, and that the CH_3F molecule is oriented in the opposite direction from the other three methyl halides. The classification by phase behavior is consistent with that expected by SIOM based on the molecular ADK model. Moreover, the directions of the detected molecules are consistent with those expected by the SIOM, whereas the large-amplitude parts (halogen atoms for CH_3I, CH_3Br, and CH_3Cl, and the methyl moieties for CH_3F) were located on the backward side and ionized electrons were removed backward at $\phi = 0$ when the optical electric field maximum pointed toward the detector. Therefore, it is reasonable to conclude that SIOM based on the molecular ADK model is the main process occurring in the phase-sensitive ionization of the four methyl halides induced by a phase-controlled $\omega + 2\omega$ field [47].

5.4.4 OCS Molecule Investigated by Nanosecond $\omega + 2\omega$ Laser Fields

Finally, we investigated the dependence of pulse-duration [48]. We investigated OCS molecules by nanosecond phase-controlled $\omega + 2\omega(1064 + 5032\,\text{nm})$ pulses generated by the Nd:YAG laser with an intensity of $5.0 \times 10^{12}\,\text{W/cm}^2$ and a pulse duration of 10 ns [48].

When OCS molecules were irradiated with the nanosecond $\omega + 2\omega$ pulses, single-charged OC^+, S^+, and parent OCS^+ were detected in the TOF mass spectrum. The breaking of the forward–backward symmetry was clearly observed in the TOF spectrum. The forward peak of the OC^+ ions was more dominant than the backward peak at $\phi = 0$. This result indicates that the OC^+ ions were preferentially ejected toward the detector at $\phi = 0$ when the electric field maximum pointed toward the detector. Conversely, the backward peak of the S^+ ions was more dominant than the forward peak at $\phi = 0$. This behavior is reversed by changing ϕ from 0 to π. A clear periodicity of 2π with considerably large contrast was observed in the I_f/I_b ratio for the OC^+ and the S^+. The phase dependencies between the OC^+ and the S^+ cations were completely out of phase with each other. This result demonstrates that oriented molecules were detected while discriminating the head–tail order of the molecules [48]. The selectivity of the oriented molecules reached 86% ($I_f/I_b = 5.9$) from OC^+ and 75% ($I_f/I_b = 3$) from S^+. *Therefore, we have experimentally confirmed that SIOM induced by directionally asymmetric tunneling ionization is free from laser wavelength constraint and observed universally in a vast range of pulse durations in the femtosecond-to-nanosecond regime.* However, many other studies concerning the interaction between molecules and intense nanosecond laser fields have confirmed that molecules can be dynamically aligned (while not discriminating the head–tail order of molecules) through the interaction between nonresonant laser fields and *induced* dipoles [1]. Therefore, it is reasonable to expect that an intense nanosecond $\omega + 2\omega$ laser field can induce SIOM in dynamically aligned molecules, rather than in randomly oriented molecules, during the laser pulse [48].

5.5 Summary

We have investigated the interaction between gas-phase molecules with asymmetric structure and intense (10^{12-13}W/cm^2) phase-controlled $\omega + 2\omega$ pulses with an asymmetric waveform. We observed phase-sensitive ionization related to molecular orientation induced by the intense phase-controlled $\omega + 2\omega$ pulses. Ionized molecules were oriented with discrimination of head–tail order, which is impossible to achieve with a monochromatic laser field with a symmetric waveform. The direction of oriented molecules can be easily flipped by changing the relative phase difference $(0, \pi)$. We have experimentally demonstrated that, as a consequence of directionally asymmetric TI, SIOM induced by phase-controlled $\omega + 2\omega$ laser fields reflects the asymmetric geometry of the HOMO structure. The present experiments were performed under the condition of Keldysh parameter $\gamma \sim 2$, which can be categorized as an intermediate region between the TI region and the MPI region. While the molecular ADK theory is quantitatively valid only in the region of $\gamma < 1$, the theory seems to be applicable for quantitative discussions on SIOM in the present study. SIOM can be achieved through discrimination of the wavefunction in the space domain by the enhancement of nonlinear interaction between the asymmetric laser fields and the asymmetric HOMO structure. SIOM is free of laser wavelength constraints and is observed over a wide range of pulse durations in the femtosecond-to-nanosecond regime. Furthermore, SIOM is free of the constraints of size, weight, and polarity of molecules, and this is an advantage compared to DMO, with which it is difficult to orient large, heavy molecules that require large torques at practical laser intensities, and with which it is impossible to orient nonpolar molecules with asymmetric structures.

Acknowledgements The author thanks M. Tachiya, T. Nakanaga, F. Ito, N. Saito, H. Nonaka, and S. Ichimura. This work was supported by the Fund for Young Researchers from the Ministry of Education, Culture, Sports, Science and Technology (MEXT); the Mitsubishi Foundation; the Sumitomo Foundation; the Precursory Research for Embryonic Science and Technology (PRESTO) program from Japan Science and Technology (JST); and a Grant-in-Aid for Young Scientists (A) and (B) from the Japan Society for the Promotion of Science (JSPS).

References

1. H. Stapelfeldt, T. Seideman, Aligning molecules with strong laser pulses. Rev. Mod. Phys. **75**, 543 (2003), and references therein
2. F. Krausz, M. Ivanov, Attosecond physics. Rev. Mod. Phys. **81**, 163 (2009), and references there in
3. L.V. Keldysh, Ionization in the field of a strong electromagnetic wave. Sov. Phys. JETP **20,** 1307 (1965)
4. F.H.M. Faisal, Multiphoton absorption of laser photons by atoms. J. Phys. B **6**, L89 (1973)
5. H.R. Reiss, Effect of an intense electromagnetic field on a weakly bound system. Phys. Rev. A **22**, 1786 (1980)
6. A.M. Perelomov, V.S. Popov, M.V. Terent'ev, Ionization of atoms in an alternating electric field. Sov. Phys. JETP **23**, 924 (1966)
7. M.V. Ammosov, N.B. Delone, V.P. Krainov, Tunnel ionization of complexatoms and of atomic ions in an alternating electromagnetic field. Sov. Phys. JETP **64**, 1191 (1987)
8. M. Uiberacker, T.h. Uphues, M. Schultze, A.J. Verhoef, V. Yakovlev, M.F. Kling, J. Raushenberger, N.M. Kabachnik, H. Schröder, M. Lezius, K.L. Kompa, H.G. Muller, M.J.J. Vrakking, S. Hendel, U. Kleineberg, U. Heinzmann, M. Drescher, F. Krausz, Attosecond real-time obasevation of electron tunneling in atoms. Nature **446**, 627 (2007)

9. X.M. Tong, Z.X. Zhao, C.D. Lin, Theory of molecular tunneling ionization. Phys. Rev. A **66**, 033402 (2002)
10. C.D. Lin, X.M. Tong, Dependence of tunneling ionization and harmonic generation on the structure of molecules by short intense laser pulses. J. Photochem. Photobio. A **182,** 213 (2006)
11. J. Muth-Böhm, A. Becker, F.H.M. Faisal, Suppressed molecular ionization for a class of diatomics in intense femtosecond laser fields. Phys. Rev. Lett. **85**, 2280 (2000)
12. A.S. Alnaser, S. Voss, X.M. Tong, C.M. Maharjan, P. Ranitovic, B. Ulrich, T. Osipov, B. Shan, Z. Chang, C.L. Cocke, Effects of molecular structure on ion disintegration patterns in ionization of O_2 and N_2 by short laser pulses. Phys. Rev. Lett. **93**, 113003 (2004)
13. A.S. Alnaser, C.M. Maharjan, X.M. Tong, B. Ulrich, P. Ranitovic, B. Shan, Z. Chang, C.D. Lin, C.L. Cocke, I.V. Litvinyuk, Effects of orbital symmetries in dissociative ionization of molecules by few-cycle laser pulses. Phys. Rev. A **71**, 031403(R) (2005)
14. P.B. Corkum, Plasma perspective on strong-field multiphoton ionization. Phys. Rev. Lett. **71**, 1994 (1993)
15. D.N. Fittinghoff, P.R. Bolton, B. Chang, K.C. Kulander, Observation of nonsequential double ionization of helium with optical tunneling. Phys. Rev. Lett. **69**, 2642 (1992)
16. B. Walker, B. Sheehy, L.F. DiMauro, P. Agostini, K.J. Schafer, K.C. Kulander, Precision measurement of strong field double ionization of helium. Phys. Rev. Lett. **73,** 1227 (1994)
17. A. Tablebpour, S. Larochelle, S.L. Chin, Non-sequential and sequential double ionization of NO in an intense femtosecond Ti:sapphire laser pulse. J. Phys. B **30**, L245 (1997)
18. A. Baltuška, T.h. Udem, M. Ulberacker, M. Hentschel, E. Goulielmakis, C.h. Gohle, R. Holzwarth, V.S. Yakovlev, A. Scrinzi, T.W. Hänsch, F. Krausz, Attosecond control of electronic processes by intense light fields. Nature **421**, 611 (2002)
19. M. Shapiro, P. Brumer, *Principles of the Quantum Control of Molecular Processes* (Wiley, New York, 2003)
20. M. Dantus, V.V. Lozovoy, Experimental coherent control of physicochemical processes. Chem. Rev. **104**, 1813 (2004)
21. G. Kurizki, M. Shapiro, P. Brumer, Phase-coherent control of photocurrent directionality in semiconductors. Phys. Rev. B **39**, 3435 (1989)
22. Y.Y. Yin, C. Chen, D.S. Elliott, A.V. Smith, Asymmetric photoelectron angular distributions from interfering photoionization processes. Phys. Rev. Lett. **69**, 2353 (1992)
23. Z-M. Wang, D.S. Elliott, Determination of the phase difference between even and odd continuum wave functions in atoms through quantum interference measurements. Phys. Rev. Lett. **87**, 173001 (2001)
24. E. Dupont, P.B. Corkum, H.C. Liu, M. Buchanan, Z.R. Wasilewski, Phse-controlled currents in semiconductors. Phys. Rev. Lett. **74**, 3596 (1995)
25. A. Hache, Y. Kostoulas, R. Atanasov, J.L.P. Hughes, J.E. Sipe, H.M. van Driel, Observation of coherently controlled photocurrent in unbiased, bulk GaAs. Phys. Rev. Lett. **78**, 306 (1997)
26. E. Charron, A. Giusti-Suzor, F.H. Mies, Coherent control of isotope separation in HD^+ photodissociation by strong fields. Phys. Rev. Lett. **75**, 2815 (1995)
27. E. Charron, A. Giusti-Suzor, F.H. Mies, Coherent control of photodissociation in intense laser fields. J. Chem. Phys. **103**, 7359 (1995)
28. B. Sheehy, B. Walker, L.F. DiMauro, Phase control in the two-color photodissociation of HD^+. Phys. Rev. Lett. **74**, 4799 (1995)
29. M.R. Thompson, M.K. Thomas, P.F. Taday, J.H. Posthumus, A.J. Langley, F.J. Frasinski, K. Codling, One and two-colour studies of the dissociative ionization and Coulomb explosion of H_2 with intense Ti:sapphire laser pulses. J. Phys. B **30**, 5755 (1997)
30. D. Ray, F.He.S. De, W. Cao, H. Mashiko, P. Ranitovic, K.P. Singh, I. Znakovskaya, U. Thumm, G.G. Paulus, M.F. Kling, I.V. Litvinyuk, C.L. Cocke, Ion-energy dependence of asymmetric dissociation of D_2 by a two-color laser field. Phys. Rev. Lett. **103**, 223201 (2009)
31. K.J. Schafer, K. Kulander, Phase-dependent effects on multiphoton ionization induced by a laser field and its second harmonic. Phys. Rev. A **45**, 8026 (1992)
32. N.B. Baranova, H.R. Reiss, B. Ya. Zel'dovich, Multiphoton and tunnel ionization by an optical field with polar asymmetry. Phys. Rev. A **48**, 1497 (1993)

33. D.W. Schumacher, F. Weihe, H.G. Muller, P.H. Bucksbaum, Phase dependence of intense field ionization: a study using two colors. Phys. Rev. Lett. **73**, 1344 (1994)
34. A.D. Bandrauk, S. Chelkowski, Asymmetric electron-nuclear dynamics in two-color laser fields: laser phase directional control of photodissociation in $H_2{}^+$. Phys. Rev. Lett. **84**, 3562 (2000)
35. S. Chelkowski, M. Zamojski, A. D. Bandrauk, Laser-phase directional control of photofragments in dissociative ionization of $H_2{}^+$ using two-color intense laser pulses. Phys. Rev. A **63**, 023409 (2001)
36. M.J.J. Vrakking, S. Stolte, Coherent control of molecular orientation. Chem. Phys. Lett. **271**, 209 (1997)
37. T. Kanai, H. Sakai, Numerical simulation of molecular orientation using strong, nonresonant, two-color laser fields. J. Chem. Phys. **115**, 5492 (2001)
38. S. Guérin, L.P. Yatsenko, H.R. Jauslin, O. Faucher, B. Lavorel, Orientation of polar molecules by laser induced adiabatic passage. Phys. Rev. Lett. **88**, 233601 (2002)
39. K. Oda, M. Hita, S. Minemoto, H. Sakai, All-optical molecular orientation, Phys. Rev. Lett. **104**, 213901 (2010)
40. C.M. Dion, A.D. Bandrauk, O. Atabek, A. Keller, H. Umeda, Y. Fujimura, Two-frequency IR laser orientation of polar molecules. Numerical simulation for HCN. Chem. Phys. Lett. **302**, 215 (1999)
41. M. Muramatsu, M. Hita, S. Minemoto, H. Sakai, Field-free molecular orientation by an intense nonresonant two-color laser field with a slow turn on and rapid turn off. Phys. Rev. A **79**, 011403 (2009)
42. R. Tehini, D. Sugny, Field-free molecular orientation by nonresonant and quasiresonant two-color laser pulses. Phys. Rev. A **77**, 023407 (2008)
43. S. De, I. Znakovskaya, D. Ray, F. Anis, Nora G. Johnson, I.A. Bocharova, M. Magrakvelidze, B.D. Esry, C.J. Cocke, I.V. Litvinyuk, M.F. Kling, Field-free orientation of CO molecules by femtosecond two-color laser fields. Phys. Rev. Lett. **103**, 153002 (2009)
44. H. Ohmura, T. Nakanaga, M. Tachiya, Coherent control of photofragment separation in the dissociative ionization of IBr. Phys. Rev. Lett. **92**, 113002 (2004)
45. H. Ohmura, T. Nakanaga, Quantum control of molecular orientation by two-color laser fields. J. Chem. Phys. **120**, 5176 (2004)
46. H. Ohmura, N. Saito, M. Tachiya, Selective ionization of oriented nonpolar molecules with asymmetric structure by phase-controlled two-color laser fields. Phys. Rev. Lett. **96**, 173001 (2006)
47. H. Ohmura, F. Ito, M. Tachiya, Phase-sensitive molecular ionization induced by a phase-controlled two-color laser field in methyl halides. Phys. Rev. A **74**, 043410 (2006)
48. H. Ohmura, M. Tachiya, Robust quantum control of molecular tunneling ionization in the space domain by phase-controlled laser fields. Phys. Rev. A **77**, 023408 (2008)
49. H. Ohmura, N. Saito, H. Nonaka, S. Ichimura, Dissociative ionization of a large molecule studied by intense phase-controlled laser fields. Phys. Rev. A **77**, 053405 (2008)
50. J. Lavancier, D. Normand, C. Cornaggia, J. Morellec, H.X. Liu, Laser-intensity dependence of the multielectron ionization of CO at 305 nm and 610 nm. Phys. Rev. A **43**, 1461 (1991)
51. M.J. DeWitt, R.J. Levis, Observing the transition from a multiphoton-dominated to a field-mediated ionization process for polyatomic molecules in intense laser fields. Phys. Rev. Lett. **81**, 5101 (1998)
52. H.R. Reiss, Inherent contradictions in the tunneling-multiphoton dichotomy. Phys. Rev. A **75**, 031404 (2007)
53. T.K. Kjeldsen, C.Z. Bisgaard, L.B. Madsen, H. Stapelfeldt, Role of symmetry in strong-field ionization of molecules. Phys. Rev. A **68**, 063407 (2003);
54. T.K. Kjeldsen, C.Z. Bisgaard, L.B. Madsen, H. Stapelfeldt, Influence of molecular symmetry on strong-field ionization: Studies on ethylene, benzene, fluorobenzene, and chlorofluorobenzene. Phys. Rev. A **71**, 013418 (2005)

55. M.J. Frisch, G.W. Trucks, H.B. Schlegel, G.E. Scuseria, M.A. Robb, J.R. Cheeseman, J.A. Montgomery Jr., T. Vreven, K.N. Kudin, J.C. Burant, J.M. Millam, S.S. Iyengar, J. Tomasi, V. Barone, B. Mennucci, M. Cossi, G. Scalmani, N. Rega, G.A. Petersson, H. Nakatsuji, M. Hada, M. Ehara, K. Toyota, R. Fukuda, J. Hasegawa, M. Ishida, T. Nakajima, Y. Honda, O. Kitao, H. Nakai, M. Klene, X. Li, J.E. Knox, H.P. Hratchian, J.B. Cross, C. Adamo, J. Jaramillo, R. Gomperts, R.E. Stratmann, O. Yazyev, A.J. Austin, R. Cammi, C. Pomelli, J.W. Ochterski, P.Y. Ayala, K. Morokuma, G.A. Voth, P. Salvador, J.J. Dannenberg, V.G. Zakrzewski, S. Dapprich, A.D. Daniels, M.C. Strain, O. Farkas, D.K. Malick, A.D. Rabuck, K. Raghavachari, J.B. Foresman, J.V. Ortiz, Q. Cui, A.G. Baboul, S. Clifford, J. Cioslowski, B.B. Stefanov, G. Liu, A. Liashenko, P. Piskorz, I. Komaromi, R.L. Martin, D.J. Fox, T. Keith, M.A. Al-Laham, C.Y. Peng, A. Nanayakkara, M. Challacombe, P.M.W. Gill, B. Johnson, W. Chen, M.W. Wong, C. Gonzalez, J.A. Pople, Gaussian 03, Revision C.02, (Gaussian, Inc., Wallingford CT, 2004)

Chapter 6
High Harmonic Generation from Aligned Molecules

Ruxin Li, Peng Liu, Pengfei Wei, Yuexun Li, Shitong Zhao, Zhinan Zeng, and Zhizhan Xu

Abstract The recent progress in the study of high-order harmonic generation (HHG) from aligned molecules at SIOM is reviewed. We identify the laser intensity dependence of HHG from aligned CO_2 molecules. The modulation inversion of harmonic yield with respect to molecular alignment can be altered dramatically by fine tuning the intensity of driving laser pulse for harmonic generation. The angular distribution of harmonic intensities is measured, and the results can be modeled by employing the strong-field approximation including ground state depletion factor.

For improving the alignment degree of molecules, we present an active control scheme – the rule of slope – to either enhance or suppress the molecular alignment by the first laser pulse. The underlying physics has been revealed both numerically and analytically.

6.1 Introduction

High-order harmonic generation (HHG) from atoms and molecules has been intensively explored in high-field laser physics owing to its application in producing ultrafast coherent extreme ultraviolet (XUV) radiation and even attosecond pulses [1–5]. In the strong-field limit, HHG is well explained by the three-step model: active electrons first tunnel through the potential barrier, are then accelerated and turned back by laser fields to recombine with the parent ions emitting high-energy photons [6]. Based on the underlying physics of HHG, the electronic dynamics of atoms can be probed in extreme high temporal resolution [7].

Comparing to atom, molecule has additional degrees of freedom, i.e., rotational and vibrational motions, which offer extra parameters as useful tools to control the

---

R. Li (✉), P. Liu, P. Wei, Y. Li, S. Zhao, Z. Zeng, and Z. Xu
State Key Laboratory of High Field Laser Physics, Shanghai Institute of Optics and Fine Mechanics, Chinese Academy of Sciences, Shanghai, 201800, China
e-mail: ruxinli@mail.shcnc.ac.cn, peng@siom.ac.cn, personinjoy@163.com, yuexunli@siom.ac.cn, zst@siom.ac.cn, zhinan_zeng@siom.ac.cn, zzxu@mail.shcnc.ac.cn

K. Yamanouchi et al. (eds.), *Progress in Ultrafast Intense Laser Science VII*, Springer Series in Chemical Physics 100, DOI 10.1007/978-3-642-18327-0_6,

HHG characteristics [8, 9]. Through the interaction of molecules and an ultrafast strong laser field, the coherent evolution of molecular rotational levels results in the periodic alignment of molecules in laboratory frame. Such a field-free alignment has potential applications in investigating all the angular distribution phenomena in molecular frame [10–16]. According to the strong-field approximation, the harmonic generation is determined by the molecular ground state, while the electron wavefunction can be approximated by a plane wave. Therefore, from the angular distribution of harmonic generation, the tomographic images of the highest occupied molecular orbital (HOMO) of nitrogen molecule has been reconstructed [12]. It is also noted that the alignment and orientation of molecules are fundamentally important for such studies.

Recently, progresses have been made in the studies of molecular harmonic generation and nonadiabatic molecular alignment by double pulses, which are addressed in this chapter.

6.2 Laser Intensity Dependence of Quantum Interference Effect in HHG

HHG from CO_2 represents an interesting new feature of molecular alignment dependence. As shown by Kanai et al., the harmonic yields for the 17th to 29th orders from aligned CO_2 molecules exhibit *inverted* modulation versus the field ionization and alignment parameter $<\cos^2\theta>$ [17]. This modulation inversion was attributed to the interference of recombination electrons from the two oxygen atoms in CO_2 molecule. Vozzi et al. reported further that the two center interference effect can be controlled by changing the ellipticity of the driving laser pulse [18]. They showed that the *inverted* modulation of harmonic signal appears for the 21st to 39th harmonic orders, while the modulation of the 41st to 49th orders harmonic emission is the same as that of molecular alignment. The difference in harmonic orders with inverted modulation cannot be explained by the constructive and destructive interference. By taking into account the ground state depletion effect on HHG, Le et al. proposed theoretically that the harmonic yield inversion can be varied by changing driving laser intensity [19]. This may explain the inversion for different harmonic orders but an experimental verification of this model is strongly desired.

To explore the underlying physics in the angular dependence of high harmonic emission from aligned molecules, the HHG from aligned CO_2 molecules as a function of the driving laser intensity is investigated. For the first time, it has been found that by changing the laser intensity, the modulation of harmonic yield from aligned molecules can be varied and even inverted with respect to molecular alignment. Our experimental results are compared with the numerical calculations employing the strong-field approximation model including ground state depletion effect. The role of laser intensity as a new parameter to control the angular dependence of HHG in aligned molecules is revealed.

6.2.1 Experimental Setup

The experiments are performed using a Ti:sapphire-based chirped pulse amplification laser system (Spectral-Physics, TSA-25), which produce 50 femtoseconds (fs) laser pulses of 10 Hz at 800 nm center wavelength. The output is split into two beams, one used as the pump pulse (for aligning molecules) and the other as the probe pulse (for driving HHG from molecules). The polarization of the pump laser beam is the same as the probe beam. The two beams were collinearly focused with a 30 cm focal length lens onto a pulsed supersonic molecular beam located in a high vacuum interaction chamber. The laser focus was about 1 mm downstream of a 0.25 mm diameter nozzle orifice. Stagnation pressure of CO_2 gas (99.998%) was around 2 bar, leading to a rotational temperature of several tens of Kelvin. The spot size of pump laser beam crossing with molecules was measured to be 110 μm (FWHM), and the laser field intensity was estimated to be $2.1 \times 10^{13}\,\mathrm{W/cm^2}$. The probe laser energy was adjustable by using a half-wave plate and a high extinction film polarizer. At the interaction position, the spot size of the probe beam was measured to be 144 μm (FWHM). The HHG spectra were detected by a homemade flat-field grating spectrometer equipped with a soft-x-ray CCD camera (Princeton Instruments, PI: SX 400).

6.2.2 Inverted Modulation of Harmonics by Laser Intensities

When a CO_2 molecule is irradiated by a short laser pulse whose duration ($\tau_{on} =$ 50 fs) is much shorter than the molecular rotational period, $\tau_{on} \ll 2\pi/\omega = 42.7$ ps for CO_2, nonadiabatic field-free alignment is achieved by the excitation of a rotational wave packet $\psi(t) = \sum_{J,M} A_{J,M}(t)|J,M\rangle$. The time evolution of the wave packet can be calculated through solving the time-dependent Schrödinger equation (TDSE), and the time-dependent alignment parameter $<\cos^2\theta>(t)$ is obtained by the evolved wave function of the rotational states. (Details are given in Sect. 6.3 of this chapter.) In the calculation, the initial rotational temperature is taken to be 80 K and the result is shown in Fig. 6.1 (dashed line, right axis).

The experimentally measured 23rd harmonic intensity as a function of pump-probe delay time is also represented in Fig. 6.1 (solid line, left axis). The probe laser intensity was $2.4 \times 10^{14}\,\mathrm{W/cm^2}$. It is evident that the modulation of harmonic signal is reversely matched with that of the molecular alignment parameters $<\cos^2\theta>$ exactly. This is consistent with the experimental results of Kanai et al. and Vozzi et al. that the harmonic signal reversely matches the alignment parameter $<\cos^2\theta>$ [17, 18]. The phenomena were regarded as the evidence of interference of the recombining electrons originated from two oxygen atoms.

We measured the 25th harmonic yield at around half revival of molecular alignment under four different probe laser intensities: $1.6 \times 10^{14}\,\mathrm{W/cm^2}$, $1.9 \times 10^{14}\,\mathrm{W/cm^2}$, $2.1 \times 10^{14}\,\mathrm{W/cm^2}$, and $2.4 \times 10^{14}\,\mathrm{W/cm^2}$. The results are

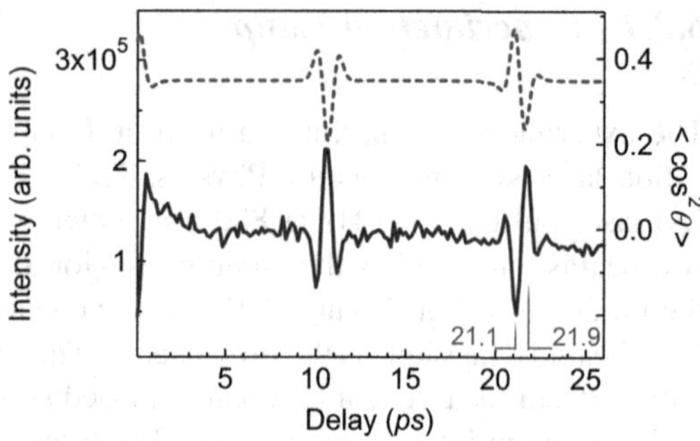

Fig. 6.1 Measured 23rd harmonic yield (*solid line*) from aligned CO_2 molecules and calculated alignment parameter $< \cos^2 \theta >$ (*dashed line*) as functions of the pump-probe delay time. The temporal modulation of the harmonic intensity reversely matches the alignment [16]

shown in Fig. 6.2a–d, respectively. As we can see, the modulation of the 25th order harmonic yield is completely inverted as the laser intensity increases to $2.1 \times 10^{14}\,\mathrm{W/cm^2}$ (Fig. 6.2c) and the inversion can survive at a higher intensity as $2.4 \times 10^{14}\,\mathrm{W/cm^2}$. It should be noted that the change in probe laser intensity does not induce significant variation of molecular alignment during the HHG process. We thus demonstrate that the angular dependence of harmonic emission with respect to molecular alignment can be varied dramatically and even inverted by tuning the probe laser intensity.

The observation is interesting in light of the fact that two-center interference effect is only determined by the alignment angle and the deBroglie wavelength of recombining electrons. As Le et al. [19] have shown in the calculation, the 33rd harmonic intensity has a reversed modulation as the evolution of molecular alignment at the driving laser intensity of $2 \times 10^{14}\,\mathrm{W/cm^2}$, and it changes to be the same as that of molecular alignment when the laser intensity decreases to $1 \times 10^{14}\,\mathrm{W/cm^2}$. Although in our experiments the 33rd harmonic is too weak to be distinguished, the results shown in Fig. 6.2 are in principle consistent with the calculated results.

From Fig. 6.1, one can also note that the 23rd harmonic intensity from the maximal aligned CO_2 increases over 35% (2.2:1.6 in arbitrary unit) around the half revival. The strongest pulse energy is estimated to be 2.81×10^{-11} J comparing to the 2.05×10^{-11} J at the time delay between the revivals. In previous studies, for the purpose of enhancing harmonic yields, methods of phase-matching and quasi phase-matching schemes [20, 21] have been developed to be able to increase the pulse energy to the order of microjoules in X-ray region, which indicates a much higher conversion efficiency than in our experiments (10^{-8} versus 10^{-6}). However, the enhancement of harmonic emission from aligned molecules can serve as a complementary mechanism to increase the coherent attosecond X-ray output.

6.2.3 Angular Distribution of Harmonics from CO_2

We further measured the harmonic intensities as a function of the polarization angle between the aligning and the driving laser field at the fixed pump-probe delay of

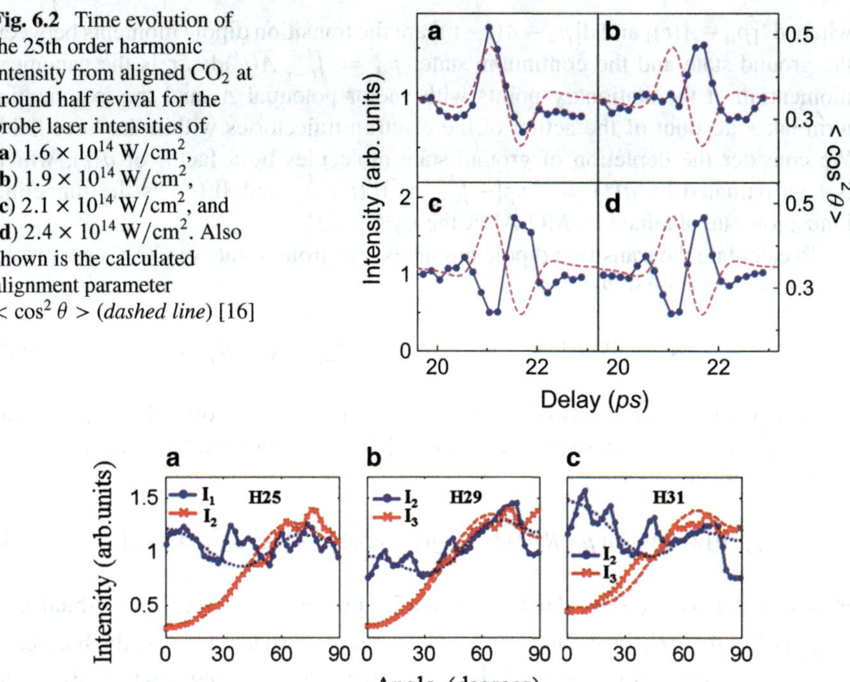

Fig. 6.2 Time evolution of the 25th order harmonic intensity from aligned CO_2 at around half revival for the probe laser intensities of (**a**) $1.6 \times 10^{14}\,\mathrm{W/cm^2}$, (**b**) $1.9 \times 10^{14}\,\mathrm{W/cm^2}$, (**c**) $2.1 \times 10^{14}\,\mathrm{W/cm^2}$, and (**d**) $2.4 \times 10^{14}\,\mathrm{W/cm^2}$. Also shown is the calculated alignment parameter $< \cos^2\theta >$ (*dashed line*) [16]

Fig. 6.3 Lineouts of harmonic orders 25, 29, and 31 that exhibit different angular distributions at different driving laser intensities [49]

21.1 ps corresponding to the alignment condition. The results for three driving laser intensities ($I_1 = 1.8 \times 10^{14}\,\mathrm{W/cm^2}$, $I_2 = 2.3 \times 10^{14}\,\mathrm{W/cm^2}$ and $I_3 = 2.8 \times 10^{14}\,\mathrm{W/cm^2}$) are shown in Fig. 6.3a–c, respectively. From Fig. 6.3a–c, one can see that the angular dependence of the 25th to 33rd harmonic emission can be changed and even inverted by enhancing the driving laser intensity.

6.2.4 *Strong-Field Approximation of Molecules*

To clarify the role of laser field intensity, we performed numerical simulation using an extended Lewenstein's strong-field approximation model including the ground state ionization effect [22]. In this model, the harmonic spectrum from molecular dipole moment in the time domain is calculated by using

$$x(t) = \mathrm{i}\int_0^{\infty} \mathrm{d}\tau \left(\frac{\pi}{\epsilon + i\tau/2}\right)^{3/2} \mathrm{d}^*[p_{\mathrm{st}} - A(t)]\mathrm{e}^{-\mathrm{i}S(p_{st},t,\tau)}$$
$$\times E(t-\tau)\cdot \mathrm{d}[p_{\mathrm{st}} - A(t-\tau)]a^*(t)a(t-\tau) + c.c., \qquad (6.1)$$

where $d^*[p_{st} - A(t)]$ and $d[p_{st} - A(t-\tau)]$ are the transition dipole moments between the ground state and the continuum state, $p_{st} = \int_{t-\tau}^{t} A(t')dt'/\tau$ is the canonical momentum at the stationary points with vector potential $\boldsymbol{A}$, and the exponential term takes account of the action of the electron trajectories within the laser field. We consider the depletion of ground state molecules by a factor of $a(t)$, which is approximated by $a(t) = \exp[-\int_{-\infty}^{t} dt' W(t')/2]$, and $W(t')$ is the tunneling ionization rate obtained by MO-ADK theory [19,23].

To calculate the transition dipole moments, the ground state of CO_2 is expressed by atomic $2p_y$ orbital $\Phi_{2p_{y'}}$ as

$$\Psi_{\pi_g}(x) \propto [\Phi_{2p_{y'}}(x + R/2) - \Phi_{2p_{y'}}(x - R/2)], \tag{6.2}$$

where R is the distance between oxygen atoms, and the $2p_{y'}$ orbital was expressed by $\sim \exp(-\alpha|x|)$ type basis function [24–26]. The time-dependent dipole transition moment is given by

$$d_{\pi_g}(p) \propto [2\mathrm{i} \sin(p \cdot R/2) \mathrm{d}_{2p_{y'}}(p) - \cos(p \cdot R/2) \tilde{\Phi}_{2p_{y'}}(p) R], \tag{6.3}$$

in which the $\mathrm{d}_{2p_{y'}}(p)$ is the atomic dipole moment from the $2p_{y'}$ orbital and $\tilde{\Phi}_{2p_{y'}}(p)$ is the $2p_{y'}$ wavefunction in the momentum space. Finally, the harmonic spectra as a function of the angle between molecular axis and laser polarization, $S(\theta)$, are obtained through the Fourier transform of the transition dipole moment.

To calculate the harmonic intensities from aligned CO_2 molecules, the time-dependent alignment distribution of molecules has to be taken into account. At field-free evolving time t, the distribution of aligned molecular axis is expressed as $P(\theta, t) = |\langle \psi(t)|\psi(t)\rangle|^2$, in which $\psi(t)$ is obtained by solving the TDSE. The harmonic intensity evolution is given by integrating the product of $S(\theta)$ and $P(\theta, t)$,

$$f(t) \propto \int S(\theta) P(\theta, t) \sin\theta \mathrm{d}\theta. \tag{6.4}$$

To compare with the experimental results, we first calculate the angular distribution $S(\theta)$ of the 25th harmonics using different driving laser intensities from $1.2 \times 10^{14}\,\mathrm{W/cm^2}$ to $2.4 \times 10^{14}\,\mathrm{W/cm^2}$. The molecular alignment distributions $P(\theta, t = 21.1\,\mathrm{ps})$ and $P(\theta, t = 21.9\,\mathrm{ps})$ are obtained from $\psi(t)$ by solving the TDSE using the experimental conditions. At these two delay times, CO_2 molecules are mostly aligned along the laser polarization direction ($\theta = 0°$) and perpendicular to laser polarization ($\theta = 90°$), respectively. Based on (6.4), the calculated harmonic yields of $f(t = 21.1\,\mathrm{ps})$ and $f(t = 21.9\,\mathrm{ps})$ are plotted in Fig. 6.4a. It is evident that the harmonic yields are clearly modulated by the laser intensity. With the laser intensity in the range from $2.4 \times 10^{14}\,\mathrm{W/cm^2}$ to $1.4 \times 10^{14}\,\mathrm{W/cm^2}$, the normalized 25th order harmonic intensity at $t = 21.9\,\mathrm{ps}$ is stronger than at $t = 21.1\,\mathrm{ps}$ (i.e., the harmonic intensity modulation is inverted with the molecular alignment). When the laser intensity decreases to $1.4 \times 10^{14}\,\mathrm{W/cm^2}$ and lower, the harmonic intensities at the two delays are reversed back to be matched with the modulation of molecular alignment. The calculated results reproduce well the experimental results shown in Fig. 6.3, except a small difference in the absolute value of laser intensity, where the inversion starts to occur. A more rigid model for the HHG in CO_2 may help to reduce this quantitative difference between the experimental and calculated results.

For comparison, we also calculate the harmonic intensities $f(t = 21.1\,\text{ps})$ and $f(t = 21.9\,\text{ps})$ without including the ground-state depletion factor $a(t)$ and the results are shown in Fig. 6.4b. It is shown that the normalized harmonic signals at $t = 21.1$ ps (the molecules are aligned to laser polarization direction) are stronger than those at $t = 21.9$ ps (the anti-aligned case) in the whole range of different probe laser intensities. However, in the experimental results shown in Fig. 6.2, we observed this modulation only when the laser field intensity is reduced to $1.6 \times 10^{14}\,\text{W/cm}^2$. Clearly, the calculation without considering the ground state depletion cannot reproduce the experimental observation of the intensity modulation inversion of the 25th order harmonic. The above calculated results indicate the necessity of including the ground state depletion factor into SFA model in the calculation of harmonic emission from aligned molecules.

The above analysis indicates that ground state depletion might be an important factor responsible for the observed laser intensity dependence of the angular distribution of harmonic emission from aligned molecules. However, a further investigation is still necessary for understanding how the probe laser intensity affects the HHG from aligned molecules.

From (6.4), we know that to control the angular distribution of HHG one could also adjust the molecular alignment distribution $P(\theta)$. The recent progress in 3D molecular alignment is expected to play a role in controlling HHG [27–29]. Experiments on higher order harmonics with finely adjusted driving laser intensity may lead to a more comprehensive understanding of angular dependence of harmonics and therefore a better control of HHG.

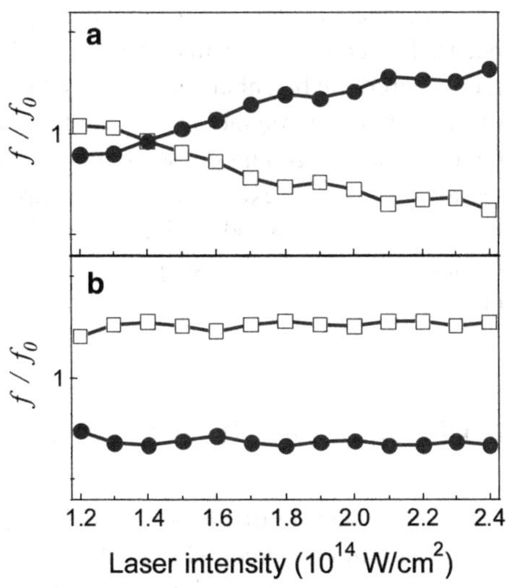

Fig. 6.4 Calculated normalized intensities (f/f_0) of the 25th order harmonic emission at two pump-probe delays: 21.1 ps (*open squares*) and 21.9 ps (*solid circles*) as functions of the probe laser intensities from $1.2 \times 10^{14}\text{W/cm}^2$ to $2.4 \times 10^{14}\,\text{W/cm}^2$. The calculations were carried out by (**a**) considering ground state depletion factor and (**b**) not considering ground state depletion factor [16]

6.3 Molecular Alignment

Irradiated by an ultrafast laser field ($\sim 10^{13}\,\mathrm{W/cm^2}$), molecules can be partially aligned along the polarization direction of the laser pulse. Alignment means an increased probability distribution along the polarization axis of the field, whereas orientation requires, in addition, the same (or apposite) direction as the polarization vector. There are two ways to align molecular ensembles – adiabatically and nonadiabatically, among which nonadiabatic alignment of the molecules attracts much attention for being able to study the molecules in a field-free alignment condition [10–16].

For most applications, it is important to select the optimal laser field so that the molecules are aligned to a degree ($< \cos^2\theta >$) that is as high as possible. For a given pulse duration, the degree of alignment can be improved by increasing the laser field intensity. However, the improvement is often limited by the saturation of molecular alignment and the maximum intensity that can be applied to the molecules without ionizing it [30, 31]. Spectral phase shaping method provides a passive control means of improving the alignment beyond the limit [32]. Accurate optimization of aligning laser pulse can be realized by this self-learning evolutionary algorithm, with which the degree of alignment can be maximized. However, this method is time consuming and needs the phase shaping instrument to apply it in experiments.

An alternative and more efficient way of achieving an enhanced degree of alignment is to employ a train of laser pulses. As the alignment is resulted from the in-phase overlapping of spherical harmonics in the time evolution of rotational wave packet, introducing the second laser pulse may modify the rotational wave packet and thereby increase the alignment degree. To strongly enhance the degree of alignment of linear molecules, Averbukh et al. proposed a specially designed sequence of pulses [33]. A two- and a three-pulse excitation schemes were also employed to provide an effective and robust molecular alignment method by Leibscher et al. [34, 35]. The enhancement of the multi-pulse alignment using this scheme has also been achieved in the recent experiments [36–38]. As shown in the reports, the alignment degree can be enhanced by applying the second laser pulse at the rising edge of the alignment degree induced by the first laser pulse, or at the full revival time T_{rev}. On the contrary, as a model scenario of coherent control of quantum states, the rotational wave packet excited by the first laser pulse can be annihilated by the second laser pulse when introduced at the half revival time [39–42]. However, a general method that relates the two-pulse alignment degree to their delay times is highly desired.

6.3.1 Rotational Wave Packet and Molecular Alignment

When an ensemble of linear molecules is illuminated by an ultrashort laser pulse whose duration is much shorter than the molecular rotational period, each initial rotational state in the Boltzmann distribution expands to a wave packet through nonresonant impulsive Raman excitation. After the aligning pulse, molecular rotational wave packet is a coherent superposition of rotational eigenstates excited from the initial state $|J_i M_0\rangle$ (J, M labeling the rotational and the magnetic quantum numbers) and undergoes field-free evolution,

$$\Psi(t) = \sum_J \tilde{A}_J^{J_i} \mathrm{e}^{-\mathrm{i}E_J t} |JM_0\rangle, \tag{6.5}$$

where $\tilde{A}_J^{J_i} = A_J^{J_i} \exp(\mathrm{i}\delta_J^{J_i})$ is the complex amplitude of the transition from J_i to J, with the real amplitude $A_J^{J_i}$ and the phase $\delta_J^{J_i}$, and E_J is the eigenenergy of the state $|JM_0\rangle$ [43].

The time evolution of the wave packet can be calculated through solving the TDSE

$$\mathrm{i}\frac{\partial \psi(\theta,t)}{\partial t} = [BJ^2 + H_{\mathrm{int}}]\psi(\theta,t), \tag{6.6}$$

in which the θ is defined by the angle between the laser field polarization and molecular axis, and the $\boldsymbol{BJ}^2$ is the rotational energy operator. The Hamiltonian of the angle-dependent AC stark shift is

$$H_{\mathrm{int}} = -\frac{1}{2}\alpha_{\perp}\varepsilon^2 - \frac{1}{2}\Delta\alpha\varepsilon^2\cos^2\theta, \tag{6.7}$$

where the $\Delta\alpha = \alpha_{\parallel} - \alpha_{\perp}$ and $\alpha_{\parallel}, \alpha_{\perp}$ are the polarizabilities along and perpendicular to the molecular axis. Finally, the time-dependent alignment parameter $< \cos^2\theta >$ is obtained from the calculated rotational wavefunction of the molecules.

The alignment parameter is described as the time-dependent average

$$\langle\cos^2\theta\rangle(t) = \sum_{J_i} \rho_{J_i} \sum_{M_0=-J_i}^{J_i} \langle\cos^2\theta\rangle_{J_i M_0}(t), \tag{6.8}$$

where ρ_{J_i} is the Boltzmann weight of the J_i state, and

$$\begin{aligned} &\langle\cos^2\theta\rangle_{J_i M_0}(t) \\ &\quad = \sum_J (A_J^{J_i})^2 \langle JM_0|\cos^2\theta|JM_0\rangle + \sum_J 2A_J^{J_i} A_{J+2}^{J_i} \langle JM_0|\cos^2\theta|J+2, M_0\rangle \\ &\quad \times \cos(\Delta\omega_{J+2,J}t + \Delta\theta_{J+2,J}) \end{aligned} \tag{6.9}$$

in which $\Delta\omega_{J+2,J} = E_{J+2} - E_J = B_0(4J+6)$ is the beat frequency of rotational states $|J+2, M_0\rangle$ and $|JM_0\rangle$, and $\Delta\theta_{J+2,J} = \delta_{J+2}^{J_i} - \delta_J^{J_i}$ is the initial phase difference between the states $|J+2, M_0\rangle$ and $|JM_0\rangle$ [40]. During the field-free evolution of the rotational wave packet, the alignment parameter demonstrates revivals due to the periodic rephasing among the beats with frequencies $\Delta\omega_{J+2,J}$, as shown in (6.9). A full revival period T_{rev} is π/B_0.

The slope of the alignment curve $\langle\cos^2\theta\rangle(t)$ is given from (6.8),

$$\mathrm{d}\langle\cos^2\theta\rangle(t)/\mathrm{d}t = \sum_{J_i}\rho_{J_i}\sum_{M_0=-J_i}^{J_i}\mathrm{d}\langle\cos^2\theta\rangle_{J_iM_0}(t)/\mathrm{d}t, \tag{6.10}$$

where

$$\mathrm{d}\langle\cos^2\theta\rangle_{J_iM_0}(t)/\mathrm{d}t = -\sum_J 2A_J^{J_i}A_{J+2}^{J_i}\langle JM_0|\cos^2\theta|J+2,M_0\rangle\Delta\omega_{J+2,J} \times\sin(\Delta\omega_{J+2,J}t+\Delta\theta_{J+2,J}). \tag{6.11}$$

It is noted that the slope is modulated by the sinusoidal term of time delay.

6.3.2 *The Rule of Slope*

In the calculation, the O_2 ($B_0 = 1.4297\,\mathrm{cm}^{-1}$, $D_e = 4.839\times10^{-6}\,\mathrm{cm}^{-1}$, and $\Delta\alpha = 1.099\,\text{Å}^3$ [44]) is aligned by two identical 50 fs laser pulses with a Gaussian temporal shape. The rotational temperature of the ensemble of molecules is chosen as 80 K according to the kinetics estimation based on usual experimental conditions. The peak intensity of the laser pulse is set to be $3.0\times10^{13}\,\mathrm{W/cm^2}$ to avoid significant ionization of the oxygen molecules.

The calculation shows that the resulted maximum alignment degree of two-pulse alignment coincides with the maximum of the slope of the $\langle\cos^2\theta\rangle$ by the first pulse, $\mathrm{d}\langle\cos^2\theta\rangle/\mathrm{d}t$. When the second laser pulse is applied at the peak position of the slope curve, the resulted alignment degree is the largest. As shown in Fig. 6.5a, the alignment degree $\langle\cos^2\theta\rangle$ by the first laser pulse is shown in the solid curve (the left axis), and its time derivative of $\langle\cos^2\theta\rangle$ is plotted in the dashed curve (the right axis). The zero point of time (t_0) is at the peak of the first pulse. One can see that the minimum of the slope is located at $t_1 = 0.50T_{\mathrm{rev}}$(5.83 ps) and the maximum is at $t_2 = 1.00T_{\mathrm{rev}}$(11.65 ps). Figure 6.5b, c show the resulted $\langle\cos^2\theta\rangle$ by introducing the second laser pulse at t_1 and t_2, respectively. As shown in Fig. 6.5b, when the second pulse applying at the time of the minimum of slope, t_1, the alignment by the first pulse is suppressed and $\langle\cos^2\theta\rangle$ becomes close to 0.33 after the second pulse. But from the Fig. 6.5c, one can see that the alignment of molecules is enhanced to 0.59 at $t = 1.25T_{\mathrm{rev}}$ (14.57 ps) by applying the second pulse at the time t_2. The antialignment of the molecules is also enhanced by the same laser condition at $t = 1.75T_{\mathrm{rev}}$ (20.39 ps), $\langle\cos^2\theta\rangle$ decreases to 0.15.

We also calculated the highest resulted alignment degree $\langle\cos^2\theta\rangle$ by adjusting the delay at around the quarter revival and the half revival. In Fig. 6.6, the $\langle\cos^2\theta\rangle$ by the first pulse and its slope are plotted in solid curve (the left axis) and the dashed curve (the right axis), respectively. From 0.11 T_{rev} to 0.64 T_{rev} (1.30 ps to 7.50 ps) that covers the quarter and the half revival, the second laser pulse is introduced and

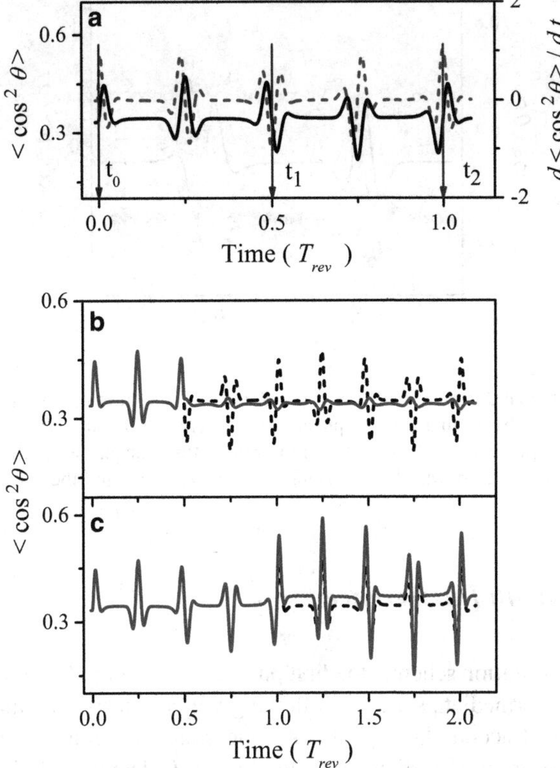

Fig. 6.5 (**a**) The alignment degree $< \cos^2\theta >$ by the first aligning pulse (the *solid curve*, the left axis) and its slope curve (the *dashed curve*, the right axis). t_0 is at the peak of the first aligning pulse. t_1 and t_2 are the minimum and maximum of the *slope curve*; (**b**) The $< \cos^2\theta >$ (the *solid curve*) by the two-pulse alignment with delay time of $t_1 = 0.50T_{rev}$; (**c**) The $< \cos^2\theta >$ (the *solid curve*) by the two-pulse alignment with delay of $t_2 = 1.0T_{rev}$ [50]

the resulted maximum and minimum alignment degree $\langle\cos^2\theta\rangle$ are plotted in the empty squared and the solid circled markers. As can be seen, the curve resembled by the maximum markers matches with the slope of $\langle\cos^2\theta\rangle$ exactly, and the minimum markers match with the slope reversely.

The calculation results agree with the previous studies in the cases of the full revival and the half revival delay times of the two laser pulses, which result in the alignment enhancement and the annihilation of rotational wave packet. Besides these two time delays, our scheme proposes a more general rule that covers all the delay times of the two laser pulses. For example, it is noted that the two pulses with the delay time of 0.24 T_{rev}, where a local maximum of the slope locates, result in the maximum alignment degree (0.59) comparable to that reached by delay time of 1.00 T_{rev}.

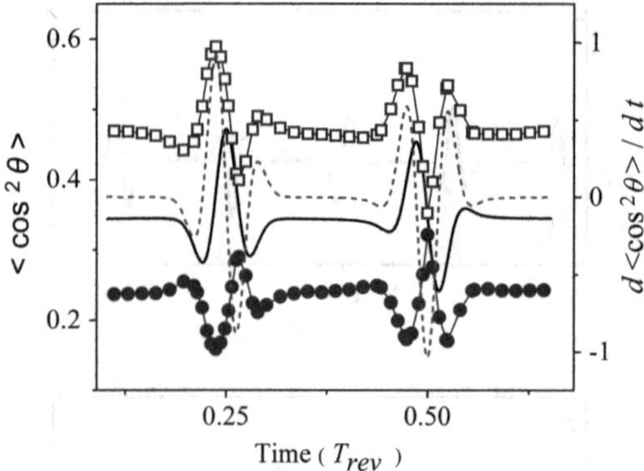

Fig. 6.6 The alignment degree $< \cos^2\theta >$ (the *solid curve*) and the slope of the $< \cos^2\theta >$ (the *dashed curve*) induced by the first aligning pulse with the field of 3.0×10^{13} W/cm^2. When applying the second aligning pulse at the delay times in the shown time range, the resulted maximum and minimum $< \cos^2\theta >$ are shown with the *empty squared markers* and the *solid circled markers* respectively [50]

6.3.3 Multi-Path Interference

In a two-pulse excitation scheme, the first pulse transfers population from the initial state J_i to the intermediate state J', following which the state J' undergoes field-free evolution and accumulates phase. A time-delayed second replica pulse then transfers population in J' further to the final state J. The probability amplitude is (M is neglected in the formula since it is conserved.)

$$\tilde{p}_J^{J_i} = \tilde{A}_{J'}^{J_i} \mathrm{e}^{-\mathrm{i}E_{J'}t} \tilde{A}_J^{J'}, \tag{6.12}$$

where $\tilde{A}_{J'}^{J_i}$ is the complex amplitude of the transition from J_i to J', with the real amplitude $A_{J'}^{J_i}$ and phase $\delta_{J'}^{J_i}$. Zero time point is set at the peak of the first pulse.

After the first pulse, the initial state expands to a rotational wave packet, so a number of intermediate levels J' contribute to the excitation from the initial state J_i to the final state J. The overall probability amplitude $\tilde{p}_J^{J_i}$ from J_i to J is a coherent superposition of all the quantum paths connecting the initial state J_i and the final state J,

$$\tilde{P}_J^{J_i}(t) = \sum_{J'} \tilde{A}_{J'}^{J_i} \mathrm{e}^{-\mathrm{i}E_{J'}t} \tilde{A}_J^{J'}. \tag{6.13}$$

When $J = J_i$, the transition probability back to the initial state is

$$|\tilde{P}_{J_i}^{J_i}(t)|^2 = \sum_{J'} (A_{J'}^{J_i})^4 + 2 \sum_{J'>J''} (A_{J'}^{J_i})^2 (A_{J''}^{J_i})^2 \cos[\Delta\omega_{J'J''}t + 2(\delta_{J'}^{J_i} - \delta_{J''}^{J_i})], \tag{6.14}$$

where $\Delta\omega_{J'J''} = E_{J'} - E_{J''}$, $J' = J'' + 2n$, $n = 1, 2, 3\ldots$. The result in (6.14) has been demonstrated in previous works on rotational wave packet reconstruction [45, 46]. Generally, the products $(A_{J''}^{J_i})^2(A_{J''+4}^{J_i})^2$, $(A_{J''}^{J_i})^2(A_{J''+6}^{J_i})^2\ldots$, which contain two or more Raman excitation steps, are much smaller than $(A_{J''}^{J_i})^2$ $(A_{J''+2}^{J_i})^2$. Therefore, it is legitimate to only consider the interference of adjacent quantum paths involving the intermediate states J' and $J' + 2$. As a result, the (6.14) turns to

$$|\tilde{P}_{J_i}^{J_i}(t)|^2 \approx \sum_J (A_J^{J_i})^4 + 2\sum_J (A_J^{J_i})^2(A_{J+2}^{J_i})^2 \cos(\Delta\omega_{J+2,J}t + 2\Delta\theta_{J+2,J}), \tag{6.15}$$

where $\Delta\theta_{J+2,J} = \delta_{J+2}^{J_i} - \delta_J^{J_i}$ is the initial phase different between the state J and the excited state $J + 2$. Equation (6.15) can also be rewritten as

$$\begin{aligned} |\tilde{P}_{J_i}^{J_i}(t)|^2 \approx &\sum_J (A_J^{J_i})^4 + 2\sum_J (A_J^{J_i})^2(A_{J+2}^{J_i})^2 \\ &\times \sin(\Delta\omega_{J+2,J}t + \Delta\theta_{J+2,J} + \Delta\delta_{J+2,J}), \end{aligned} \tag{6.16}$$

where $\Delta\delta_{J+2,J} = \Delta\theta_{J+2,J} + \pi/2$. In previous theoretical and experimental works, it has been found that the dominant initial phase differences are $-\pi/2$ in the impulsive approximation [45–47]. So we neglect $\Delta\delta_{J+2,J}$. Consequently, (6.16) becomes

$$|\tilde{P}_{J_i}^{J_i}(t)|^2 \approx \sum_J (A_J^{J_i})^4 + 2\sum_J (A_J^{J_i})^2(A_{J+2}^{J_i})^2 \sin(\Delta\omega_{J+2,J}t + \Delta\theta_{J+2,J}), \tag{6.17}$$

from which one can see that the constructive or destructive interference occurs among different transition paths when the time delay varies. Note that such population probability is modulated by the same sinusoidal term as in (6.11).

After taking into account the thermal distribution of all molecular rotational levels, the total transition probability for molecules being back to the original states after two-pulse excitation is matched reversely with the slope of the alignment degree by the first aligning pulse, which confirms the rule of slope.

6.3.4 Selective Population Transition

The two extreme cases of the two-pulse alignment are when the time delay of the two pulses is in one revival and a half revival, under which conditions, the alignment is either maximal enhanced or suppressed by the second pulse. To further understand this scenario, a theoretical analysis is given to reveal the underlying physics.

The transition amplitude corresponding to the Raman frequency $\omega_J = E_{J+2} - E_J = B_0(4J+6)$ (E_J is the eigenvalue of the Jth state and B_0 is the rotational constant.) is proportional to

$$A(\omega_J) \propto \int_{-\infty}^{\infty} E(\omega)E^*(\omega - \omega_J)\mathrm{d}\omega, \tag{6.18}$$

where $E(\omega)$ is the spectrum of the aligning laser pulse [48]. For the impulsive Raman process, transition occurs for all pairs of photons with the frequency difference of ω_J. Consider a laser field composing of two replicas of an FTL pulse temporally separated by ΔT,

$$E_d(t) = E_s(t - \Delta T/2) + E_s(t + \Delta T/2), \tag{6.19}$$

where $E_d(t)$ describes the electric field of double pulses in time domain and $E_s(t)$ the laser field of a single pulse. The spectrum of this pulse pair is

$$E_d(\omega) = 2E_s(\omega)\cos(\Delta T\omega/2), \tag{6.20}$$

where $E_s(\omega)$ is the spectrum of a single FTL pulse. From (6.20), one can find that the spectrum of a pulse pair can be understood as the spectrum of a single FTL pulse with an amplitude modulation whose period is related to the temporal separation of two pulses.

Substitute (6.20) into (6.18), one gets

$$\begin{aligned} A_d(\omega_J) &= \int_{-\infty}^{\infty} E_d(\omega)E_d^*(\omega - \omega_J)\mathrm{d}\omega \\ &= 2\int_{-\infty}^{\infty} E_s(\omega)E_s^*(\omega - \omega_J)\cos[(\omega - \omega_J/2)\Delta T]\mathrm{d}\omega l. \\ &\quad +2\int_{-\infty}^{\infty} E_s(\omega)E_s^*(\omega - \omega_J)\cos(\Delta T\omega_J/2)\mathrm{d}\omega \end{aligned} \tag{6.21}$$

Generally, the delay between two pulses is much longer than the laser period,$\omega \gg 1/\Delta T$, so the first term of (6.21) is an integral over a fast oscillated function, which is approximately zero. Therefore, only the second term of (6.21) remains,

$$A_d(\omega_J) \approx 2A_s(\omega_J)\cos(\Delta T\omega_J/2). \tag{6.22}$$

It is noted that $A_d(\omega_J)$ is modulated by ΔT.

When $\Delta T = T_{\mathrm{rev}}/2$, $\Delta T\omega_J/2 = (J + 3/2)\pi$, then we have $A_d(\omega_J) \approx 0$ for all rotational quantum numbers. The net Raman transition is suppressed by the pulse pair. As a result, the alignment signal disappears. In the previous study on alignment suppression, the "zero-effect pulse pair" operator is proven to induce no net excitation [40]. Our analysis of the suppressed Raman transition also leads to the same conclusion. On the contrary, when $\Delta T = T_{\mathrm{rev}}$, $\Delta T\omega_J/2 = (2J + 3)\pi$,

$A_d(\omega_J) \approx -2A_s(\omega_J)\forall J$ reaches to the maximum. This means that the second pulse, when applied at a full revival period after the peak of the first pulse, results in the wave packet most strongly broadened in J space, which increases the degree of alignment.

Taking the rotational level of $|30\rangle$ ($J = 3$; $M = 0$, labeling the rotational and the magnetic quantum numbers) as an initial state, we calculate the time evolution of the population of related rotational levels during the interaction of the laser pulse and molecules. Figure 6.7a, c are the same and showing the population of the rotational levels of $|10\rangle, |30\rangle, |50\rangle$ and $|70\rangle$ during the first pulse. As can be seen, the population on the level of $|30\rangle$ is partially transferred mainly to two adjacent levels, $|10\rangle$ and $|50\rangle$, during the first pulse. Figure 6.7b, d shows the evolution of rotational levels as the second pulse applied at $T_{\text{rev}}/2$ and T_{rev}, respectively. As shown in Fig. 6.7a, the second pulse, when applied at $T_{\text{rev}}/2$, transfers the population in levels of $|10\rangle$ and $|50\rangle$ back to the initial state $|30\rangle$. As shown in (6.9), the alignment signal is the result of the interference of all the beat signals of adjacent states with $\Delta J = \pm 2$. As the population is transferred back to the initial state, there is no Raman excitation after the aligning laser pulses, resulting in the disappearance of the beats that could otherwise synthesize the alignment signal. Consequently, the molecular

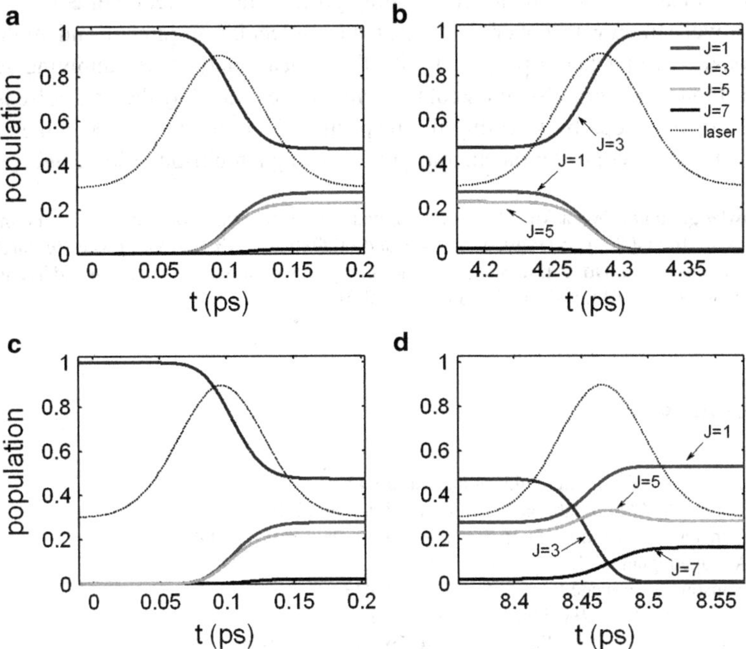

Fig. 6.7 The evolution of population for rotational eigenstates, excited from the initial state $|30\rangle$, during the aligning laser field. (**a**) and (**c**) are the same showing the population evolution during the first pulse for the eye convenience. (**b**) the population evolution during the second pulse when it is applied at $T_{\text{rev}}/2$. (**d**) the population evolution during the second pulse when it is applied at T_{rev}. The *dotted line* represents the laser envelope [51]

alignment disappears. In contrast, when the second pulse is applied at T_{rev}, as shown in Fig. 6.7d, the population in the initial state $|30\rangle$ is transferred further to the rotational levels $|10\rangle$ and $|50\rangle$ and even to the newly populated level $|70\rangle$, leaving a more depleted initial state. The stepwise excitation from the initial state to higher J states shown in Fig. 6.7c, d shows the Raman process in temporal-resolved details. This leads to a strongly broadened rotational wave packet. Consequently, the summation of the beat amplitudes increases as the population transferred to more rotational levels. As a result, when rephasing occurs, the constructive interference among beat signals leads to the enhanced molecular alignment. The controlled coherent population transfer from the initial state $|30\rangle$ demonstrated in Fig. 6.7 exemplifies the similar processes starting from other initial states in the thermal ensemble.

6.4 Summary

We experimentally demonstrate that the harmonic emission from aligned CO_2 molecules as a function of alignment angle can be altered dramatically by fine tuning the intensity of driving laser pulses. The experimental results can be modeled by the strong-field approximation model including ground state depletion effect.

A novel strategy to coherent control the molecular alignment and molecular rotational wave packet is proposed with theoretical analysis. By applying second ultrafast intense laser pulse, molecular alignment generated by the first pulse is suppressed or enhanced by selecting the respective delay times. This scheme can be applied to active control molecular alignment using a multi-pulse laser field.

Acknowledgements We acknowledge the support from National Natural Science Foundation (Grant Nos. 10734080, 60578049, 10523003 and 60978012), 973 National Basic Research Program of China (Grant No. 2006CB806000) and Shanghai Commission of Science and Technology (Grant Nos. 06DZ22015, 0652nm005 and 07pj14091).

References

1. R. Kienberger, E. Goulielmakis et al., Nature **427**, 817 (2004)
2. P. Tzallas, D. Charalambidis et al., Nature **426**, 267 (2003)
3. Y. Nabekawa, T. Shimizu et al., Phys. Rev. Lett. **96**, 083901 (2006)
4. N.A. Papadogiannis, B. Witzel et al., Phys. Rev. Lett. **83**, 4289 (1999)
5. G. Sansone, E. Benedetti et al., Science **314**, 443 (2006)
6. P.B. Corkum, Phys. Rev. Lett. **71**, 1994 (1993)
7. A. Zair, M. Holler et al., Phys. Rev. Lett. **100**, 143902 (2008)
8. N. Hay, R. Velotta et al., J. Phys. B **35**, 1051 (2002)
9. J. Itatani, D. Zeidler et al., Phys. Rev. Lett. **94**, 123902 (2005)
10. H. Stapelfeld, T. Seideman, Rev. Modern Phys. **75**, 543 (2003)
11. I.V. Litvinyuk, K.F. Lee et al., Phys. Rev. Lett. **90**, 233003 (2003)
12. J. Itatani, J. Levesque et al., Nature **432**, 867 (2004)
13. R.A. Bartels, T.C. Weinacht et al., Phys. Rev. Lett. **88**, 013903 (2002)

14. V. Kalosha, M. Spanner et al., Phys. Rev. Lett. **88**, 103901 (2002)
15. R. Velotta, N. Hay et al., Phys. Rev. Lett. **87**, 183901 (2001)
16. P. Liu, P.F. Yu et al., Phys. Rev. A **78**, 015802 (2008)
17. T. Kanai, S. Minemoto et al., Nature **435**, 470 (2005)
18. C. Vozzi, F. Calegari et al., Phys. Rev. Lett. **95**, 153902 (2005)
19. A.-T. Le, X.M. Tong et al., Phys. Rev. A **73**, 041402 (2006)
20. E. Takahashi, Y. Nabekawa et al., Phys. Rev. A **61**, 021802 (2002)
21. T. Popmintchev, M. Chen et al., PNAS **106**, 10516 (2009)
22. M. Lewenstein, P. Balcou et al., Phys. Rev. A **49**, 2117 (1994)
23. X.M. Tong, Z.X. Zhao et al., Phys. Rev. A **66**, 033402 (2002)
24. T. Kanai, S. Minemoto et al., Phys. Rev. Lett. **98**, 053002 (2007)
25. X.X. Zhou, X.M. Tong et al., Phys. Rev. A **72**, 033412 (2005)
26. X.X. Zhou, X.M. Tong et al., Phys. Rev. A **71**, 061801 (2005)
27. J.J. Larsen, K. Hald et al., Phys. Rev. Lett. **85**, 2470 (2000)
28. K.F. Lee, D.M. Villeneuve et al., Phys. Rev. Lett. **97**, 173001 (2006)
29. S.V. Simon, K. Vinod et al., Phys. Rev. Lett. **99**, 143602 (2007)
30. M. Machholm, J. Chem. Phys. **115**, 10724 (2001)
31. F. Rosca-Pruna, M.J.J. Vrakking, Phys. Rev. Lett. **87**, 153902 (2001)
32. E. Hertz, A. Rouzée et al., Phys. Rev. A **75**, 031403 (2007)
33. I.Sh. Averbukh, R. Arvieu, Phys. Rev. Lett. **87**, 163601 (2001)
34. M. Leibscher, I.Sh. Averbukh et al., Phys. Rev. Lett. **90**, 213001 (2003)
35. M. Leibscher, I.Sh. Averbukh et al., Phys. Rev. A **69**, 013402 (2004)
36. C.Z. Bisgaard, M.D. Poulsen et al., Phys. Rev. Lett. **92**, 173004 (2004)
37. C.Z. Bisgaard, S.S. Viftrup et al., Phys. Rev. A **73**, 053410 (2006)
38. K.F. Lee, I.V. Litvinyuk et al., J. Phys. B **37**, L43 (2004)
39. M. Spanner, E.A. Shapiro et al., Phys. Rev. Lett. **92**, 093001 (2004)
40. M. Renard, E. Hertz et al., Phys. Rev. A **72**, 025401 (2005)
41. S. Fleischer, I.Sh. Averbukh et al., Phys. Rev. Lett. **99**, 093002 (2007)
42. K.F. Lee, E.A. Shapiro et al., Phys. Rev. A **73**, 033403 (2006)
43. P.W. Dooley, I.V. Litvinyuk et al., Phys. Rev. A **68**, 023406 (2003)
44. K.J. Miller, J. Am. Chem. Soc. **112**, 8543 (1990)
45. H. Hasegawa, Y. Ohshima, Phys. Rev. Lett. **101**, 053002 (2008)
46. H. Hasegawa, Y. Ohshima, Proc. of SPIE, **7027**, 70271F (2008)
47. V. Renard, M. Renard et al., Phys. Rev. A **70**, 033420 (2004)
48. D. Meshulach, Y. Silberberg, Phys. Rev. A **60**, 1287 (1999)
49. P. Wei, P. Liu et al., Phys. Rev. A **79**, 053814 (2009)
50. Y. Li, P. Liu et al., Chem. Phys. Lett. **475**, 183 (2009)
51. S. Zhao, P. Liu et al., Chem. Phys. Lett. **480**, 67 (2009)

Chapter 7
New Methods For Computing High-Order Harmonic Generation and Propagation

J.A. Pérez-Hernández, C. Hernandez-García, J. Ramos, E. Conejero Jarque, L. Plaja, and L. Roso

Abstract Due to its nonperturbative character, the theoretical modelization of strong field phenomena is a challenging aspiration. In this chapter, we shall consider the problem of high-order harmonic generation and propagation, and review some recent proposals that conform an alternative approach to the standard procedures. In particular, the semiclassical description of the single-atom response can be nowadays replaced to include the full quantum description. Also, the limits of the Strong-Field Approximation can be extended to include the influence of the strong field on the ground state. These two aspects allow for a new procedure, here referred to as SFA+, for calculating the high-order harmonic generation spectrum, which is demonstrated to improve the quantitative accuracy and to recover, for instance, the correct dependence of the harmonic yield with the laser wavelength. On the contrary, the problem of harmonic propagation has also been tackled recently from a new perspective: the combination of SFA+ methods with a Discrete Dipole approach. This latter strategy is not based on the differential wave equation for the fields, but on its integral version, and finds some advantages with respect to the usual approximations (slowly varying envelopes, paraxial, etc).

7.1 Introduction

The progress of laser technology in the past two decades has led to an unprecedented development of intense sources. From a fundamental viewpoint, the interaction of ultraintense lasers with atoms constitutes a paradigmatic example of nonperturbative physics. Besides the practical interest of the phenomena associated, there is

J.A. Pérez-Hernández (✉) and L. Roso
Centro de Láseres Pulsados CLPU, E-37008 Salamanca, Spain
e-mail: joseap@usal.es, roso@usal.es

C. Hernandez-García, J. Ramos, E.C. Jarque, and L. Plaja
Departamento de Física Aplicada, Universidad de Salamanca, E-37008 Salamanca, Spain
e-mail: carloshergar@usal.es, jarp@usal.es, enrikecj@usal.es, lplaja@usal.es

K. Yamanouchi et al. (eds.), *Progress in Ultrafast Intense Laser Science VII*,
Springer Series in Chemical Physics 100, DOI 10.1007/978-3-642-18327-0_7,

a fundamental appeal in developing theoretical methods to describe this problem, since the experimental validation is often possible in small laboratories throughout the world. For the same reason, the interplay between theory experiment makes this field specially dynamical. The theoretical description of an atomic electron submitted to an intense electromagnetic field corresponds to the integration of the time-dependent Schrödinger equation (TDSE), which has to be done numerically. Nowadays, however, the exact 3D solution is only feasible for one- and two-electron systems, taking some tens of minutes in the first case, while being a formidable task for supercomputers in the second. Even for the single electron, the exact solution of propagation equations involving macroscopic targets, and therefore a huge number of atoms, is far from the present and near-future computing capabilities. In this scenario, the development of approximated models becomes mandatory. Among them, the S-matrix formulation combined with the Strong-Field Approximation constitutes an accredited strategy to approach the problem.[1]The first studies in this direction were aimed to the computation of ionization rates and the description of the photoelectron spectrum [1–3], but these techniques were progressively extended to treat new phenomenology, as the multielectron ionization [4] or the harmonic generation [5–7]. In this latter case, a *standard* approach combines SFA with a saddle point method to compute the harmonic spectra very efficiently in terms of computing time. As a result of the saddle point integration, the standard theory offers a semiclassical description in terms of electronic trajectories, which constitute an extraordinary tool for the physical understanding of high-order harmonic generation. The fundamental process arising from this description [8, 9] consists of the rescattering of an ionized electron with the parent ion. High-order harmonics are generated by the dipole acceleration associated with the transition between the free electron and the fundamental atomic state, during the rescattering event. The resulting harmonic spectra are characterized by a plateau structure of similar harmonic intensities followed by an abrupt cut-off. According to the standard model, the extension of this spectral plateau is determined by the maximum kinetic energy of the free electron upon rescattering, which follows the simple law $I_p + 3.17U_p$ (I_p being the ionization energy, and $U_p = q^2E^2/4m\omega^2$ the ponderomotive energy). The standard model also succeeds in predicting the mode locking of the highest order harmonics, chirps, and modulation of the yields with the intensity [10]. However, despite these achievements, recent studies on the scaling of the harmonic yield with wavelength point out departures between the predictions of this model and the exact TDSE [11–13]. We have recently shown that, for the quantitative improvement of the model predictions, the SFA has to be relaxed to incorporate the field-induced dynamics into the ground state, at least during the rescattering event [14]. This new approach is also particular as it does not resort to the saddle-point approximation, therefore including the full quantum description of the harmonic generation process. In the following, we shall refer to this new strategy as SFA+.

[1] For a review of these techniques, see H. R. Reiss, in Progress in Ultrafast Intense Laser Science, 1 (Springer, 2008).

To be applicable to the experimental realm, the modeling of a harmonic generation has to incorporate also the interference and propagation effects derived from the macroscopic size of the targets. This is usually done [15–17] by means of the numerical integration of the wave equation, where the source term is evaluated from the single-atom computations described above. Even for the case of low pressure gas targets, the phase and intensity variations of the laser beam in space (which is often considered a gaussian beam) require the integration of the source term in many spatial points. This is an extraordinary load in terms of computing time, which can be tackled in two different ways: either the source term is computed exactly from the TDSE for some set of intensities corresponding to a sample of the field amplitudes in the gaussian beam, and the spatial phase shift is introduced as an *ad hoc* phase factor proportional to the harmonic order [15], or the source term is computed at every point of a spatial grid, needing therefore a fast model to compute the single-atom harmonics. For this latter case, the standard model (employing the saddle point approximation in temporal and the momentum space integrals) is the standard choice. We have, however, developed a propagation code within the SFA+ approach. Although in principle SFA+ computations are slower than those from the standard saddle point-SFA, our approach to propagation is based on the integral solution of the wave equation and the Discrete Dipole Approximation (DDA), which is faster than the numerical integration [18].

At the end of this chapter, we shall review these two combined strategies, SFA+ and DDA, to develop a novel approach to the propagation problem. In the following sections, we will derive the SFA+ approach from the S-matrix formalism and discuss some results in comparison with the exact integration of the TDSE; next, we will expose the DDA approach to propagation and give some results for the angular distribution of the harmonic radiation far field.

7.2 Computing High-Order Harmonic Generation Within the SFA+

Let us start considering the description of an isolated atom in interaction with an intense electromagnetic field. In the single-active electron approximation, and assuming a nucleus with infinite mass, the system dynamics is described by the time-evolution of the electron wavefunction according to the Schrödinger equation,

$$i\hbar\frac{\partial}{\partial t}|\psi(t)\rangle = [H_a + V_i(t)]\,|\psi(t)\rangle, \tag{7.1}$$

where $H_a = p^2/2m - Zq^2/r$ is the atomic Hamiltonian ($q = -|e|$, Z the atomic number) and $V_i(t) = -(q/mc)A(t)p_z + q^2/(2mc^2)A^2(t)$, as we assume a linearly polarized field and the dipole approximation. The exact integral solution can be written in terms of propagators as

$$-\mathrm{i}|\psi(t)\rangle = G_a^+(t,t_0)|\psi(t_0)\rangle + \frac{1}{\hbar}\int_{t_0}^{t} \mathrm{d}t' G^+(t,t')V_i(t')G_a^+(t',t_0)|\psi(t_0)\rangle, \tag{7.2}$$

G^+ being the Green's function of the whole problem, and G_a^+ the one associated with the field-free case. Let us now consider a splitting of the Hilbert space into two subspaces, one for the bound states of the atom, Q, and the other for the continuum, P. In association, we define the corresponding projectors $\hat{Q}$ and $\hat{P}$. By definition, $\hat{Q}+\hat{P}=1$, $\hat{Q}\hat{P}=\hat{P}\hat{Q}=0$, $\hat{Q}^2=\hat{Q}$, $\hat{P}^2=\hat{P}$, and $|\psi(t_0)\rangle=\hat{Q}|\psi(t_0)\rangle$, $|\psi(t_0)\rangle$ being the initial bound state (note that this does not imply that the subspace Q contains only one state). Also, by definition, $\left[H_a,\hat{Q}\right]=0$, while we will assume that the electrons promoted to the continuum have no possibility to recombine, therefore $\left[H,\hat{P}\right]\simeq 0$, in accordance with the SFA.

Imposing these definitions, (7.2) leads to two coupled equations (one for the bound part of the wavefunction and other for the free part)

$$\begin{aligned} -\mathrm{i}\hat{Q}|\psi(t)\rangle &= G_a^+(t,t_0)|\psi(t_0)\rangle \\ &\quad + \frac{1}{\hbar}\int_{t_0}^{t} \mathrm{d}t' \hat{Q}G^+(t,t')\hat{Q}V_i(t')\hat{Q}\, G_a^+(t',t_0)|\psi(t_0)\rangle \end{aligned} \tag{7.3}$$

$$-\mathrm{i}\hat{P}|\psi(t)\rangle = \frac{1}{\hbar}\int_{t_0}^{t} \mathrm{d}t' \hat{P}G^+(t,t')\hat{P}V_i(t')\hat{Q}\, G_a^+(t',t_0)|\psi(t_0)\rangle. \tag{7.4}$$

The second term in the *rhs* of (7.3) describes the non-ionizing excitation of the ground-state at t' and its subsequent evolution until t in the combined influence of the atom and the field. Therefore, (7.3) describes the evolution of the bound part of the wavefunction as the superposition of the bare (field free) evolution and the bound-state excitations (here referred to as field-dressing). The SFA consists of setting $\hat{Q}V_i(t')\hat{Q}=0$ in (7.3), considering that the field interaction leads invariably to ionization, together with the former condition $\left[H,\hat{P}\right]=0$, which prevents the recombination of an ionized state. Thus, in considering $\hat{Q}V_i(t')\hat{Q}\neq 0$ in (7.3), we soften the constraints of standard SFA. We shall, therefore, refer to this approach as SFA+.

Although the harmonic conversion efficiency is typically orders of magnitude smaller than unity, the use of intense lasers ensures a sufficient number of radiated photons to allow for a classical description of the harmonic field. The single-atom radiation, therefore, is proportional to the dipole acceleration (see Sect. 7.4) $a(t)=\langle\psi(t)|\hat{a}|\psi(t)\rangle$, that can be evaluated according to the Ehrenfest theorem, $\hat{a}=-(q/m)\partial V_c/\partial z$ (V_c being the Coulomb potential, $-Zq^2/r$ in our case). The higher-frequency harmonics correspond to the most energetic photons, thus involving the higher energy transitions, i.e. free to bound. Therefore, we may approximate

$$a(t)\simeq\langle\psi(t)|\hat{Q}\hat{a}\hat{P}|\psi(t)\rangle + c.c. = a_b(t)+a_d(t)+c.c., \tag{7.5}$$

where a_b and a_d are two interfering contributions to the total acceleration, associated with transitions between the continuum to the bare ground state or to its field-dressing (not considered in the SFA), respectively [14],

$$a_b(t) = \frac{1}{\hbar}\int_{t_0}^{t} dt_1 \langle\phi_0|G_a^-(t_0,t)\hat{a}\hat{P}G^+(t,t_1)\hat{P}V_i(t_1)\hat{Q}G_a^+(t_1,t_0)|\phi_0\rangle \quad (7.6)$$

$$a_d(t) = \frac{1}{\hbar^2}\int_{t_0}^{t} dt_2 \langle\phi_0|G_a^-(t_0,t_2)\hat{Q}V_i(t_2)\hat{Q}G^-(t_2,t)\hat{Q}\hat{a} \times \int_{t_0}^{t_2} dt_1 \hat{P}G^+(t,t_1)\hat{P}V_i(t_1)\hat{Q}G_a^+(t_1,t_0)|\phi_0\rangle, \quad (7.7)$$

where we have already defined the initial bound state by the atomic ground state $|\phi_0\rangle$.

We now must give a form to the operators $\hat{P}V_i(t_1)\hat{Q}$, $\hat{P}G^+(t,t_1)\hat{P}$, $\hat{Q}V_i(t_2)\hat{Q}$ and $\hat{Q}G^-(t,t_2)\hat{Q}$. The operators $\hat{P}V_i(t_1)\hat{Q}$ and $\hat{P}G^+(t,t_1)\hat{P}$ in (7.4), (7.6), and (7.7) can be evaluated according to the standard procedure in SFA: First, we consider a planewave basis, $\{\mathbf{k}\}$, for the subspace defined by $\hat{P}$, therefore

$$\hat{P} \simeq \int d\mathbf{k}|\mathbf{k}\rangle\langle\mathbf{k}|. \quad (7.8)$$

Each planewave evolves as a Volkov wave of momentum $\mathbf{p} = \hbar\mathbf{k}$ under the influence of the electromagnetic field. Therefore,

$$\hat{P}V_i(t_1)\hat{Q} \simeq \int d\mathbf{k} V_i(\mathbf{k},t_1)|\mathbf{k}\rangle\langle\mathbf{k}|\hat{Q} \quad (7.9)$$

with $V_i(\mathbf{k},t_1) = -(q/mc)A(t_1)k_z + q^2/(2mc^2)A^2(t_1)$. In addition, we have

$$\hat{P}G^+(t,t_1)\hat{P} = -\mathrm{i}\frac{C_F}{r^n}\exp\left[-(\mathrm{i}/\hbar)\int_{t_1}^{t} d\tau\hat{P}H(\tau)\hat{P}\right], \quad (7.10)$$

where we have introduced the Coulomb factor [19] $C_F/r^n = \left(2Z^2/n^2E_0r\right)^n$, (for Hydrogen $n = 1, Z = 1$) and with

$$\hat{P}H(\tau)\hat{P} \simeq \int d\mathbf{k}\,\epsilon(\mathbf{k},\tau)|\mathbf{k}\rangle\langle\mathbf{k}|, \quad (7.11)$$

where $\epsilon(\mathbf{k},\tau) = \hbar^2k^2/2m - (q/mc)A(\tau)k_z + q^2/(2mc^2)A^2(\tau)$. With this definitions, the bare state contribution to the acceleration, (7.6), can be written as $a_b(t) = \int d\mathbf{k}a_b(\mathbf{k},t)$, where

$$a_b(\mathbf{k},t) = -\frac{\mathrm{i}}{\hbar}C_F\int_{t_0}^{t} dt_1 \mathrm{e}^{\mathrm{i}\epsilon_0(t-t_1)/\hbar}\mathrm{e}^{-\mathrm{i}\frac{1}{\hbar}\int_{t_1}^{t}\epsilon(\mathbf{k},\tau)d\tau}\langle\phi_0|\hat{a}|\mathbf{k}\rangle V_i(\mathbf{k},t_1)\langle\mathbf{k}|r^{-n}|\phi_0\rangle. \quad (7.12)$$

The evaluation of the operators of the field-dressing part, (7.7), is less straightforward in the general case. However, some simplifications arise for the study of harmonic generation. High-order harmonics are generated through transitions from continuum to bound states that take place during the process of rescattering of the ionized electron with the parent ion. Instead of assuming this process as instantaneous, let us consider that the harmonic generation at time t is triggered by the collision of the ionized electron taking place over the small time interval, $t - \delta t_s$ to t. Assuming the bound state wavefunction to be the ground state at the beginning of this interval, (7.3) predicts its evolution during the temporal lapse of rescattering, by setting the lower limit t_0 of the time integral to the initial time $t - \delta t_s$,

$$-\mathrm{i}\hat{Q}|\psi(t)\rangle \simeq G_a^+(t,t_0)|\phi_0\rangle + \frac{1}{\hbar}\int_{t-\delta t_s}^{t} \mathrm{d}t' \hat{Q}G^+(t,t')\hat{Q}V_i(t')\hat{Q}\, G_a^+(t',t_0))|\phi_0\rangle, \tag{7.13}$$

and

$$a_d(t) \simeq \frac{1}{\hbar^2}\int_{t-\delta t_s}^{t} \mathrm{d}t_2 \langle\phi_0|G_a^-(t_0,t_2)\hat{Q}V_i(t_2)\hat{Q}G^-(t_2,t)\hat{Q}\hat{a} \times \int_{t_0}^{t_2} \mathrm{d}t_1 \hat{P}G^+(t,t_1)\hat{P}V_i(t_1)\hat{Q}G_a^+(t_1,t_0)|\phi_0\rangle. \tag{7.14}$$

The dynamics of the bound excitations during the time lapse δt_s is given by the operator $\hat{Q}G^-(t_2,t)\hat{Q}$ which is, in turn, a function of the total Hamiltonian $H(t) = H_a + V_i(t)$. The rescattering event is defined by the overlap of the free electron wavefunction with the coordinate origin, where the potential singularity is located. With this definition, for the most energetic electrons, the scattering time lapse can be evaluated as [14]

$$\delta t_s \simeq (3\pi/2\omega_0)\sqrt{|\epsilon_0|/3.17U_p}, \tag{7.15}$$

where U_p is the ponderomotive energy. For large U_p, this time lapse is small enough to approximate the time-dependent operator $V_i(t)$ in (7.13) and (7.14) by its time average over δt_s

$$\Delta_s = \langle V_i(t)\rangle = (1/\delta t_s)\int_{t-\delta t_s}^{t} \left[-(q\hbar/mc)A(\tau)k_z + (q^2/2mc^2)A^2(\tau)\right]\mathrm{d}\tau, \tag{7.16}$$

where k_z is a relevant momentum of the state, that can be evaluated as [14]

$$k_z = -\frac{2}{\hbar}\sqrt{mU_p}\frac{\sin\omega_0\delta t_s}{\omega_0\delta t_s}\left[1 - \sqrt{1 - \frac{1}{6}\left(1 + \frac{\epsilon_0}{U_p} + \frac{\sin 2\omega_0\delta t_s}{2\omega_0\delta t_s}\right)\left(\frac{\sin\omega_0\delta t_s}{\omega_0\delta t_s}\right)^{-2}}\right]. \tag{7.17}$$

Therefore, $\hat{Q}G^{-}(t_2,t)\hat{Q}$ can be approximated by $\mathrm{i}\exp[\mathrm{i}(\epsilon_0+\Delta_s)(t-t_2)]$, where we assume that the mean energy of the ground state after interacting with $\hat{Q}V_i(t_2)\hat{Q}$ is $\simeq \epsilon_0$. On the contrary, we should look for a simplified form of the interaction operator $\hat{Q}V_i(t_2)\hat{Q}$. A first choice would be to replace it by the time-averaged form Δ_s, but it proves not to be a sufficiently accurate approximation. We shall, therefore, consider the factorization $V_i(t) = H(t) - H_a \simeq p^2/2m + \Delta_s - H_a$. The Coulomb term in $H(t)$ is neglected as we assume the ground-state excitations to have minimal probability near the origin. With these approximations, and after some algebra [14], the dressing contribution to the acceleration can be written in terms of the bare contribution as

$$a_d(\mathbf{k},t) \simeq -\left[1+\frac{k^2/2m-\epsilon_0}{\Delta_s}\right]a_b(\mathbf{k},t). \tag{7.18}$$

Therefore, the total acceleration is given by

$$a(t) = -\int \mathrm{d}\mathbf{k}\frac{\hbar^2k^2/2m-\epsilon_0}{\Delta_s}a_b(\mathbf{k},t) + c.c. \tag{7.19}$$

Note that the opposite sign of a_d relative to a_b, see (7.18), leads to the destructive interference between the bare and dressing contributions to the acceleration. This is a main result of the SFA+ approach, as the standard SFA considers only the bare contribution, a_b. The degree of interference changes with the laser parameters and affects the harmonic yield. In particular, SFA+ gives a proper account of the scaling of the harmonic intensities with the laser wavelength [14].

Finally, the time integral leading to $a_b(\mathbf{k})$ can be computed very effectively numerically without recurring to the saddle-point approximation and, thus, retaining the full quantum description of the process. This is done by integrating the set of (uncoupled) one-dimensional differential equations, each associated with a particular Volkov wave $\mathbf{k}$, that result from differentiating (7.12)

$$\frac{\mathrm{d}}{\mathrm{d}t}a_b(\mathbf{k},t) = \frac{\mathrm{i}}{\hbar}\left[\epsilon_0-\epsilon(\mathbf{k},t)\right]a_b(\mathbf{k},t) - \frac{\mathrm{i}}{\hbar}C_F\langle\phi_0|\hat{a}|\mathbf{k}\rangle V_i(\mathbf{k},t)\langle\mathbf{k}|r^{-n}|\phi_0\rangle. \tag{7.20}$$

7.3 SFA+ Deliverables

Let us now turn to some particular outcomes of the SFA+ model. We shall first show some comparisons of the harmonic spectra computed with our SFA+ model with the standard SFA and with the exact solution of the 3D TDSE. Afterward, as our procedure is fully quantum (not semiclassical), we will analyze the different momentum contributions to the total harmonic yield, and discuss paths for harmonic generation that fall behind the standard semiclassical description. For the rest of this chapter, we shall consider Hydrogen as atomic species. The models, however, are universal and can be applied to any species as long as the single electron approximation remains valid, and the matrix elements of the ground to the plane-wave continuum transitions and acceleration can be evaluated. The applicability of the Coulomb factor in [19], derived for the ionization process, to the problem of harmonic generation in species different than Hydrogen is currently under study.

7.3.1 Quantitative Description of the Harmonic Spectra

To gain insight into the validity of SFA+, we show in Fig. 7.1 the results for the computations of the harmonic spectra for three different laser parameters. The plots in the figure correspond to the direct outcome from the Fourier transform of the dipole acceleration in the different approaches, i.e. no relative shift or scalings have been used. The result of the exact computation of the 3D TDSE is plotted in blue lines, while the result of the SFA+ computation is plotted in filled green areas, and the orange line corresponds to the computation in the SFA approach, i.e. not including ground-state dressing, $a_d(t)$. We note that here the SFA computations are carried on using (7.20), i.e. without resort to the saddle point method. The possible issues derived from this latter method, and from the associated semiclassical description, fall out of the scope of this chapter. However, it has been recently tested to be a fairly good approach, for laser parameters deep in the tunnel ionization regime [20]. The choice of the laser parameters in Fig. 7.1 is done to show as much variety as possible. Therefore, the field intensities are changed from 1.58×10^{14} W/cm^2 in part (a) and (b) to 9×10^{13} in (c), wavelengths correspond to (a) 800 nm, (b) 1,200 nm, and (c) 1,600 nm, the field envelopes are 9-cycle trapezoidal (half-cycle turn-on and turn-off) in (a) and 4-cycle sine squared in (b) and (c). More comparisons including, for instance, carrier-envelope phase offsets can be found in earlier references [14,21]. As a general trend, the SFA+ method offers a more accurate description of the high-order harmonic plateau than the SFA, and for a wider frequency interval. For the longer pulses (Fig. 7.1a), the effect of ionization shows up, and our SFA+ model gives harmonic yields slightly higher than the TDSE. The implementation of ionization in our model is planned for the future, resorting to the ADK [22] rates in a similar fashion as it is done in the standard saddle point SFA approaches. We may notice that both models, SFA and SFA+, become progressively inaccurate for the lowest harmonics. These correspond to low-energy transitions, involving excited states in the atom, and are not fully accounted into the dressing term of SFA+, while are ruled out completely in the standard SFA. A rough estimation is to consider the law $|\epsilon| + 3.17U_p$ valid also for the maximum frequency reachable from excited states. In such case, the influence of the excited states extends in the harmonic spectrum up to a maximum energy $|\epsilon_0| - |\epsilon|$ below the cut-off frequency. Therefore, one can consider that the frequency window extending $|\epsilon_0| - |\epsilon|$ below the cut-off is produced solely by the transitions involving the ground state. For the case of Hydrogen, this window is of about 10 eV. Therefore, for the 800 nm case of Fig. 7.1a, this defines a region window of validity for the SFA and SFA+ approaches from the cut-off at the 29th harmonic, down to the 23rd harmonic. The figure seems

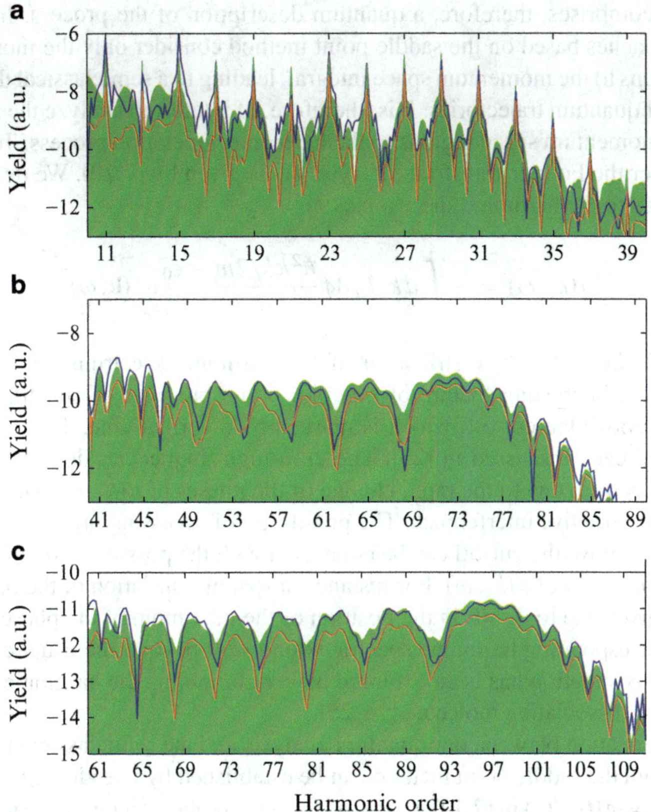

Fig. 7.1 Comparison of the harmonic spectra computed from the exact numerical integration of the TDSE (*blue line*), the SFA model (*orange line*), and SFA+ model (*green filled area*) for different laser parameters: (**a**) 9-cycle pulse of trapezoidal shape, with half period turn-on and turn-off and constant intensity in between, with intensity $\simeq 1.58 \times 10^{14}$ W/cm^2 and wavelength 800 nm, (**b**) 4-cycle $\sin^2$ envelope with intensity $\simeq 1.58 \times 10^{14}$ W/cm^2 and wavelength 1,200 nm, and (**c**) 4-cycle $\sin^2$ envelope with intensity $\simeq 9 \times 10^{13}$ W/cm^2 and wavelength 1,600 nm

to corroborate our discussion, as the description of the harmonic spectrum by the SFA+ model in this window is excellent. We shall note, also, that this window estimation is quite conservative, as the model description can be accurate in a wider spectral region if the population of the excited states is small, for instance for longer wavelengths (Fig. 7.1b, c).

7.3.2 Momentum Space Contributions to the Harmonic Spectra

The SFA+ model computes the harmonic spectrum from the superposition of the independent contributions at every point in the momentum space, see (7.12) and

(7.19). It comprises, therefore, a quantum description of the process. In contrast, SFA approaches based on the saddle point method consider only the most relevant contributions to the momentum space integral, leading to a semiclassical description in terms of quantum trajectories. It is, therefore, of interest to analyze the role of the different momentum space regions to the harmonic generation process. To this end, we consider the Fourier transform of $a_b(\mathbf{k}, \mathbf{t})$, as given by (7.20). We define the k_z contribution to the harmonic spectrum as

$$a(k_z, \omega) = -\int dk_\rho k_\rho d\phi \frac{\hbar^2 k^2/2m - \epsilon_0}{\Delta_s} a_b(\mathbf{k}, \omega). \tag{7.21}$$

Figure 7.2 shows the k_z distribution of the harmonic spectrum for the case of Fig. 7.1b. An interesting feature of this plot is the presence of spectral contributions well above the cut-off, which is marked by a vertical line. This is a general behavior, already discussed in [23]. The ultra-high frequencies do not show up in the total spectrum due to the rapid change of the phases of $a(k_z, \omega)$ with k_z, which leads to a destructive interference. The possibility of accessing high-harmonic generation well above the cut-off can be connected with the possibility of changing the phase variations in of $a(k_z, \omega)$. For instance, a spatial translation of the rescattering potential gives rise to a shift in the position of the maximum of the phase parabola, and the corresponding harmonic spectra extend well beyond the usual cut-off frequency. This situation has been explored before computing the harmonic yield of a double-well dissociating molecule [24, 25].

The connection between our quantum analysis and the semiclassical trajectories derived from the saddle point method can be established by introducing an artificial time window $g(t_1 - t_w)$ in (7.12), so that $V_i(\mathbf{k}, t_1)$ is replaced by $g(t_1)V_i(\mathbf{k}, t_1)$. $g(t_1)$ is defined to be nonzero at a small time interval around some particular t_w [23]. The

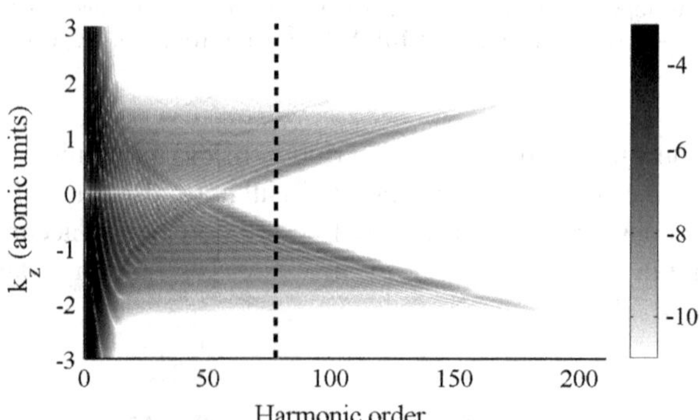

Fig. 7.2 Momentum space, k_z, distribution of the spectral contributions to the harmonic radiation, corresponding to the case of Fig. 7.1b. The cut-off frequency of the integrated spectrum is shown as *vertical dashed line*. The harmonic intensities are plotted with gray tones using logarithmic scale, and in arbitrary units

harmonics generated by this modified interaction correspond to the contribution to the total harmonic spectrum of the electrons ionized during this restricted temporal window centered at t_w. In the limit of small windows, it corresponds to the almost instantaneous release of a wavepacket to the continuum, which is then driven by the field as a free electron. According to the Ehrenfest theorem, the mean value of the position operator of this wavepacket will follow the trajectory of a classical electron released at time t_w. Note, however, that the description of this event in our model is still quantum mechanical, and therefore takes fully into account the spread of the wavefunction during its excursion through the continuum.

Figure 7.3 shows the resulting dipole acceleration computed from our SFA+ model with a temporal window of 1/8 of the laser period, centered at t_w near the field maximum. The trajectory corresponding to a classical electron born at time t_w at $z = 0$, with zero velocity, is shown with a black dashed line. The harmonic burst at the recollision of the electron with the parent ion is labeled as (b). Note, however, that there are two more harmonic radiation events, which are not taken into account by the semiclassical description: (a) radiation of the electron leaving the parent ion (way-out generation) and, (b) radiation that takes place as the electron trajectory approaches the parent ion but in absence of a recolliding classical trajectory (close-up radiation). These two latter processes can be explained satisfactorily considering the quantum spread of the wavefunction, which overlaps the ion location even though its coordinate mean value is far from it. The part of the wavefunction near the ion still feels a strong Coulomb potential, and therefore suffers the acceleration that triggers the harmonic radiation.

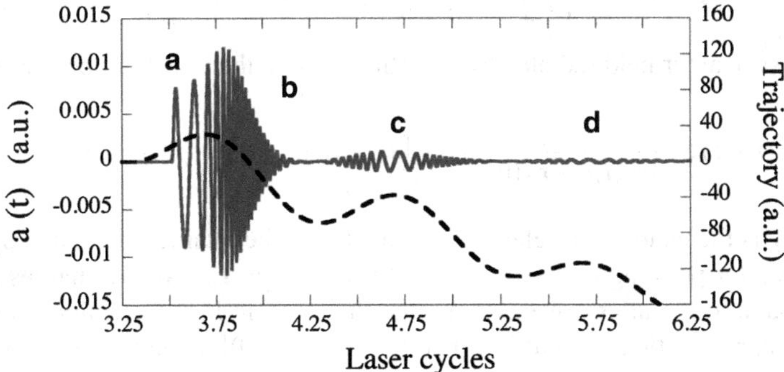

Fig. 7.3 Contribution to the total dipole acceleration of the electrons ionized during a time lapse of 1/8 of period near the field maximum. The *dashed line* shows the corresponding trajectory of a classical electron starting at the same instant of time at the coordinate origin with zero velocity. The different events for harmonic radiation labeled as (**a**) way-out, (**b**) rescattering, and (**c**) two events of close-up radiation. The laser pulse is modeled by an eight-cycle pulse of $\lambda = 800$ nm with intensities 3.5×10^{14} W/cm^2

7.4 Computation of the High-Order Harmonic Propagation

Even for the case of low-pressure targets, the harmonic spectra observed in the experiment differs substantially from the single atom case. Besides the issues connected with the spectral efficiency of the detectors, etc., the fundamental physical modifications come from three different facts. In the first place, a focused laser beam is inhomogeneous and, therefore, the atoms located at different places in the target respond to different field parameters (intensity and phase). Second, the target radiated far field is the interference of the single-atom contributions, which are radiated at different spatial points and that reach the detector at different retarded times. Finally, the target itself responds to the radiated field as it propagates, as described by the index of refraction. A faithful comparison of the theory with the experiments, therefore, should include these macroscopic effects by means of a proper description of the harmonic propagation.

The general problem of propagation amounts to solve the wave equation for the electric field $\mathbf{E}$,

$$\nabla^2\mathbf{E} - \frac{1}{c^2}\frac{\partial^2}{\partial t^2}\mathbf{E} = \frac{4\pi}{c^2}\frac{\partial}{\partial t}\mathbf{J}, \tag{7.22}$$

where $\mathbf{J}$ is the current density. Most of the approaches to high-order harmonic propagation are based on the numerical solution of this equation. In our case, we will start from a different point of view, considering the formal integral solution [26]: $\mathbf{E}(\mathbf{r},t) = \mathbf{E}_0(\mathbf{r},t) + \mathbf{E}_i(\mathbf{r},t)$, where $\mathbf{E}_0(\mathbf{r},t)$ is the laser field, as it propagates in vacuum, and $\mathbf{E}_i(\mathbf{r},t)$ is the total field radiated by the accelerated charges in the target,

$$\mathbf{E}_i(\mathbf{r},t) = -\frac{1}{c^2}\int d\mathbf{r}'\frac{1}{|\mathbf{r}-\mathbf{r}'|}\left[\frac{\partial}{\partial t'}\mathbf{J}(\mathbf{r}',t')\right]_{t'=t-|\mathbf{r}-\mathbf{r}'|/c}. \tag{7.23}$$

The transversal far field radiated by the jth charge in the target can be written as,

$$\mathbf{E}_i^j(\mathbf{r}_d,t) = \frac{1}{c^2}\frac{q_j}{|\mathbf{r}_d-\mathbf{r}_j(0)|}\mathbf{s}_d\times\left[\mathbf{s}_d\times\mathbf{a}_j(t-|\mathbf{r}_d-\mathbf{r}_j(0)|/c)\right], \tag{7.24}$$

where $\mathbf{a}_j$ is the charge's acceleration, evaluated at the retarded time, and $\mathbf{s}_d$ is the unitary vector pointing to a virtual detector located at $\mathbf{r}_d$. As the charges are initially bound to the atoms, we assume that they remain in the vicinity at all times (dipole approximation) so that $|\mathbf{r}-\mathbf{r}_j(t)| \simeq |\mathbf{r}-\mathbf{r}_j(0)|$. Although this expression is only valid for radiation emission in vacuum, it is found to be a good approximation for the high-order harmonic radiation [18], as the propagation wavevector for the qth harmonic becomes $k_q \simeq qk_0$ [15]. Therefore, the global field radiated by the target can be written as the superposition of the individual charge contributions $\mathbf{E}_i(\mathbf{r},t) = \sum_{j=1}^{N}\mathbf{E}_i^j(\mathbf{r},t)$. Although the target refractive index for low pressure targets is effectively the unit, it is essential to include the correction at the fundamental frequency. For our case of interest, the fundamental field is affected mainly by the presence of free charges resulting from ionization, so its wavenumber is shifted from the vacuum value as $k_1^2 = k_0^2(1-\omega_p^2/\omega_0^2)$, ω_p being the plasma frequency

[26]. In a typical experimental situation, this plasma contribution amounts to a small dephase at the fundamental frequency. However, it has a relevant effect in the harmonic generation, as the wavenumber of the generated qth order harmonic is approximately qk_1, and it has a particular impact in the phase matching of the harmonics, $\Delta k_q = k_q - qk_1 \simeq q\omega_p^2/2\omega_0^2$.

The description of the plasma-induced phase-matching has to take into account the target inhomogenities and, also, the time dependence of the ionization process. Therefore, we shall compute the spatial phase of the fundamental field as

$$\int_{-\infty}^{z} k_1(x,y,\xi)\mathrm{d}\xi \simeq k_0 z - \frac{2\pi e^2}{{\omega_0}^2 m} P_f(\mathbf{r},t) \int_{-\infty}^{z} N(x,y,\xi)\mathrm{d}\xi, \tag{7.25}$$

being $N(x,y,z)$ the atomic density distribution and $P_f(\mathbf{r},t)$ the probability for ionized electrons that we calculate as follows

$$P_f(\mathbf{r},t) = 1 - \mathrm{e}^{-\int_{-\infty}^{t} w_{\mathrm{ADK}}(\mathbf{r},\tau)\mathrm{d}\tau}, \tag{7.26}$$

where $w_{\mathrm{ADK}}(\mathbf{r},t)$ is the ionization rate, calculated from the Amosov–Delone–Krainov (ADK) formula for Hydrogen [22]

$$w_{\mathrm{ADK}}(\mathbf{r},t) = \frac{2e^2}{\pi}\left[\frac{3}{\pi|E(\mathbf{r},t)|}\right]^{1/2} \mathrm{e}^{-\frac{2}{3|E(\mathbf{r},t)|}}. \tag{7.27}$$

7.4.1 Discrete-Dipole Approach

In principle, our integral approach demands the evaluation of the dipole acceleration associated with each charge in the target. The total field is afterward computed as a superposition of the yield of each of these elementary sources. Even though (7.20) and (7.25) offer a fast method for the computation of the elementary dipoles, the evaluation at all charges in the target becomes an unreasonable goal, even for the case of low pressure targets (with about 10^{18} atm/cm^3). We shall, therefore, resort to a Discrete Dipole approach [27], where the interaction volume is discretized in a set of small macroscopic cells. The size of the cells is defined according to the following restrictions (a) each cell must include a sufficiently large number of physical charges to approximate their density by a continuous distribution, while (b) it should be small enough to approximate the fundamental field in it as a plane wave. Instead of defining the cells with definite limits, we replace their geometrical shape by a localized gaussian distribution of equivalent size

$$g(\mathbf{r}) = C\mathrm{e}^{-\mathbf{r}^2/\sigma^2}, \tag{7.28}$$

where C is the normalization factor, σ is related to the diameter of the cell d as $d = 2\sigma\sqrt{ln2}$, and $\mathbf{r}$ is a coordinate on the local reference frame with origin at the cell's center. The field radiated by the macroscopic cell can be computed assuming the local plane wave approximation for the fundamental, and integrating the radiated field over the gaussian distribution, for every frequency component of the dipole acceleration, following (7.23). Finally, the far field radiation from the single macroscopic cell located at the point r_j of the target can be evaluated as [18]

$$\mathbf{E}_{i,j}(\mathbf{r}_d,\omega) \propto N(\mathbf{r}_j)\, \mathbf{s}_d \times \left[\mathbf{s}_d \times \mathbf{a}(\mathbf{r}_j,\omega)\right] \mathrm{e}^{\mathrm{i}\omega \frac{|\mathbf{r}_d-\mathbf{r}_j|}{c}} F(\theta_{d,j},\omega)), \quad (7.29)$$

where $F(\theta_{d,j},\omega)$ is defined as

$$F(\theta_{d,j},\omega) \propto e^{\frac{1}{2}\frac{\omega^2}{c^2}\sigma^2(1-\cos\theta_{d,j})}, \quad (7.30)$$

$\theta_{d,j}$ being the angle between the detector position and the local field propagation vector at the cell j. Note that the field radiated by the cell, (7.29), corresponds to the field of a single radiator located at the cell's center $\mathbf{r_j}$, modulated by a form factor $F(\theta_{d,j},\omega)$ that takes into account the interfering contributions of the rest of the radiators in the macroscopic cell. Figure 7.4 shows the angular distribution of the radiation spectra (a) for a single radiator and (b) for a cell of diameter 1 micron. Note that, as a result of the macroscopic size of the cell, the global matching of the field emitted by the different points in the cell leads to a constructive interference restricted along the propagation direction of the fundamental field.

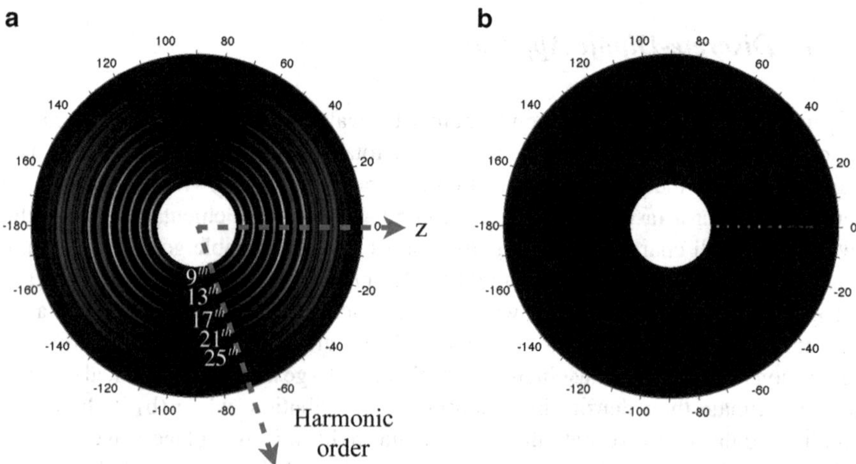

Fig. 7.4 Angular distribution of the radiation spectra (**a**) for a single radiator and (**b**) for a cell of 1 micron diameter. The radial axis corresponds to the high-order harmonic order, angles are represented in degrees, and the z-axis corresponds to the propagation direction of the local plane wave

7.4.2 Results

We shall consider a low-pressure Hydrogen jet interacting with an 8-cycle 800 nm laser pulse of peak intensity 1.57×10^{14} W/cm^2 (see Fig. 7.5). We assume a fundamental field of the form

$$\mathbf{E}_1(\mathbf{r},t) = A_0\eta(z-ct)U(\mathbf{r})\mathrm{e}^{-\mathrm{i}\left(\int_{-\infty}^{z} k_1(x,y,\xi)\mathrm{d}\xi-\omega t\right)} \tag{7.31}$$

with A_0 the peak field amplitude, $\eta(z-ct)$ a $\sin^2$ temporal envelope, 8 cycles long, and $U(\mathbf{r})$ a gaussian profile [28]

$$U(\mathbf{r}) = \frac{W_0}{W(z)}\mathrm{e}^{-\frac{\rho^2}{W^2(z)}}\mathrm{e}^{\mathrm{i}\frac{\omega}{c}\frac{\rho^2}{2R(z)}+\mathrm{i}\zeta(z)} \tag{7.32}$$

with a beam waist $W_0 = 30\mu$m (Rayleigh length 7.1 mm). The spatial integral in the exponent of (7.31) is computed using (7.25). The gas jet, directed along the x-axis (perpendicular to the field propagation), is modeled by a gaussian distribution along the y and z dimensions, and a constant profile along its axial dimension, x,

$$N(y,z) = N_0\mathrm{e}^{-\frac{(y-y_c)^2}{2\sigma_y^2}}\mathrm{e}^{-\frac{(z-z_c)^2}{2\sigma_z^2}}, \tag{7.33}$$

where N_0 is the maximum gas density over the interacting volume, and (y_c, z_c) are the coordinates of the beam axis center with respect to the laser beam focal point (located at the origin). We have chosen gas beam half widths of $\sigma_y = \sigma_z = 400\mu m$. The angular distribution of the far field is shown in Fig. 7.6, for two different positions of the gas jet: $x_c = y_c = 0$, (a) $z_c = -2.5$ mm (jet before the laser focus) and (b) $z_c = +2.5$ mm (jet after the laser focus). Due to the spatial symmetry of the laser beam, a test atom located at the center of the gas beam feels the same intensity in both cases. In addition, despite the Gouy phase is opposite, it does not have a

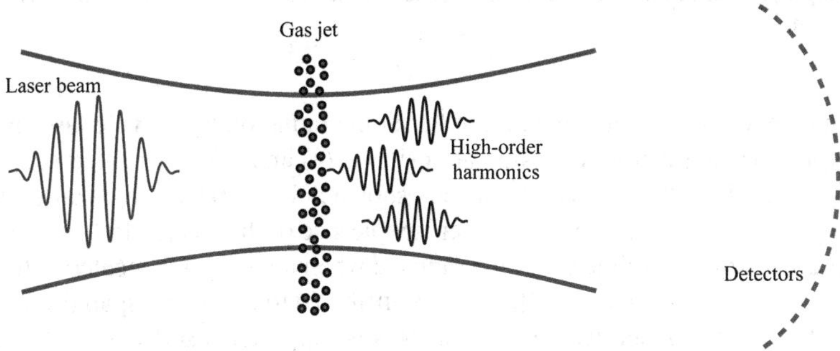

Fig. 7.5 Interaction geometry of a typical experiment of high-order harmonic generation by low-pressure gases

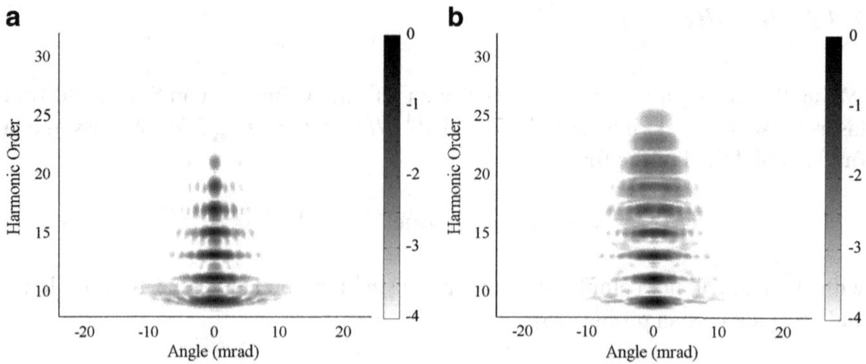

Fig. 7.6 Angular distribution of the radiated spectra for two symmetrical positions of the gas jet relative to the laser focal point: (**a**) $z_c = -2.5$ mm and (**b**) $z_c = 2.5$ mm. The *gray scale* corresponds to the logarithm of the harmonic intensities, in arbitrary units

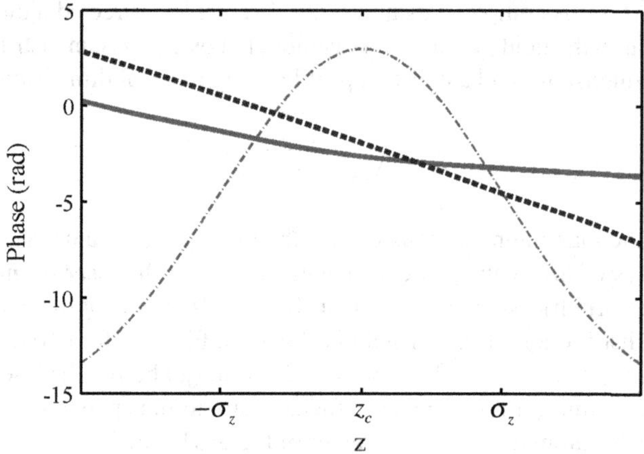

Fig. 7.7 Phase distribution of the 21st harmonic through the target, along the laser propagation axis. The *dashed line* corresponds to the gas jet located before the laser focus and the *solid line* to the target positioned after the focus. As a reference, the *dashed-dot line* shows the gas density distribution

fundamental effect for the single atom harmonic radiation by 8-cycle laser pulses. Therefore, the single atom emission at positions (a) and (b) is expected to be very similar. Thus, the differences in the spectra shown in Fig. 7.6 have to be attributed to the propagation effects through the macroscopic size of the gas jet. In fact, the complex angular structure of the spectrum radiated when the gas jet is located before the focus is a known phenomena [29], which is attributed to the strong spatial variations of the phase of the generated harmonics. This is supported also by our calculation, as shown in Fig. 7.7, where we plot the phase distribution of the generated 21st harmonic along the gas jet in both situations. The slower variation of the phase for the target positioned after the focus is, therefore, responsible for the smoother angular distribution of the radiated spectra. The reason behind the differences in the phase behavior can be explained in terms of the cancelation of its main contributions: the intrinsic phase of the high-order harmonic generation process, the phase of the gaussian beam (in particular the Gouy phase), and the phase induced by the free electron component. If the target is located after the focus, these three terms tend to cancel each other, while they tend to reinforce in the case before the focus. Note also that the position of the harmonic cut-off is also affected by the propagation through the target, the most drastic reduction corresponding to the target placed before the focus. This sensibility of the spectral cut-off to the propagation has been already reported in [30].

Acknowledgements We acknowledge support from Spanish Ministerio de Ciencia e Innovación through the Consolider Program SAUUL (CSD2007-00013) and Research project FIS2009-09522, from Junta de Castilla y León through the Program for Groups of Excellence (GR27) and from the EC's Seventh Framework Programme (LASERLAB-EUROPE, grant agreement nątranslated 228334). We also acknowledge support from the Centro de Laseres Pulsados, CLPU, Salamanca, Spain.

References

1. L.V. Keldysh, Zh. Eksp. Teor. Fiz. **47**, 1945–1956 (1964) [Sov. Phys. JETP **20** 1307–1314 (1965)]
2. F.H.M. Faisal, J. Phys. B **6**, L89–L92 (1973)
3. H.R. Reiss, Phys. Rev. A **22,** 1786–1813 (1980)
4. A. Becker, L. Plaja, P. Moreno, M. Nurhuda, F.H.M. Faisal, Phys. Rev. A **64**, 023408 (2001)
5. M. Lewenstein, Ph. Balcou, M. Yu. Ivanov, A. L'Huillier, P.B. Corkum, Phys. Rev. A **49**, 2117 (1994)
6. M. Lewenstein, P. Salières, A. L'Huiller, Phys. Rev. A **52**, 4747–4754 (1995)
7. W. Becker, A. Lohr, M. Kleber, M. Lewenstein, Phys. Rev. A **56** 645 (1997)
8. K. Schafer, B. Yang, L. DiMauro, K. Kulander, Phys. Rev. Let. **70**, 1599 (1993)
9. P.B. Corkum, Phys. Rev. Lett. **71**, 1994–1997 (1993)
10. A. Zaïr, M. Holler, A. Guadalini, F. Shapper, J. Biegert, L. Gallmann, U. Keller, A.S. Wyatt, A. Monmayrant, I.A. Walmsley, E. Cornier, T. Auguste, J.P. Caumes, P. Salières, Phys. Rev. Lett. **100**, 143902-1-4 (2008)
11. J. Tate, T. Auguste, H.G. Muller, P. Salières, P. Agostini, L.F. Di Mauro, Phys. Rev. Lett. **98**, 013901-1-4 (2007)
12. K. Schiessl, K.L. Ishikawa, E. Person, J. Burgdörfer, Phys. Rev. Lett. **99**, 253903-1-4 (2007)
13. M.V. Frolov, N.L. Manakov, A.F. Starace, Phys. Rev. Lett. **100**, 173001-1-4 (2008)
14. J.A. Pérez-Hernández, L. Roso, L. Plaja, Optic. Express **17**, 9891(2009)
15. A. L'Huillier, Ph. Balcou, Phys. Rev. A **46**, 2778 (1992)
16. A. L'Huillier, X.F. Li, L.A. Lompré, J. Opt. Soc. Am. B, **7**, 527 (1990)
17. T. Auguste, O. Gobert, B. Carré, Phys. Rev. A **78**, 033411 (2008)
18. C. Hernández-García, J.A. Pérez-Hernández, J. Ramos, E. Conejero Jarque, L. Roso, L. Plaja submitted. Phys. Rev. A **82**, 033432 (2010)
19. V.P. Krainov, J. Opt. Soc. Am. B, **14** 425 (1997)
20. C.C. Chirilă, M. Lein, Phys. Rev. A **80**, 013405 (2009)
21. J.A. Pérez-Hernández, L. Roso, L. Plaja, Laser Phys. **19** 1581, (2009)
22. M.V. Amosov, N.B. Delone, V.P. Krainov, Zh. Eksp. Teor. Fiz. **91**, 2008 (1986)
23. J.A. Pérez-Hernández, L. Plaja, Phys. Rev. A **76** 023829 (2007)
24. P. Moreno, L. Plaja, L. Roso, Phys. Rev. A, **55**, R1593 (1997)

25. R. Numico, P. Moreno, L. Plaja, L. Roso, J. Phys B At. Mol. Opt. Phys, **31**, 4163 (1998)
26. J.D. Jackson, *Classical Electrodynamics* (Wiley, New York, 1999)
27. E.M. Purcell, C.R. Pennypacker, Astrophys. J. **186**, 705 (1973)
28. B.E.A. Saleh, M.C. Teich, *Fundamentals of Photonics* (Wiley, New York, 2007)
29. P. Salieres, A. L'Huillier, M. Lewenstein, Phys. Rev. Lett. **74**, 3776 (1995)
30. A. L'Huillier, M. Lewenstein, P. Salieres, Ph. Balcou, Phys. Rev. A **48**, R3433 (1993)

Chapter 8
On the Generation of Intense Isolated Attosecond Pulses by Many-Cycle Laser Fields

Paris Tzallas, Emmanouil Skantzakis, Jann E. Kruse, and Dimitrios Charalambidis

Abstract Real-time observation of ultrafast dynamics in all states of matter requires temporal resolution on the atomic unit of time (24.189 asec) (1 asec = 10^{-18} s). Tools for tracking such ultrafast dynamics are ultrashort light pulses. During the last decade, continuous efforts in ultrashort pulse engineering led to the development of light pulses width duration close to the atomic unit of time. Attosecond (asec) pulses have been synthesized by broadband coherent extreme ultraviolet (XUV) radiation generated by the interaction of gases or solids with an intense IR fs pulse. *Asec* pulse trains can be generated when the medium interacts with many-cycle driving IR fs laser fields. In this case, a broadband XUV frequency comb is emitted from the medium. The Fourier synthesis of a part of the comb results in an *asec* pulse train. Isolated *asec* pulses are generated when the medium is forced to emit XUV radiation only during few cycles of the driving laser field. This leads to the emission of a broadband quasicontinuum XUV radiation. The Fourier synthesis of the continuum part of the spectrum results in an isolated *asec* pulse. For the realization of studies of ultrafast dynamics, intense *asec* pulses are preferable. If the pulses are intense enough to induce a nonlinear process in a target system, they can be used for ultrafast dynamic studies in an XUV pump-probe configuration. Although trains of intense *asec* pulses are commonly produced nowadays, the generation of intense isolated *asec* pulses remains a challenge.

Here, we review a recently developed approach for the generation of intense *asec* pulses using high-peak-power many-cycle laser fields. The approach is based on

---

P. Tzallas (✉)
Foundation for Research and Technology–Hellas, Institute of Electronic Structure and Laser, PO Box 1527, GR-711 10 Heraklion, Crete, Greece
e-mail: ptzallas@iesl.forth.gr

E. Skantzakis, J.E. Kruse, and D. Charalambidis
Foundation for Research and Technology–Hellas, Institute of Electronic Structure and Laser, PO Box 1527, GR-711 10 Heraklion, Crete, Greece
e-mail: chara@iesl.forth.gr
and
Department of Physics, University of Crete, PO Box 2208, GR71003 Heraklion, Crete, Greece
e-mail: skanman@iesl.forth.gr, jkruse@iesl.forth.gr, chara@iesl.forth.gr

K. Yamanouchi et al. (eds.), *Progress in Ultrafast Intense Laser Science VII*,
Springer Series in Chemical Physics 100, DOI 10.1007/978-3-642-18327-0_8,

controlling, with *asec* precession, the response of the atomic dipole to an external many-cycle driving field in such a way as to emit an isolated *asec* XUV burst. This approach has been implemented by using the inteferometric polarization gating (IPG) technique. The bandwidth of the generated XUV radiation is large enough to enable the synthesis of isolated XUV pulses with durations of a few hundred *asec*.

The technique paves the way for the generation of intense isolated *asec* pulses, tuneable in duration and frequency, for carrier-envelope phase (CEP) variation studies of many-cycle driving fields, and it offers exciting opportunities for multiphoton XUV-pump-XUV-probe experiments.

8.1 Introduction

In the last decade, dedicated efforts in the development of ultrashort radiation pulses led to the breakthrough of light pulses width duration in the *asec* timescale. Gas and recently solid targets are the nonlinear (NL) media mainly used in laser-driven *asec* sources [1–4]. *Asec* pulses have been achieved through phase locking of broadband coherent XUV radiation generated by the interaction of gases and solids with intense IR fs pulses. Trains of *asec* pulses are generated, when the NL medium interacts with many-cycle driving IR fs (>5 fs) laser fields. In this case, a broadband XUV frequency comb is emitted by the medium. The Fourier synthesis of a selected part of the frequency comb results in the formation of an *asec* pulse train. For gas media, the generation of *asec* pulse trains is well established [5–10], and is essentially understood in the framework of the three-step model [11, 12]. The generation of *asec* pulse trains in solid media has also been recently demonstrated experimentally [13, 14], while the coherent wake emission (CWE) [15–18] and relativistic oscillating mirror (ROM) [19–24] models successfully describe the process at IR laser intensities $<10^{18}$ W/cm^2 and $>10^{18}$ W/cm^2, respectively. In both, gases and solids targets, a burst of continuum XUV radiation is emitted periodically. In gases, the process is repeated twice per laser cycle and in solids once. As the process is periodic, the emitted spectrum consists of a superposition of coherent continua, which in the time domain is equivalent to a train of *asec* pulses. Using high-power many-cycle laser pulses with duration >20 fs, intense *asec* pulse trains have been generated and already used for the study of NL phenomena in the XUV spectral region [25–30]. They thus fulfill the requirements for their temporal characterization, on the basis of second-order autocorrelation techniques [7,9] and open the road toward XUV-pump-probe experiments.

Isolated *asec* pulses are generated when the medium is steered to emit XUV radiation only during few cycles of the driving laser field. Thus, the emitted spectrum is a broadband quasicontinuum in the XUV. The Fourier synthesis of the continuum part of the spectrum results in an isolated *asec* pulse. Such pulses have been generated so far only from gas media by few-cycle laser pulses [31–34]. In the spirit of the three-step model, if the process is confined to a single revisit of the core by the driven electron, a single continuum is emitted in the form of an isolated pulse. In mathematical terms, the Fourier synthesis of a broad continuum corresponds in the time domain to a single temporal occurrence, whereas a discrete spectrum leads to a repetitive process. Thus, the emission of a single coherent continuum is a crucial requirement for single *asec* pulse generation. Separable XUV continuum emission ensues only during a small fraction of optical cycle of the driving laser field with the highest amplitude. Separation may occur either by selecting the highest energy part of the spectrum [35], or through the temporal modulation of the laser field's ellipticity confining the linear polarization (and thus the XUV emission) at the pulse center [36]. In both cases, the stabilization of the relative carrier-envelope phase (CEP) is highly pertinent [35–38]. In contrast to *asec* trains, isolated *asec* pulses of relatively low energy [31–34] are generated in gases by means of low energy, $\leq$5 fs long laser pulses. For the generation of isolated *asec* pulses, alternative approaches based on polarization gating [39] of low-energy few-cycle pulses [40, 41] and on collective effects in harmonic generation process by sub-20 fs pulses [42, 43] have been recently proposed and implemented. Such isolated *asec* pulses are, so far, less intense and contain notably less photons per pulse than *asec* pulse trains. The highly desirable generation of intense single *asec* pulses remains a challenge. Toward this goal, two different approaches are currently considered: either by developing high-power few-cycle laser systems or, starting with existing high-peak-power many-cycle laser pulses, by isolating half a laser period, during which intense XUV radiation is emitted. The route toward the generation of intense *asec* pulses has recently opened up through the successful implementation of the inteferometric polarization gating (IPG) technique in the generation of quasicontinuum XUV radiation using high power $\sim$50 fs laser pulses [36, 44–46].

Here, we review successive developments in a novel approach, leading to the generation of intense coherent continuum broadband XUV radiation utilizing conventional high peak-power many-cycle laser pulses. By controlling the ellipticity of a many-cycle driving laser field, we can force the medium to emit essentially once and thus an isolated XUV burst. The approach has been realized exploiting the IPG technique. The implementation of the technique in gas phase media is presented here theoretically and experimentally. Analytical expressions for the Ellipticity Modulated (ElMo) many-cycle driving fields coming out of an IPG device are given, while single atom quantum mechanical calculations demonstrate the applicability of the technique in generating intense isolated *asec* pulses. In conjunction with modeling, the experimental observation of intense coherent XUV radiation continua, emitted by the interaction of ellipticity modulated 50 fs driving laser fileds with gases is detailed. The bandwidth of the generated XUV radiation is broad enough to enable the synthesis of isolated XUV pulses with durations of a few hundred *asec* and energy of the order of tens of nJ.

8.2 Generation of Trains and Isolated Attosecond Pulses

8.2.1 Generation of Asec Pulse Trains

The principle of *asec* pulse train generation in gases is shown schematically in Fig. 8.1. Briefly, by focusing an intense linearly polarized (LP) many-cycle

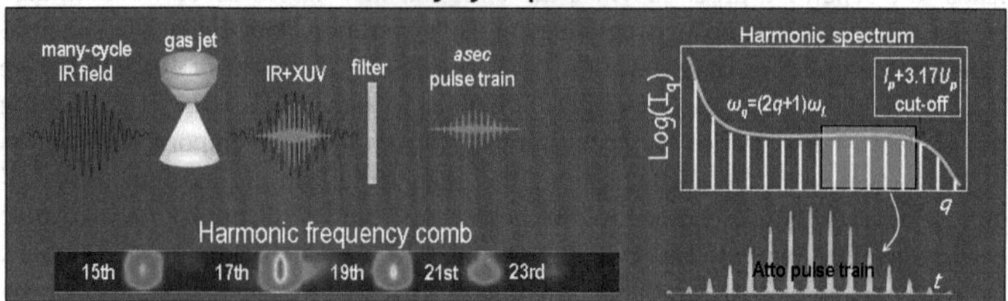

Fig. 8.1 Schematic of *asec* pulse train generation by a many-cycle process. By focusing a many-cycle LP driving laser field in a gas-phase medium XUV radiation is emitted in the direction of the driving laser field. The spectrum of the XUV radiation contains well-confined odd harmonic frequencies. By using the proper metal filter a part of the spectrum can be selected. The temporal profile of the selected spectrum may then have the shape of an *asec* pulse train. The contour iso-intensity color plot shows a typical harmonic spectrum, which has been recorded by using a 50 fs long driving pulse

Ti:Sapphire fs laser beam into a gas phase medium, the medium is nonlinearly driven to emit short-wavelength radiation in the propagation direction of the laser field. The emitted spectrum consists of odd harmonics of the driving frequency and presents a characteristic behavior:

1. The amplitude throughout the first harmonic orders rapid decreases.
2. When the harmonic photon energy becomes equal or larger than the ionization energy of the medium, its conversion efficiency becomes and remains constant up to highest-order harmonics, forming the *plateau* region of the spectrum.
3. Then the harmonic amplitude drops fast in the *cutoff* region.

When fulfilling the appropriate phase-matching conditions (atomic response and macroscopic response) leading to phase locking between the harmonics and, by using a filter, selecting a group of harmonics blocking at the same time the fundamental pulse, an *asec* pulse train is formed with the individual pulses in the train being separated by half of the laser period. The physical process of above-threshold high-order harmonic generation by LP many-cycle infrared intense laser pulses is descried by the well-known classical three-step model in its classical [11] or quantum mechanical version [12]. The process is governed by the recombination of electrons ejected into the continuum and driven back toward the core upon reversal of the LP driving field. Within a fraction of half the laser period, the electron may revisit the parent ion to recombine and emit a burst of continuum XUV radiation. The periodic repetition of the process leads to the formation of an *asec* pulse train. For reasons of symmetry, the periodic motion of the electron results in an XUV frequency comb consisting only of odd harmonic peaks. Such a source can be considered a high-intensity XUV source, since it may induce NL XUV processes and thus can be used in NL autocorrelation diagnostics or XUV-pump-XUV-probe applications [7, 9, 47, 48].

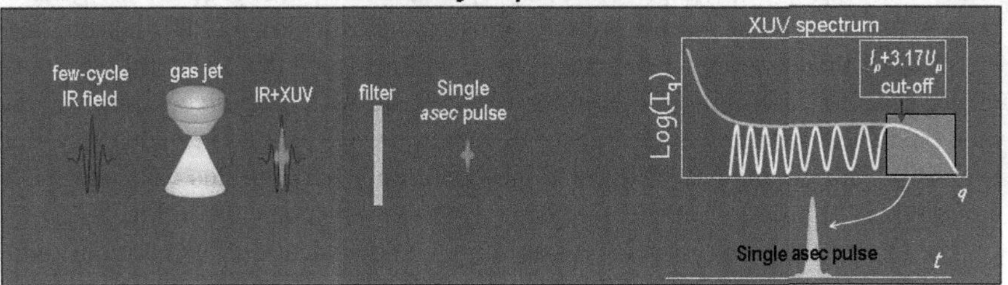

Fig. 8.2 Schematic representation of the isolated *asec* pulse generation process by a few-cycle driving field. By focusing a few-cycle LP driving laser field in a gas phase medium the XUV radiation is emitted in the propagation direction of laser beam. The spectrum of the XUV radiation is quasicontinuous with a continuum part in the cutoff region. By using the proper metal filter the continuum part of the spectrum can be selected. The temporal profile of the selected spectrum may then have the shape of an isolated *asec* pulse

8.2.2 Generation of Isolated Asec Pulses

Rigorously, isolated *asec* pulses are generated when XUV emission is restricted to occur within half a cycle of the driving field. The NL medium is then emitting a coherent continuum; however, isolated pulse generation is also feasible if the emission is restricted to few cycles. Indeed, few-cycle driving laser fields are commonly used for the generation of isolated *asec* pulses. The emitted XUV spectrum is then quasicontinuum, with a pure continuum part in the cutoff region, which is generated by the half cycle of the laser field with the highest amplitude, that is by the central half cycle (Fig. 8.2).

Selecting the cutoff spectral region by a thin metal filter is equivalent to selecting part of one burst of the few-pulse train. An isolated *asec* pulse is thus transmitted through the filter. These types of *asec* sources have so far low pulse energy due to the limited power delivered by the few-cycle laser systems.

8.3 Generation of Isolated *Asec* Pulses by Using Polarization Gating Approaches

8.3.1 Polarization Gating Approach

There are two possible solutions to the challenging task of generating high-energy isolated *asec* pulses: Either to develop high-peak-power few-cycle laser systems or to use the already commercially available high-peak-power many-cycle laser systems. The generation of isolated *asec* pulses by many-cycle driving fields requires the development of a pulse picker in the *asec* timescale. Since the pulse selection from an *asec* pulse train is electronically impossible due to the limited speed of electronics, it has to be implemented by optical means. This requires the generation of a temporal gate, with a width τ_g close to half the driving laser period, near the peak of the envelope of the laser pulse, within which the XUV emission will take place. By controlling the width τ_g of the gate with *asec* precession, we can manipulate the dynamics of the quasi-free electron wave packets in such a way, that it revisits and recombines with the atomic core only once. In this case, the XUV emission is confined to a single *asec* burst as shown schematically in Fig. 8.3.

The idea of the gate formation is based on the dependence of the harmonic generation on the ellipticity of the driving laser field [39]. The harmonic intensity I_q rapidly drops with the ellipticity ε of the driving laser field [39, 49]. According to lowest order perturbation theory [50, 51], the dependence of the harmonic intensity from the harmonic order q and ellipticity ε is given by

$$I_q = ((1-\varepsilon^2)/(1+\varepsilon^2))^{(q-1)}.$$

For linear polarization, where $\varepsilon = 0$, I_q is maximum ($I_q^{\max}$) while for circular polarization $I_q(\varepsilon = 1) \approx 0$. There is a value of ellipticity, known as ellipticity threshold (ε_{th}), above which $I_q < I_q^{\max}/2$. This threshold can be used to define the beginning and the end of the gate, thus creating an XUV emission switch. Within the interval during which the gate is open (τ_g) the XUV emission is taking place. During this interval, the ellipticity ε, varies from $\varepsilon = 0$ to $\varepsilon = \varepsilon_{\text{th}}$. Utilizing a half-cycle temporal gate in the driving field, an ElMo driving laser field is formed with circular polarization at its tails and linear at the peak of the pulse envelope. In this case, the XUV emission can be confined to a single XUV *asec* burst with a continuous spectrum.

8.3.2 Wave-Plates Polarization Gating Approach

Polarization gating has been successfully implemented to produce broadband continuous XUV radiation [40], as well as the shortest ever XUV pulse [33], using 5 fs long CEP stabilized laser pulses and a very convenient experimental setup simply based on two wave-plates (Fig. 8.4).

Two delayed and partially overlapping pulses with perpendicular polarization produced by the first birefringent plate synthesize a pulse with changing ellipticity ε that reaches unity at the center. The second wave-plate ($\lambda/4$) reverses the circularly polarized field at the center to be LP and leaves the field circularly or elliptically polarized elsewhere. However, this method is essentially applicable only to few-cycle driving pulses. This is because the gate width produced by this setup [52–55] reads

$$\tau_g \approx (\varepsilon_{\text{th}}\tau_L^2)/(\Delta t \ln 2), \tag{8.1}$$

where τ_L is the initial pulse width and Δt the delay between the two pulses. This relation implies that to maintain the gate width narrow while increasing the pulse width, the delay Δt has to be increased to very large values, leading to an overlap

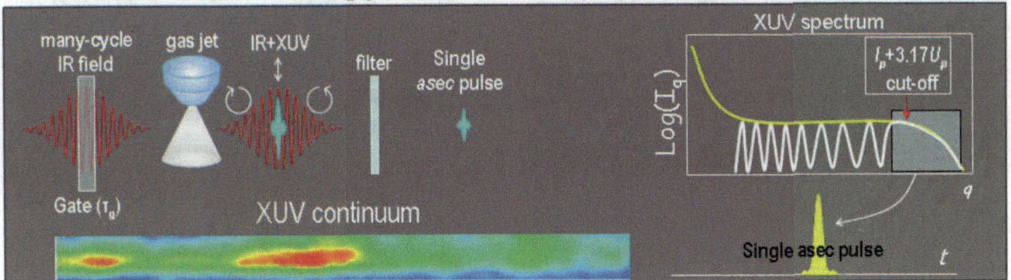

Fig. 8.3 Schematic representation of the generation of an isolated *asec* pulse applying the Polarization Gating (PG) concept using a many-cycle ElMo driving field. By placing a few cycles wide gate at the peak of the many-cycle driving field envelope, the XUV emission is confined to few XUV bursts. The spectrum of the XUV radiation is quasicontinuous with a continuum part in the cutoff region. By using the proper filter the continuum part of the spectrum can be selected. The temporal shape of the selected spectrum can be that of an isolated *asec* pulse. The contour iso-intensity color plot shows a typical quasicontinuum spectrum, which has been recorded by using an Interferometric Polarization Gating (IPG) device descried in Sect. 8.3.3.2 and 50 fs long driving pulse [36]

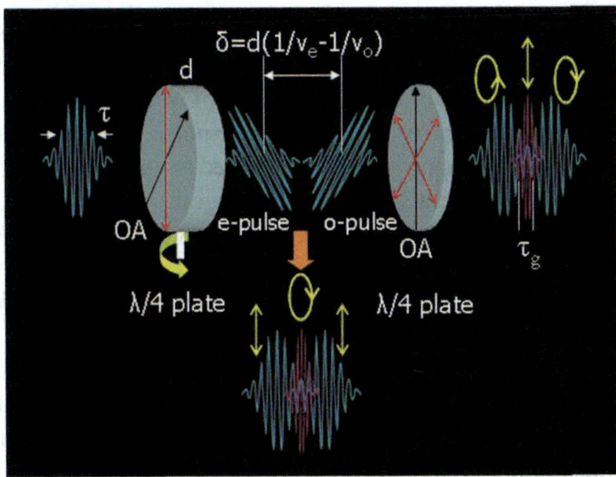

Fig. 8.4 WP-PG approach based on pair of $\lambda/4$ wave-plates. The laser pulse is passing through a multiple-order $\lambda/4$ wave-plate with its optical axis (OA) at 45° relative to the field polarization to produce two delayed pulses with perpendicular polarizations. This forms an ellipticity modulated field with circular polarization at its center. A second $\lambda/4$ wave-plate reverses the circular polarization to linear forming the gate

of only the very far pulse edges and thus to a very low conversion efficiency. For $\tau_g = 5$ fs and typical 40–60 fs long high-peak power pulse, (8.1) results required Δt delays ranging between 70 and 160 fs, imposing practically no overlap between the two pulses (Fig. 8.5a).

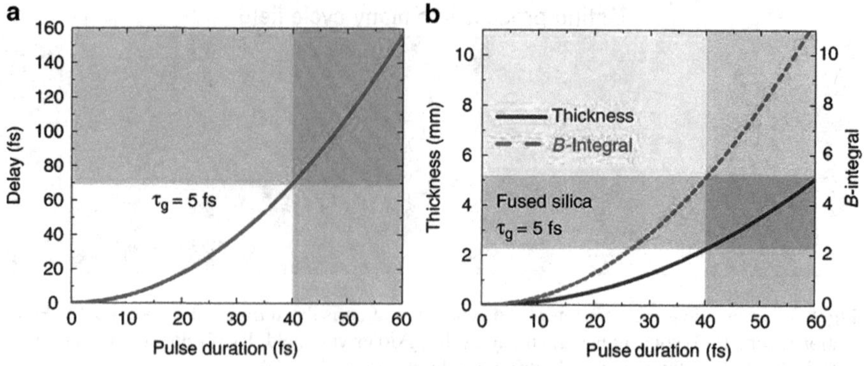

Fig. 8.5 (**a**) Delay between the two pulses in the WP-PG approach required for a gate width of 5 fs as a function of the pulse width. For widths between 40 and 60 fs, this delay is about twice the pulse width. (**b**) Plate thickness (*blue solid line*) and corresponding B-integral (*red dashed line*) for an intensity of 1 TW/cm^2 in the WP-PG approach required for a gate width of 5 fs as a function of the pulse width. For widths between 40 and 60 fs, the B-integral values range from about 5 to 11, i.e. above the tolerance threshold [45]

The restriction of the method to few-cycle pulses, which at present have limited energy content (less than 1 mJ at 5 fs, with the exception of the ∼8 fs high-power system of Vrije Universitat [56] and 8 fs, 100 mJ of Max-Planck-Institut für Quantenoptik [57]) is prohibitive for energetic *asec* pulse generation. An additional drawback of the wave-plate approach, when applied to many-cycle high-peak-power pulses is that it inevitably leads to large B-integral values exceeding by far the safety threshold. The situation is illustrated in Fig. 8.5b. For 40–60 fs long pulses, the first fused silica wave-plate thickness has to be 2–5 mm to achieve a 5 fs gate. For an intensity of 1 TWcm^{-2}, the B-integral corresponding to this thickness ranges between 5 and 11, much larger than the allowed threshold (Fig. 8.5b).

Approaches based on polarization gating [39] of low-energy few-cycle pulses [40, 41] and on collective effects in harmonic generation processes by using sub-25 fs pulses [42, 43] have also been recently proposed and implemented. Such isolated *asec* pulses are, so far, less intense and contain notably less photons than *asec* pulse trains.

The highly desirable generation of intense single *asec* pulses remains a challenge. The route toward the generation of high-energy *asec* pulses has recently opened up through the successful implementation of the IPG technique in the generation of quasicontinuum XUV radiation using high power ∼50 fs laser pulses [36, 44]. This will be the main issue of the next sections of this article. Also, approaches similar to the IPG, based on two color driving laser fields and called Double Optical Gating (DOG) [41] and Generalized Double Optical Gating (GDOG) [58] approaches, have been recently implemented for the generation of isolated *asec* pulses by ∼30 fs long IR driving laser fields [59].

8.3.3 Interferometric Polarization Gating Approach

In the IPG approach, the generated bandwidth is broad enough to enable the synthesis of isolated XUV pulses with pulse durations of a few hundreds of *asec*. For a large number of existing high-peak-power laser installations, the present approach opens up exciting prospects for the generation of intense isolated *asec* pulses. In particular, it is applicable without any conceptual modification to the XUV generation from laser–plasma interactions at relativistic intensities, aiming at the generation of isolated *asec* pulses with unprecedented XUV intensities [20, 60].

In the IPG approach, the gate width (τ_g) is substantially less sensitive to the duration of the driving field and thus it is applicable to many-cycle pulses as well, with peak power of hundreds of TW. The essence of the method, which relates to a proposal of [46], is the synthesis of an ElMo pulse from four individually controlled LP pulses of variable field amplitude, relative polarization direction and delay. The method is schematically shown in Fig. 8.6.

The four pulses originate from the same initial laser pulse. Two of them are co-propagating and delayed with respect to each other by an odd number of half laser periods, so that their overlapping parts form a destructive interference minimum. This minimum will play the role of a mould for the gate. Its steepness can be controlled by the variable delay δ and field ratio E_1/E_2. The steepness becomes maximum when the delay δ is of the order of the initial pulse duration τ_L. The delay between the other two pulses Δt is of the same order of magnitude but such that their interference is constructive. The polarization planes of the two resulting superposition fields are subsequently (or previously) rotated so as to become perpendicular. Obviously, when these two fields are recombined with zero delay with respect to each other (or with a delay equal to an integer number of laser periods) they synthesize an *ElMo* pulse, with linear polarization only in its central part. Thus,

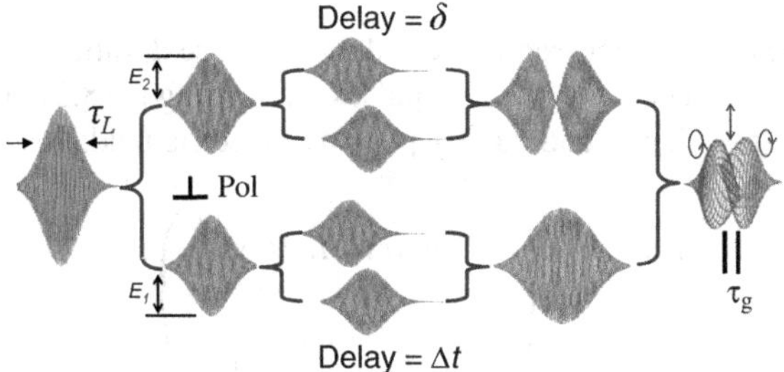

Fig. 8.6 Schematic representation of the IPG approach. The incoming pulse, with duration τ_L, is split into two with perpendicular polarizations and different amplitudes E_d and E_c. Then, each of these pulses is split again into two. Two of the appropriately delayed (Δt) pulses interfere constructively, while the other two with delay δ form a destructive interference minimum. The polarization planes of the superposition are mutually perpendicular. The two waveforms are superimposed forming an ElMo pulse with linear polarization in its central part. τ_g is called "gate width" and corresponds to the temporal window where the polarization is almost linear

the gate width τ_g can be less than half a cycle of the driving field, by controlling the parameters: (a) the ratio E_1/E_2 of the field amplitudes entering the two interferometers and (b) the delays, δ and Δt, between the two pulses in the pulse pairs. This can be done by using a Double Michelson Interferometer (DMI) [36] or a Double Mach-Zender (DMZ) [44] arrangement, which is described in detail in Sect. 8.3.1.1. The main difference between these two arrangements is that the energy content of the driving laser field within the τ_g in DMZ is a factor of 2 larger compared to DMI on the cost of reducing the control of the amplitudes of the four fields, to the control of the amplitudes of two pairs of fields.

8.3.3.1 Theoretical Treatment in the IPG Approach

In DMZ arrangement where $\tau_L \approx \delta = \Delta t$, an analytical expression for the gate width τ_g can be derived [44]. Assuming a Gaussian temporal pulse profile, the constructively (E_c) and destructively (E_d) interfering fields, coming out of the DMZ-IPG device, can be written as

$$\begin{aligned} \vec{E}_c(t) &= \hat{y} E_{c0}\left(\exp\left[-\left(\frac{t-\delta/2}{\tau_L/\sqrt{2\ln 2}}\right)^2\right]\cos(\omega_L t)\right. \\ &\quad \left. + \exp\left[-\left(\frac{t+\delta/2}{\tau_L/\sqrt{2\ln 2}}\right)^2\right]\cos(\omega_L t)\right) \\ \vec{E}_d(t) &= \hat{x} E_{d0}\left(\exp\left[-\left(\frac{t-\delta/2}{\tau_L/\sqrt{2\ln 2}}\right)^2\right]\sin(\omega_L t)\right. \\ &\quad \left. - \exp\left[-\left(\frac{t+\delta/2}{\tau_L/\sqrt{2\ln 2}}\right)^2\right]\sin(\omega_L t)\right), \end{aligned} \tag{8.2}$$

where ω_L is the laser carrier frequency and τ_L the pulse duration, $\delta = \Delta t \approx \tau_L$ are the delays introduced between the two pulses. According to (8.2), the ellipticity for the two superimposed, perpendicularly polarized electric fields reads,

$$\begin{aligned} \varepsilon(t) &= \tan\left(\frac{1}{2}\sin^{-1}\left(\frac{2\left|\vec{E}_c(t)\right|\left|\vec{E}_d(t)\right|}{\left|\vec{E}_c(t)\right|^2 + \left|\vec{E}_d(t)\right|^2}\right)\right) \\ &= \tan\left(\frac{1}{2}\sin^{-1}\left(\frac{2(2^{8\delta t/\tau_L^2} - 1)\lambda}{\lambda^2(1 - 2^{4\delta t/\tau_L^2})^2 + (1 + 2^{4\delta t/\tau_L^2})^2}\right)\right), \end{aligned}$$

where $\lambda = E_{d0}/E_{c0} = E_2/E_1$ is the electric field amplitude ratio (E_1 and E_2 are the field amplitudes of Fig. 8.6). By setting $\varepsilon = \varepsilon_{th}$ in the above equation, the τ_g

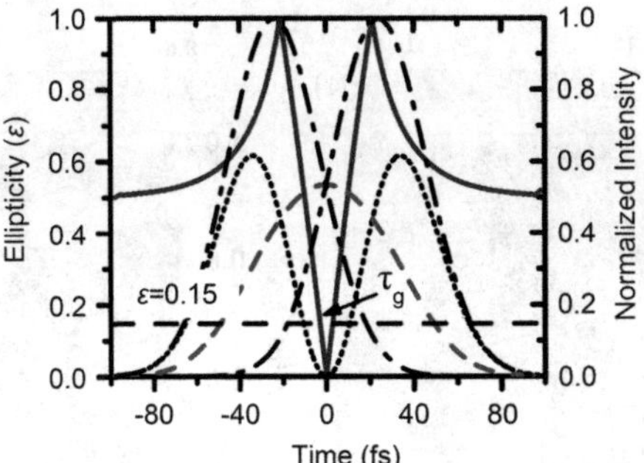

Fig. 8.7 *Solid blue line*: Ellipticity of the polarization-modulated pulse used for the generation of the continuum XUV radiation. *Black dashed–dotted line*: normalized intensity of the two pulses, which partially overlap to form a pulse with a destructive interference minimum in its central part (*black dotted line*). *Red dashed line*: intensity distribution of the pulse showing a constructive interference maximum. The threshold ellipticity $\varepsilon_{\text{th}} = 15\%$ and the time gate width $\tau_{\text{g}} \sim 5\,\text{fs}$ are shown

becomes,

$$\tau_{\text{g}} = \frac{\tau_L^2 \log_2(A)}{2\delta}, \tag{8.3}$$

where $A = \left(\frac{-2\lambda\sqrt{1-B^2}+B(\lambda^2-1)}{B-2\lambda+B\lambda^2}\right)$ and $B = \sin(2\tan^{-1}(\varepsilon_{\text{th}}))$. Figure 8.7 shows the calculated ellipticity of the polarization modulated pulse and the relative intensities of the two perpendicularly polarized fields at the output of the device for a 50 fs long laser pulse.

The contour plot of Fig. 8.8a shows the τ_{g} as a function of δ and the intensity ratio $I_{\text{c}0}/I_{\text{d}0} = 1/\lambda^2$ for a 50 fs long pulse and $\varepsilon_{\text{th}} = 0.15$. An important parameter, which has to be calculated to estimate the conversion efficiency of the XUV generation process is the energy content of the driving laser field within the gate. The intensity (I_{g}) of the driving laser field within τ_{g} is given by

$$I_{\text{g}} = \frac{\tau_L}{\tau_{\text{g}}} \frac{\int_{-\tau_{\text{g}}/2}^{\tau_{\text{g}}/2} \left|\vec{E}_{\text{c}}(t)\right|^2 \text{d}t}{\int_{-\infty}^{\infty} \left|\vec{E}_{\text{in}}(t)\right|^2 \text{d}t} I_{\text{in}} = \frac{\tau_L}{\tau_{\text{g}}} \frac{\frac{1}{E_{\text{c}0}^2} \int_{-\tau_{\text{g}}/2}^{\tau_{\text{g}}/2} \left|\vec{E}_{\text{c}}(t)\right|^2 \text{d}t}{\frac{1}{R_C} \int_{-\infty}^{\infty} \left|\vec{E}_{\text{c}}(t)\right|^2 \text{d}t + \frac{\lambda^2}{1-R_C} \int_{-\infty}^{\infty} \left|\vec{E}_{\text{d}}(t)\right|^2 \text{d}t} I_{\text{in}},$$

where $R_c = 1/(\lambda + 1)$ is the percentage of the $\vec{E}_{\text{c}}(t)$ field (R_c is the reflectivity of the beam splitter BS3 in Fig. 8.12) at the output of the DMZ-IPG device and E_{in}, I_{in} are the total field amplitude and intensity of the laser pulse entering into the DMZ-IPG, respectively. The contour plot in Fig. 8.8b depicts the intensity ratio $I_{\text{g}}/I_{\text{in}}$ as a function of δ and $I_{\text{c}0}/I_{\text{d}0}$.

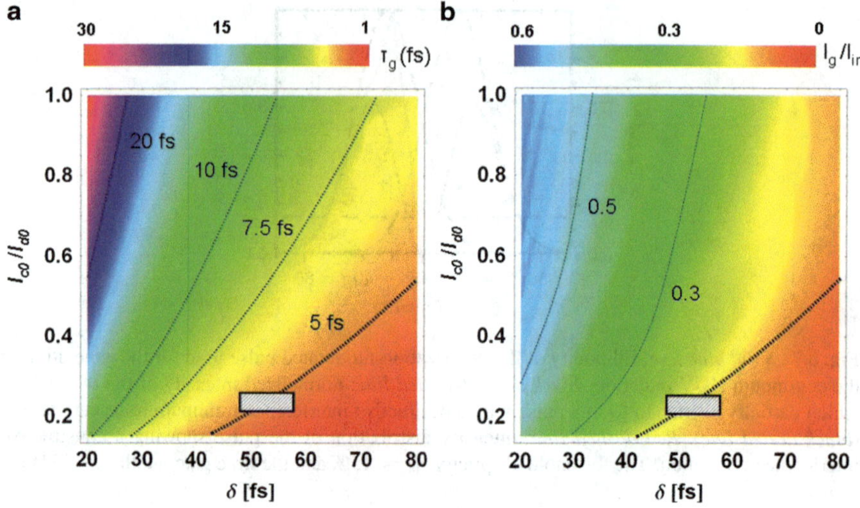

Fig. 8.8 (**a**) τ_g as a function of the delay δ and the intensity ratio $I_{c0}/I_{d0} = 1/\lambda^2$ for a 50 fs pulse and $\varepsilon_{th} = 0.15$ in DMZ arrangement. (**b**) Intensity ratio I_g/I_{in} as a function of δ and I_{c0}/I_{d0}. In both graphs, the *black solid line* depicts the area where $\tau_g \sim 5$ fs and the *shaded area* indicates the parameter range of $I_{c0}/I_{d0} = 0.25$ and $\delta = 50$ fs used

For the calculation of the spectrum emitted by an ElMo pulse, the quantum mechanical three-step model [12,61] has been solved. In the model, the saddle point method has been used and the spectrum was obtained by calculating the acceleration of the single-atom dipole moment. The expression for the single-atom dipole moment is [12]

$$x(t_r) = \mathrm{i}\int_0^{t_r} \mathrm{d}t_i \int \mathrm{d}^3 p \times [\mathrm{d}^*(\boldsymbol{p} - \boldsymbol{A}(t_r))] \times [\boldsymbol{E}(t_i) \cdot \mathrm{d}(\boldsymbol{p} - \boldsymbol{A}(t_i))] \times \mathrm{e}^{-\mathrm{i}S(p,t_r,t_i)} + c.c., \tag{8.4}$$

where the quantity S denotes the quasiclassical action that the electron experiences during its excursion in the continuum and reads as,

$$S(\boldsymbol{p}, t_r, t_i) = \int_{t_i}^{t_r} \mathrm{d}t(([\boldsymbol{p} - \boldsymbol{A}(t)]^2/2) + I_p) \tag{8.5}$$

t_r and t_i are the recombination and ionization time, respectively. In Feynman's spirit, (8.5) is an integral over all possible electron trajectories that are characterized by t_r, t_i and $\boldsymbol{p}$. The time that the electron spends in the continuum is $\tau = t_r - t_i$. In the above equations, we have introduced the canonical momentum $\boldsymbol{p} = \boldsymbol{u} + \boldsymbol{A}(t)$, with $\boldsymbol{A}(t)$ being the vector potential of the ellipticity-modulated electric field, which is given by (8.2). $\boldsymbol{u}$ is the electron velocity in the continuum and $\boldsymbol{d}$ is the dipole matrix element for bound-free transitions. As shown in [12,61], (8.4) can be solved by using the saddle-point approximation (SPA). In this approximation, the integral over all possible electron trajectories becomes a sum over the most significant trajectories, with values of t_r, t_i, and $\boldsymbol{p}_s$, which are determined by the principle of stationary action [12]. This results in a single atom dipole moment which reads as,

$$x(t_r) = \mathrm{i} \int_{-\infty}^{t_r} \mathrm{d}t_i \left(\frac{\pi}{\mu - i(t_r - t_i)/2} \right)^{3/2} [\mathrm{d}^*(\boldsymbol{p}_s - \boldsymbol{A}(t_r))] \times [\boldsymbol{E}(t_i) \cdot \mathrm{d}(\boldsymbol{p}_s - \boldsymbol{A}(t_i))] \times \mathrm{e}^{-\mathrm{i}S(\boldsymbol{p}_s, t_r, t_i)} + c.c.,$$

where μ is a positive regularization constant (associated with the effect of quantum diffusion [62]) and $\boldsymbol{p}_s = \frac{1}{t_r - t_i} \int_{t_i}^{t_r} \boldsymbol{A}(t)\,\mathrm{d}t$ is the stationary value of the momentum, which is obtained by setting $\nabla_p S(\boldsymbol{p}, t_r, t_i)|_{\boldsymbol{p}_s} = 0$.

The Fourier transform of the single-atom dipole moment can be calculated as

$$x(\omega) = \int_{-\infty}^{+\infty} x(t_r)\mathrm{e}^{\mathrm{i}\omega t_r}\,\mathrm{d}t_r \tag{8.6}$$

and the harmonic emission rate is given by [63]

$$W(\omega) = \omega^3 |x(\omega)|^2. \tag{8.7}$$

The generalized phase term of a specific frequency ω is

$$\Theta(\boldsymbol{p}_s, t_r, t_i) = \omega t_r - S(\boldsymbol{p}_s, t_r, t_i). \tag{8.8}$$

To solve (8.6), and to obtain the most significant values of t_i and t_r contributing to the spectrum, the stationary phase approximation ($\partial\Theta/\partial t_r|_{t_{\mathrm{rs}}} = \partial\Theta/\partial t|_{t_{\mathrm{is}}} = 0$) has to be applied. For a given value of the photon energy $\hbar\omega$, a series of saddle-point solutions $(\boldsymbol{p}_s, t_{\mathrm{rs}}, t_{\mathrm{is}})$ is obtained. By using this approximation to the generalized phase term of (8.8), the Fourier transform of the dipole moment, $x(\omega)$, can be transformed to [8],

$$x(\omega) = \sum_s |x_s(\omega)|\mathrm{e}^{-\mathrm{i}\Phi_s(\omega)} = \sum_s \frac{\mathrm{i}2\pi}{\sqrt{\det(\ddot{S})}} \left[\frac{\pi}{\varepsilon - i(t_{\mathrm{rs}} - t_{\mathrm{is}})/2} \right]^{3/2} \times [\mathrm{d}^*(\boldsymbol{p}_s - \boldsymbol{A}(t_{\mathrm{rs}}))] \times [\boldsymbol{E}(t_{\mathrm{is}}) \cdot \mathrm{d}(\boldsymbol{p}_s - \boldsymbol{A}(t_{\mathrm{is}}))] \times \exp[-\mathrm{i}S(\boldsymbol{p}_s, t_{\mathrm{rs}}, t_{\mathrm{is}}) + \mathrm{i}\omega t_{\mathrm{rs}}]. \tag{8.9}$$

Equation (8.9) is a coherent superposition of the contributions from the complex saddle-point solutions of $(\boldsymbol{p}_s, t_{\mathrm{rs}}, t_{\mathrm{is}})$. In (8.9), $\Phi_s(\omega)$ is the phase of the complex function of the dipole moment $x_s(\omega)$, and $\det(\ddot{S}) = \det(\ddot{\Theta})$ is the determinant of the 2×2 matrix of the second derivatives of $\Theta(\boldsymbol{p}_s, t_r, t_i)$ with respect to t_r and t_i, evaluated in correspondence of the saddle-point solutions $(\boldsymbol{p}_s, t_{\mathrm{rs}}, t_{\mathrm{is}})$. The physical interpretation of the saddle-point solutions is that photons of energy $\hbar\omega$ are

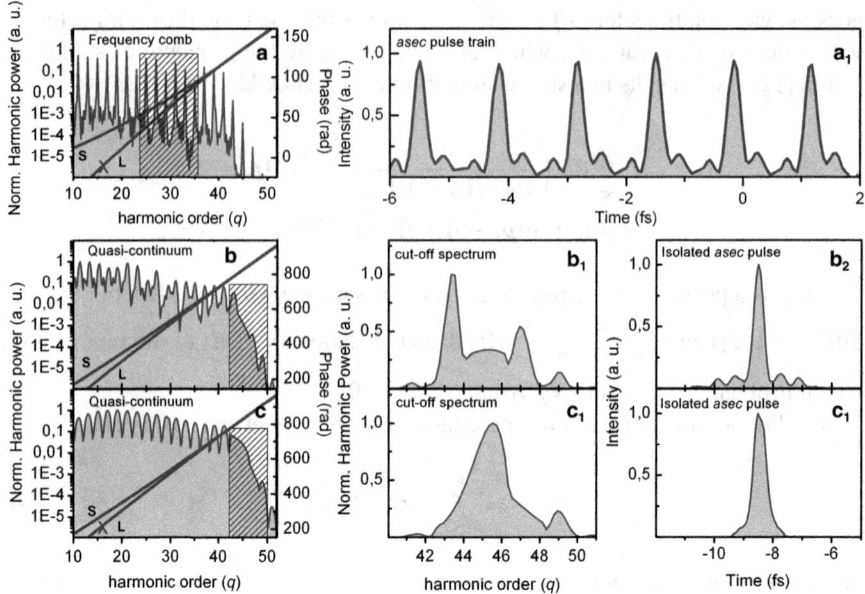

Fig. 8.9 XUV Spectra and reconstructed *asec* pulses obtained by solving the three-step quantum mechanical model [12, 61] in a single-atom interaction for a linearly polarized many-cycle 50 fs pulse and an elliptically modulated 50 fs pulse with linear polarization in a time gate of ~5 fs width. (**a**) Harmonic frequency comb calculated using both short (S) and long (L) electron trajectories. (**a**$_1$) *Asec* pulse train calculated using the spectrum (**a**) and only the phases of short (S) electron trajectories. (**b**) Quasicontinuum XUV spectrum calculated using both electron trajectories. (**b1**) Continuum spectrum of the cutoff region of (**b**) after reflection from a multilayer mirror. (**b**$_2$) Isolated *asec* pulses calculated using the spectrum of (**b**$_1$) and the phases of both electron trajectories. (**c**) Quasicontinuum XUV spectrum calculated using only the short (S) electron trajectories. (**c**$_1$) Continuum spectrum of the cu-off region of (**c**) after reflection from a multilayer mirror. (**c**$_2$) Isolated *asec* pulses calculated using the spectrum of (**c**$_1$) and the phases of the short (S) electron trajectory

emitted mostly by electrons that are set free at time t_{is}, have acquired momentum $\boldsymbol{p}_s$ in the electric field, and recombine with the nucleus at a time t_{rs}. According to this model, two interfering electron trajectories with two different flight times $\tau_q^L(I_L)$ and $\tau_q^S(I_L)$ contribute to the emission of each harmonic q, at a given driving laser intensity I_L. L and S stand for "long" and "short" electron trajectories. The phase of each frequency component ω is obtained by $\text{Re}[\Theta(\boldsymbol{p}_s, t_{\text{rs}}, t_{\text{is}})]$ and the cutoff frequency by the value of ω where $\text{Im}[\Theta(\boldsymbol{p}_s, t_{\text{rs}}, t_{\text{is}})]_{\text{Short}} = \text{Im}[\Theta(\boldsymbol{p}_s, t_{\text{rs}}, t_{\text{is}})]_{\text{Long}}$.

Figure 8.9 shows how an XUV frequency comb spectrum resulting from the interaction of atoms with a 50 fs LP laser goes over to a XUV quasicontinuum spectrum in an elliptically modulated 50 fs pulse with linear polarization in a time gate of ~5 fs width.

Figure 8.9a and b show the harmonic frequency comb and the quasicontinuum XUV spectrum calculated for the interaction of a single atom with a many-cycle LP pulse and for an ElMo 50 fs pulse with linear polarization in a time gate of ~5 fs width, respectively, by taking into account the contribution of both, the short (S) and long (L), electron trajectories. Figure 8.9c corresponds to the calculated quasicontinuum XUV spectra obtained by using an ElMo 50 fs laser pulse with linear polarization in a time gate with $\tau_g \sim 5$ fs by taking into account the contribution of the short (S) electron trajectory. The green lines in Fig. 8.9a, b, c show the calculated harmonic phases resulting from the short (S) and long (L) electron trajectories. Here, it is important to note that the difference of the two electron trajectories in the plateau spectral region disappears in the cutoff and the two trajectories degenerate to one. Figure 8.9a_1 shows a typical reconstructed *asec* pulse train obtained by taking into account the harmonic amplitudes and phases in the shaded harmonic plateau area of Fig. 8.9a. For this reconstruction and for simplicity reasons, it has been assumed that the contribution of the long trajectory is zero [64, 65], which as has been recently demonstrated is not always the case [10]. This was considered to happen when focusing the laser at the position $\sim b/2$ (where b is the confocal parameter of the laser beam) before the gas medium [64]. In case of contribution of both electron trajectories, the duration of the *asec* pulses in the train is longer than that shown in Fig. 8.9a_1 [10] (not shown here). By using the proper multilayer mirror, the cutoff part of the quasicontinuum spectra of Fig. 8.9b and c can be selected (shaded areas) as shown in Fig. 8.9b_1 and c_1, respectively. In this case, an isolated *asec* pulse is obtained (Fig. 8.9b_2 and c_2), independently of the focusing conditions, by taking into account the XUV amplitudes in the cutoff spectral region (Fig. 8.9b_1 and c_1) and their corresponding phases.

8.3.3.2 Coherent Continuum XUV Radiation Generated by a Many-Cycle Laser Field Utilizing a DMI-IPG Device

The IPG technique has been implemented utilizing two Michelson Interferometers (MI) as described in [36]. With this setup, efficient polarization gating has been demonstrated through the transition from a discrete harmonic spectrum (harmonics 15th to 23rd) to a broadband XUV continuum [36] spanning over >15 eV. An illustration of the Dual Michelson Interferometer (DMI) arrangement, the XUV generation apparatus, and the XUV detection unit is shown in Fig. 8.10.

In this approach, the laser system used is a 10 Hz Ti:sapphire, delivering 50 fs pulses, with an energy of up to 150 mJ per pulse and a carrier wavelength of 800 nm. The laser beam is split into two parts by a beam splitter (see Fig. 8.10) and the resulting two pulses enter the MI arrangements. One of the beams (first pulse) enters the first MI, introducing a time delay (δ) between the two arms, which is a multiple of the laser period, so that the output pulse possesses a constructive interference maximum at its center. The other beam (second pulse) enters the second MI, where the delay is set to a fixed value of about $\delta = \tau_L = 50$ fs forming a destructive interference at its center. The linear polarizations of the two outgoing beams are adjusted to be perpendicular to each other by a $\lambda/2$ zero-order wave-plate. The attenuator placed before the first MI, together with the variable delay of the first MI, serves to adjust the intensity distributions of the two pulses from the two MIs

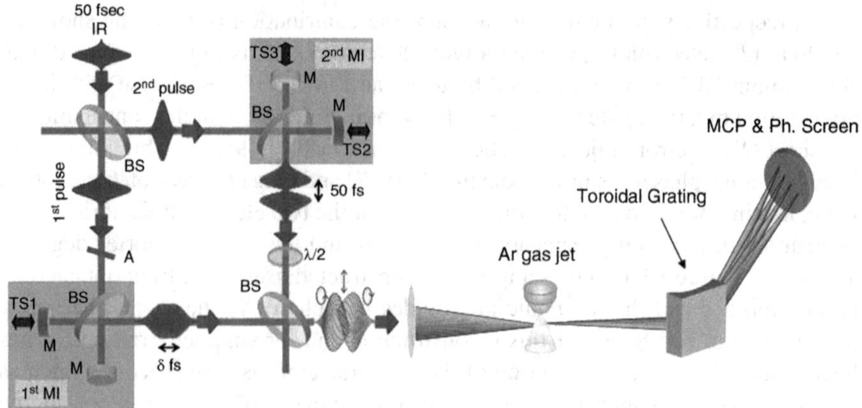

Fig. 8.10 IPG apparatus based on DMI arrangement [36]. *BS*: beam splitters. *M*: flat mirrors. *TS1,2,3*: piezoelectric translation stages. *A*: intensity attenuator. *1st and 2nd MI*: first and second Michelson Interferometers. The tailored output beam is focused into a pulsed Ar gas jet, where the harmonic generation is taking place. The harmonic radiation is monitored by an XUV toroidal grating monochromator in a grazing-incident configuration, equipped with an imaging detector, coupled to a CCD camera capable of recording single shot measurements

as to result in an optimal combination of gate width and gate energy content avoiding any secondary gate formation at the tails of the tailored pulse. On the basis of the calculated dependence of τ_g and I_g/I_{in} on δ and I_{c0}/I_{d0} (Fig. 8.8), the δ and I_{c0}/I_{d0} parameter values have been chosen to result in a gate width $\tau_g \sim 5$ fs and a ratio $I_g/I_{in} \sim 0.035$ (shaded area in Fig. 8.8). In the present case, the ratio I_g/I_{in} is smaller than $(I_g/I_{in})_{DMZ}/2 \approx 0.08$ due to the combination of the 50% reflectivity beam splitters (BS) and the attenuator (A) (see Fig. 8.10).

The tailored output beam is focused into a pulsed Ar gas jet, where the harmonic generation is taking place. The beam focus is placed before the Ar gas jet to favor the short electron trajectory [64]. The harmonic radiation is monitored by an XUV toroidal grating monochromator in a grazing-incident configuration, equipped with an imaging detector, coupled to a CCD camera capable of recording single shot spectra (Fig. 8.11).

Figure 8.11a (upper panel) shows an iso-intensity color plot on a linear scale depicting the continuum spectrum generated by an ellipticity-modulated pulse possessing almost linear polarization in a time gate of ~5 fs width. The lower panel of Fig. 8.11a shows a discrete spectrum of harmonics generated by an LP pulse. The transition from a continuum spectrum to spectrum with well-confined harmonic peaks is indicative of the transition from an isolated *asec* pulse to an *asec* pulse train when the gate is OFF. The red and blue lines in Fig. 8.11b correspond to the line out of the normalized spectra of the upper and lower panels of Fig. 8.11a, respectively. The spectra in Fig. 8.11a and b are an average of 300 shots. An interesting aspect of the results obtained is that the CEP variation is imprinted in the spectra (Fig. 8.11c) when recording single shots. They vary from continuum to semidiscrete,

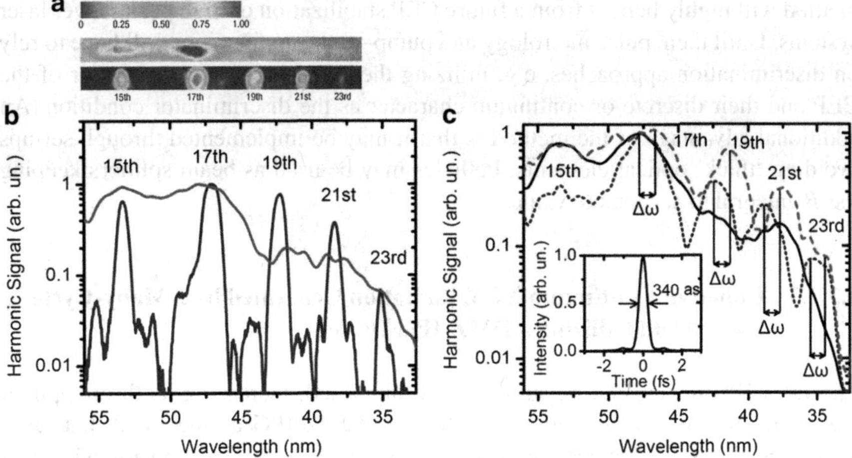

Fig. 8.11 Generation of a continuum XUV spectrum by using a DMI-IPG device [36]. (**a**) *Upper panel*: Iso-intensity color plot on a linear scale showing the continuum spectrum generated by the central part of the ElMo pulse, that is defined by the time window $\tau_g \sim 5\,\text{fs}$ (GATE ON). *Lower panel*: Iso-intensity color plot on a linear scale showing a discrete spectrum of harmonics generated by a LP pulse (GATE OFF). (**b**) The *red and blue lines* correspond to the line out of the normalized spectra of the upper and lower panels of (**a**), respectively. (**c**) Evidence of the CEP variation, reflected in the different structure of the spectra and the frequency shift ($\Delta\omega$) of the harmonic peaks. *Solid black line*: Solid normalized XUV spectrum generated during a small fraction of the highest-amplitude optical cycle of the driving laser field. *Red dashed and blue dotted lines*: normalized XUV spectra, each one generated by a driving laser field with different CEPs values. The *inset* shows the calculated temporal profile of the FTL (Fourier Transform Limit) single *asec* pulse resulting from the 55 to 32 nm spectral range of the spectrum, transmitted by a Si filter

while the semidiscrete spectra depict peak positions varying from shot to shot [36]. This behavior can be attributed to the shot-to-shot variation of the relative CEP of the laser.

Observed CEP effects in the multi-optical-cycle regime are rare [66]. Thus, the general belief is that CEP plays no role when using many-cycle pulses. The present result together with the work of [40] invalidates this assumption. The reason is that the narrow gate introduced makes the conditions equivalent to those when using few-cycle pulses. Indeed, the destructive interference minimum along with the gate phase is locked to the envelope phase (the minimum is equidistant from the positions of the peak maxima of the two interfering pulses), while the quasilinearly polarized driving field phase (in the gate) is locked to the carrier phase. So, the CEP becomes the "carrier-gate phase." The variation of the single shot harmonic spectra due to the CEP variation may serve as a rough measure of the gate width. It is worth noting that if the maximum possible gate power content is not required, by narrowing the gate width the emission may become binary, i.e. either emission of an XUV continuum or essentially no XUV emission. The emitted continuum is coherent and thus forms an ultra short isolated pulse, the duration of which remains to be measured. The method will highly benefit from a future CEP stabilization of high peak power laser systems. Until then, pulse metrology and pump–probe applications will have to rely on discrimination approaches, e.g. utilizing the XUV spectra as a monitor of the CEP and their discrete or continuum character as the discriminator condition. An additional advantage of the method is that it may be implemented through set-ups avoiding 'thick' optical elements. Pellicles may be used as beam splitters keeping the B-integral to acceptable values.

8.3.3.3 Coherent Continuum XUV Radiation Generated by a Many-Cycle Laser Field Utilizing a DMZ-IPG Device

The DMI-IPG device can be substantially simplified, increasing its throughput at the same time by a factor of 2 [44]. It is the DMZ-IPG arrangement that takes advantage of the fact that destructive interference occurs at the second arm whenever constructive interference occurs at the exit arm of a collinear interferometric setup. This modification is presented in Fig. 8.12. Indeed, when the first pulse after the beam splitter BS2 depicts a destructive interference minimum, the second one will have a constructive interference maximum. These two pulses are formed through recombination at the beam splitter BS2 of pulses appropriately delayed in the DMZ interferometer. Variable delay is here introduced by rotating the BK7 plates. The I_{c0}/I_{d0} ratio can be adjusted by the last beam splitter (BS3). The appropriate value of this ratio depends on the initial pulse duration. The 20%:80% ratio is for the 50 fs initial pulse duration. The setup presented is simpler, more stable and has a factor of two higher throughput than the DMI arrangement. The stability of the setup depends mainly on the long-term drift of the piezo delay stages. The simplicity and improvement in the stability result from the reduction of the number of the independent delay stages and the replacement of the piezo-translation stages by piezo-tilted delay stages. The only disadvantage of this setup is that the delay of all four pulses

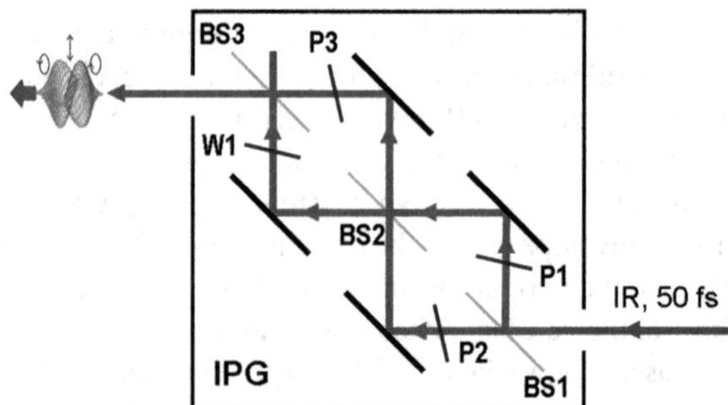

Fig. 8.12 IPG apparatus based on a DMZ arrangement [44]. *P1, P2, P3*: Delay plates. W1: $\lambda/2$ wave-plate. *BS1, BS2, BS3*: Beam splitters. BS1, BS2 are 50:50 beam splitters, while BS3 has 20% reflectivity and 80% transmission. The beam coming out of the IPG was focused by a lens into a pulsed gas jet where the XUV radiation was generated

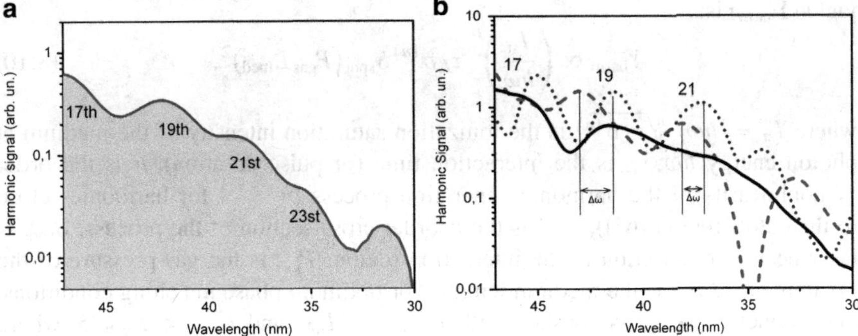

Fig. 8.13 Generation of a continuum XUV spectrum by using the DMZ-IPG device. (**a**) Continuum XUV spectrum close to the cutoff region, generated by the central part of the ElMo pulse, that is defined by the time window $\tau_g \sim 5$ fs (GATE ON). The spectrum is an average of 300 shots. (**b**) The dependence of the XUV spectrum on the CEP variation of the laser pulse is reflected in the different structure of the spectra and the frequency shift of the harmonic peaks between the single shot measurements. *Solid black line*: XUV spectrum generated during a small fraction of the highest amplitude optical cycle of the driving laser field. *Red dashed and blue dotted lines*: XUV spectra, each one generated by a driving laser field with different CEPs values

cannot be separately adjusted. The two pulses synthesizing the first field and those synthesizing the second field have the same (variable) delay; however, this is not a critical factor.

After finding the proper parameter values of I_{c0}/I_{d0} and $\delta, \Delta t$ for $\tau_g \sim 5$ fs, measurements as those in Fig. 8.11 have been also performed by using the DMZ arrangement (Fig. 8.13). The incoming pulse is split into two by a 50:50 beam splitter (BS1). The plates (P1, P2) introduce a delay δ between the pulses, which equals a multiple of the half laser period. The two pulses are recombined at the beam combiner (BS2). By properly setting the value of δ, one of the pulses after the (BS2) possesses at its center constructive, while the other one has destructive interference at its center. The plate (P3) and the $\lambda/2$ wave-plate (W1) introduce a $\pi/2$ phase difference and set perpendicular polarizations between the two fields reflected and transmitted from the (BS2). The two pulses are recombined in a 20% reflection and 80% transmission beam splitter (BS3). The results on the generation of coherent quasicontinuum XUV radiation and the CEP variation measurements (Fig. 8.13) are similar as in the case of using DMI arrangement (Fig. 8.11).

8.3.3.4 Coherent Continuum XUV Radiation in the Sub-100 nJ Range Generated by a High-Power Many Cycle Laser Field Utilizing an DMZ-IPG Device

The central challenge for the production of high intensity isolated *asec* pulses is the conversion efficiency of the XUV generation process, which has to be substantially increased. It is well known that the above-ionization-threshold harmonic generation yield (Y_{harm}) is:

$$Y_{\text{harm}} \propto \left(\frac{I_{\text{st}}}{\hbar\omega}\right)^n \tau_L \sigma^{(n)} S_{\text{spot}} (P_{\text{gas}} L_{\text{med}})^2, \tag{8.10}$$

where $I_{\text{st}} = \hbar\omega / \sqrt[n]{\tau_L \sigma^{(n)}}$ is the ionization saturation intensity of the medium at photon energy $\hbar\omega$, τ_L is the interaction time (or pulse duration), n is the order of nonlinearity of the harmonic generation process ($n \sim 4$ for harmonics close to the cutoff region [67]), $\sigma^{(n)}$ is the n-order cross-section of the process, S_{spot} is laser beam cross-section in the interaction region, P_{gas} is the gas pressure of the medium and L_{med} is the medium length. For optimum phase matching conditions, the product $P_{\text{gas}} L_{\text{med}}$ is constant with $L_{\text{med}} > 3L_{\text{abs}}$ and $L_{\text{abs}} < L_{\text{coh}}/5$, where $L_{\text{abs}} = (\sigma_{\text{ion}}^{(1)} \rho)^{-1}$ is the XUV absorption length, $L_{\text{coh}} = \pi / |\Delta \boldsymbol{k}|$ is the coherence length, $\sigma_{\text{ion}}^{(1)}$ is the single-photon ionization cross-section, ρ is the atomic density of the medium and $\Delta \boldsymbol{k} = \boldsymbol{k}_q - q\boldsymbol{k}_L$ is the wavevector difference of the harmonic q and the polarization driving fields [68–70].

Based on (a) (8.10), (b) Fig. 8.7, and (c), the measured conversion efficiency values for Xe, Ar and Ne gases in case of using 60 fs pulses [71], the energy of the emitted quasicontinuum XUV radiation generated by a many-cycle laser field using the DMZ-IPG device in a loose focusing configuration, can be estimated (Table 8.1).

In the following, we summarize the results obtained in generating high-energy coherent continuum XUV radiation exploiting the above-described DMZ-IPG device. High-energy coherent continuum XUV radiation has been generated in the spectral region around 25 nm, with a spectral width supporting single pulses of 260 as duration and energy in the range of tens of nano-joules. The radiation is generated by the interaction of Xe, Kr, or Ar gas with a high-power many-cycle laser field in a loose focusing configuration by means of a DMZ-IPG arrangement [44] (Fig. 8.14). The IPG output beam was focused 3.5 m after the lens into a pulsed gas jet, of Xe, Kr, or Ar, where the XUV radiation was generated. After the jet, an Si plate with a thin oxidized layer [72] was placed at Brewster's angle of 75° of the fundamental, reflecting the harmonics [73] toward the detection area while reducing their IR content. A 3 mm diameter aperture was placed after the Si plate transmitting the central part of the XUV and residual IR beam [74].

Table 8.1 Estimated energies of the emitted coherent quasicontinuum XUV radiation generated by a 50 fs pulse using the DMZ-IPG device in a loose focusing configuration. $E_{\text{cut-off}}$: Cutoff energy of the XUV radiation, E_{in}: Laser pulse energy entering the IPG device, D: Laser beam diameter in the interaction region, E_{g}: Driving field energy content within the τ_{g}, and E_{XUV}: energy of the continuum part of the emitted XUV radiation

DMZ-IPG ($\tau_g \sim 5$ fs)	Xe	Kr	Ar	Ne	He
I_{st} (W/cm^2)	$\sim 1.7 \times 10^{14}$	$\sim 2.7 \times 10^{14}$	$\sim 3 \times 10^{14}$	$\sim 7 \times 10^{14}$	$\sim 2 \times 10^{15}$
$E_{\text{cut-off}}$ (eV)	44	67	72	154	403
E_{in}(mJ) ($\tau_L = 50$ fs)	119	188	212	488	1375
E_{g}(mJ) ($D = 0.5$ mm)	1.9	3	3.4	7.8	22
E_{XUV}(nJ)	~38	~36	~41	~0.4	<0.4

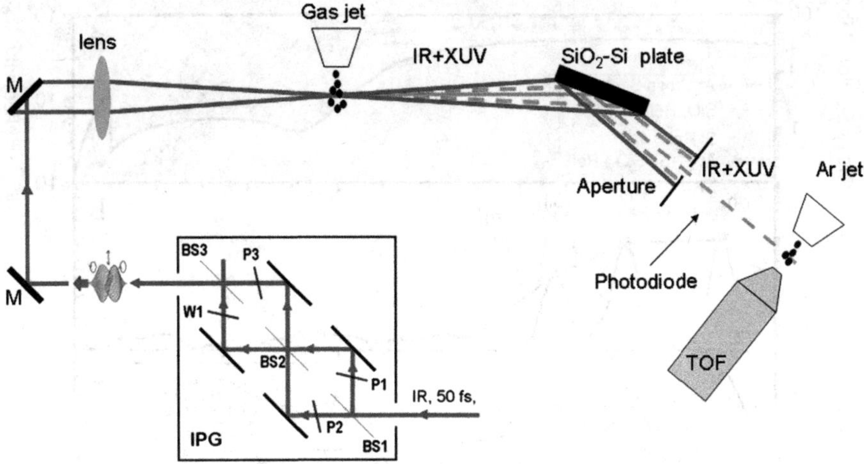

Fig. 8.14 IPG apparatus based on dual Mach–Zehnder arrangement [44]. P1, P2; W1; and BS1, BS2, BS3 are the delay plates, $\lambda/2$ wave-plates, and beam splitters, respectively. BS1, BS2 are 50/50 beam splitters, while BS3 has 20% reflectivity and 80% transmission

Figure 8.15a mimics the reflectivity of the Si plate with the oxidized surface layer (gray dotted–dashed curve) used, arbitrarily taking the average reflectivity of Si (gray dotted curve) and the SiO_2 (gray dashed curve) plate. The transmitted unfocused XUV beam was ionizing an Ar gas. The electrons produced by the interaction of Ar atoms with the XUV radiation were detected by a μ -metal-shielded time of light (TOF) spectrometer. The spectral intensity distribution of the XUV radiation was obtained by measuring the single-photon ionization photoelectron (PE) spectra induced by the XUV radiation with photon energy higher than the ionization potential (E_{IP}) of Ar ($E_{IP}(Ar) = 15.76\,eV$). The single-photon ionization cross section of Ar [75] is shown with a black solid curve in Fig. 8.15a. To obtain the XUV intensity distribution at the interaction region, the PE spectrum has been corrected with respect to the TOF collection solid angle as well as to the photoionization cross section of Ar.

Figure 8.15b shows the normalized harmonic intensity distribution generated in Xe (black solid curve) and Kr (gray solid curve) by a long laser pulse formed by the constructive interference arm of the IPG (gating off). For each of the PE traces, 400 shots were accumulated. Well-resolved harmonics up to 17th and 21st order have been recorded for Xe and Kr, respectively. Turning the gating process on, by using both arms of the IPG with relative phase $\Delta\varphi = \pi/2$ between the constructive and the destructive interference fields and choosing the parameter range such as to be in the shaded area of Fig. 8.8a and b, a quasicontinuum spectrum spanning from 80 to 15 nm has been recorded (Fig. 8.15c) and corrected (inset of Fig. 8.15c) for Xe (black solid curve) and Kr (gray filled area). In the range of 40–15 nm, the spectrum becomes a continuum and carries substantial energy only when the gating process is on. Since the laser used is not CEP stabilized, the latter result does not

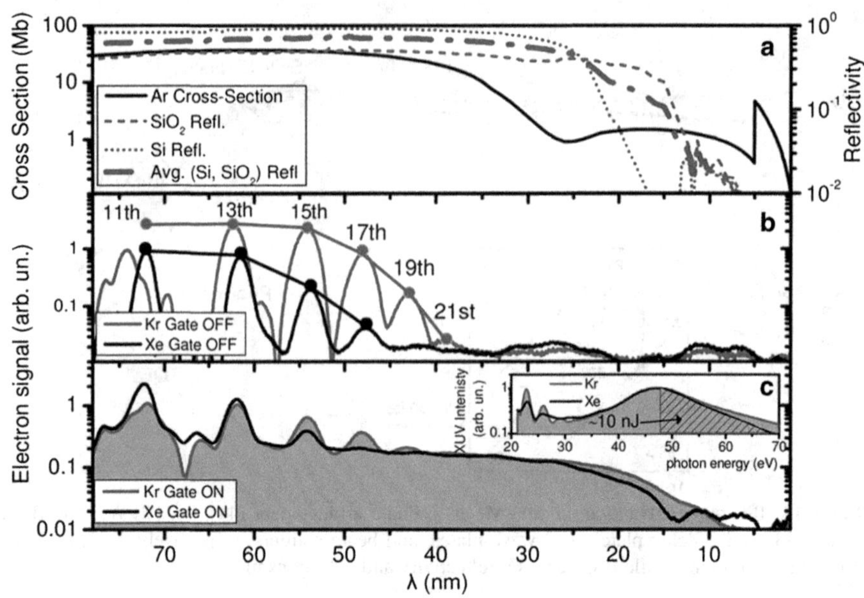

Fig. 8.15 (**a**) *Black solid line*: Single-photon ionization cross-section of Ar. *Gray dashed-doted line, gray doted line and gray dashed line*: Reflectivity of the oxidized Si, Si, and SiO_2 plate, respectively. (**b**) *Black solid line and gray solid line*: Harmonic spectrum generated in Xe and Kr, respectively, by a linear polarized pulse. (**c**) *Black solid line and gray filled area*: Quasicontinuum spectrum generated in Xe and Kr gas when a temporal gate of $\tau_g \sim 4.5$ fs duration is on. Inset in (**c**) shows the spectrum corrected for the wafer reflectivity and Ar cross-section [44]

warrant emission of isolated *asec* pulses [35]. Nevertheless, single-*asec* pulses can be isolated, by using an appropriate multilayer mirror selecting the continuum part of the radiation in the cutoff energy region (line shaded area in inset). In principle, the radiation in the cutoff region, with an energy spread of >15 eV could support single pulses with duration of 260 as. By changing the relative phase between the constructive and the destructive interference fields to $\Delta\varphi = 0$, a discrete harmonic spectrum appears. When $\Delta\varphi = \pi/2$, the intensity of the driving field in a gate $\tau_g \sim 4.5$ fs is estimated to be $\sim 5 \times 10^{14}$ W/cm^2. According to the cutoff law, and for $I_{st} \sim 1.7 \times 10^{14}$ and $\sim 3 \times 10^{14}$ W/cm^2, $E_{cut-off}$ for Xe and Kr is estimated to be ~44 and ~67 eV, respectively. $E_{cut-off}$ for Xe is fairly close to the experimental one, while for Kr it cannot be detected owing to the low reflectivity of the Si wafer and the Ar cross-section. However, the higher cutoff energy of Kr as compared with Xe is indicated in the experimental data by the observed slightly higher PE signal recorded at photon energies >50 eV. The spectra using Ar as the harmonic generation medium have also been recorded but are not shown here, as they are almost identical to those of Kr. A notable advancement achieved by the results of this work is the high photon flux of the continuum radiation. The energy measurement is performed by a calibrated XUV photodiode introduced directly after the 3 mm aperture. The energy in the range of 80–15 nm is found to be ~90 nJ, while that of the continuum (>30 eV) ~20 nJ and corresponds to a conversion efficiency of 2×10^{-5} relative to the energy of the IR field within the gate. This leads to energy of ~10 nJ in the cutoff region (shaded area in inset of Fig. 8.15c). These results are in reasonable agreement with the estimated values of Table 8.1. The energy of the continuum is determined by subtracting from the diode signal the contribution of the signal induced by photons with energy of <30 eV. The latter is estimated from the diode signal with and without gating and the corresponding PE signal. The contribution of the IR to the diode signal has been measured by switching off the harmonic production jet.

The measured ~10 nJ energy confined into temporal intervals of the order of ~300 as leads to a power of about ~20 TW that could lead to focused intensities of the order of 10^{14} W/cm^2. In a more realistic estimate, single pulses of ~260 as duration and intensities up to 10^{13} W/cm^2 can be reached by using an appropriate multilayer mirror. These conditions are sufficient for the observation of two-photon processes induced by isolated *asec* pulses, enormously facilitating *asec* scale XUV-pump-probe experiments.

8.4 Comparison Between the DMI-IPG, DMZ-IPG and WP-PG, Two-Color PG Approaches

A comparison between the IPG and wave-plates polarization gating (WP-PG) approaches is summarized in Fig. 8.16. The gate intensity content I_g/I_{in} (where I_g is the intensity of the driving laser field within τ_g and I_{in} is the total incoming intensity, respectively) drops fast with increasing pulse duration in the wave-plate method, while for both IPG set-ups the reduction is rather slow. Above 30 fs both IPG methods have a higher I_g/I_{in} ratio than the wave-plate approach, while the Mach–Zehnder setup remains superior for any pulse duration. Given the fact that currently I_{in} is much higher in many-cycle than in few-cycle pulses, both IPG approaches secure much higher I_g values and thus more intense *asec* pulses. To give an example from a 20TW system (e.g. 1 J, 40 fs), one may reach gate power content ~2TW (10 mJ, 5 fs) with low B-integral value arrangements. A necessary prerequisite for that is that the target medium is not depleted. For atomic targets ionization from the elliptically polarized part of the pulse may deplete the target. This problem may be avoided by reducing the intensity while keeping the power high. This is always possible by using looser focusing and enlarged target areas. Finally, the target depletion problem is eliminated when using solid targets as in the generation of harmonics from surface plasma for which the IPG method is highly beneficial [76].

The recently developed two-color PG approaches [41, 58, 59] have intensity content in the gate comparable to the DMZ-IPG. Although these approaches are collinear and thus more compact, a detailed study is needed to evaluate the influence of the propagation of the two-color field in the optical arrangement. Also, they

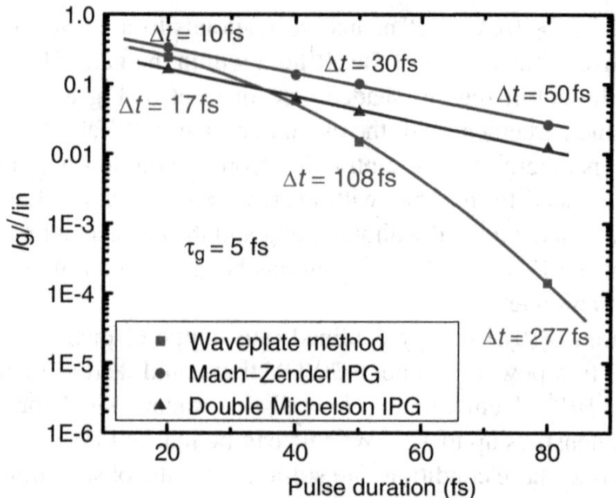

Fig. 8.16 Comparison of the gate intensity content I_g/I_{in} (where I_g is the driving field intensity content within the 5 fs gate and I_{in} is the input laser intensity) for different polarization gating setups as a function of the pulse duration. The *green, blue, and red lines* correspond to DMZ-IPG, DMI-IPG and wave-plates method, respectively. Δt in the IPG approaches correspond to the delay value between the two pulses in the pairs and the Δt in the wave-plates approach correspond to the delay value between the two perpendicular polarized pulses [45]

Table 8.2 Comparison between DMI-IPG and DMZ-IPG approaches. The dots depict the advantage of each method compared to the other. The numbers in the intensity content (I_g/I_{in}) correspond to the case of a 50 fs laser pulse. The general relation applying to any pulse duration is that $(I_g/I_{in})_{DMZ} = 2(I_g/I_{in})_{DMI}$

	Intensity content (I_g/I_{in}) in τ_g	Degrees of freedom for τ_g control
DMI	0.08	Full control of each pulse in the DMI (•)
DMZ	0.16 (•)	Control of $\delta = \tau$ and E_{c0}/E_{d0}

may suffer from limitations due to the larger the B-integral value when utilizing high-peak-power many-cycle pulses (see Fig. 8.5).

Table 8.2 summarizes the comparison between the DMI-IPG and DMZ-IPG approaches by presenting with dots the advantages of each method.

8.5 Summary

In this article, we have reviewed a recently developed novel approach, for the generation of intense coherent continuum broadband XUV radiation using high peak-power many-cycle laser pulses. Intense isolated *asec* pulses are superior compared to *asec* pulse trains for the realization of studies of ultrafast dynamics.

Studies of dynamics that are based on time domain spectroscopy rely on pump-probe approaches and thus on, at least, two-photon transitions. Thus high XUV intensity is required for the realization of XUV-pump-XUV-probe schemes. Isolated pulses greatly facilitate the experiment as they keep the two steps of the pump-probe sequence clearly separable. To the challenging quest of the generation of high-intensity isolated *asec* pulses, there are two possible solutions. Either to develop high-power few-cycle laser systems or to use the already commercially available high-power many-cycle laser systems combined with polarization gating techniques. This work focuses on the second approach. It is a new version of what is known as polarization gating, which can be applied, in contrast to the previous ones, to high-peak-power many-cycle laser pulses. The principle of this IPG technique has been demonstrated theoretically and experimentally in gas phase media. XUV bandwidths large enough to enable the synthesis of isolated XUV pulses, with durations of a few hundred *asec*, have been achieved. The energy of the emitted XUV continua is of the order of tens of nJ/pulse, the highest so far demonstrated XUV continuum energy. The performance of the technique may be greatly improved, reaching the pulse energy level of 1μ J/pulse by using larger focal lengths in the driving field and properly arranged successive gas sources [77]. An important finding, which resulted from applying the technique, is its sensitivity to the CEP of the driving pulse. The shot-to-shot emitted XUV spectra strongly depend on the CEP. The technique provides the means for the generation of intense isolated *asec* pulses tuneable in duration and frequency. It can be further used for CEP variation studies and to monitor many-cycle driving fields and offers exciting opportunities for time delay spectroscopy [78] and multiphoton XUV-pump-XUV-probe studies [79] in all states of matter.

Acknowledgements This work is supported in part by the Ultraviolet Laser Facility (ULF) operating at FORTH-IESL (contract HPRI-CT-2001-00139), the ELI research infrastructure preparatory phase program, the FASTQUAST ITN, the ATTOFEL ITN, and the FLUX program (PIAPP-GA-2008-218053) of the 7th FP.

References

1. P. Agostini, L.F. DiMauro, The physics of attosecond light pulses, J. Phys. B **67**, 813 (2004)
2. P.B. Corkum, F. Krausz, Attosecond science, Nat. Phys. **3**, 381 (2007)
3. F. Krausz, M. Ivanov, Attosecond physics, Rev. Mod. Phys. **81**, 163 (2009)
4. U. Teubner, P. Gibbon, High-order harmonics from laser-irradiated surfaces, Rev. Mod. Phys. **81**, 445 (2009)
5. N.A. Pappadogiannis et al., Observation of attosecond light localization in high order harmonic generation, Phys. Rev. Lett. **83**, 4289 (1999)
6. P.M. Paul et al., Observation of train of attosecond pulses from high harmonic generation, Science **292**, 1689 (2001)
7. P. Tzallas et al., Direct observation of attosecond light bunching. Nature **426**, 267 (2003)
8. Y. Mairesse et al., Attosecond Synchronization of High-Harmonic Soft X-rays, Science **302**, 1540 (2003)
9. Y. Nabekawa et al., Conclusive evidence of an attosecond pulse train observed with the mode-resolved autocorrelation technique, Phys. Rev. Lett. **96**, 083901 (2005)

10. J.E. Kruse et al., The challenge of attosecond pulse metrology, (arXiv:1005.0568v1)
11. P.B. Corkum, Plasma prospective on strong-field multiphoton ionization, Phys. Rev. Lett. **71**, 1994 (1993)
12. M. Lewenstein et al., Theory of high harmonic generation by low-frequency laser fields, Phys. Rev. A **49**, 2117–2132 (1994)
13. Y. Nomura et al., Attosecond phase locking of harmonics emitted from laser-produced plasmas, Nat. Phys. **5**, 124 (2009)
14. R. Hörlein et al., Temporal characterization of attosecond pulses emitted from solid-density plasmas, New J. Phys. **12**, 043020 (2010)
15. F. Quéré et al, Coherent Wake Emission of High-Order Harmonics from Overdense Plasmas, Phys. Rev. Lett. **96**, 125004 (2006)
16. A. Tarasevitch et al, Transition to the Relativistic Regime in High Order Harmonic Generation, Phys. Rev. Lett. **98**, 103902 (2007)
17. F. Quéré et al,Phase Properties of Laser High-Order Harmonics Generated on Plasma Mirrors, Phys. Rev. Lett. **100**, 095004 (2008)
18. C. Thaury et al, Coherent dynamics of plasma mirrors, Nat. Phys. **4**, 631 (2008)
19. S.V. Bulanov et al.,Interaction of an ultrashort, relativistically strong laser pulse with an overdense plasma, Phys. Plasmas **1**, 745 (1994)
20. G.D. Tsakiris et al., Route to intense single attosecond pulses, New J. Phys.**8**, 19 (2006)
21. T. Baeva et al., Theory of high-order harmonic generation in relativistic laser interaction with overdense plasma, Phys. Rev. E **74**, 046404 (2006)
22. B. Dromey et al., High harmonic generation in relativistic limit, Nat. Phys. **2**, 456 (2006)
23. S. Gordienko et al., Relativistic Doppler Effect: Universal Spectra and Zeptosecond Pulses, Phys. Rev. Lett. **93**, 115002 (2004)
24. R. Lichters et al., Short-pulse laser harmonics from oscillating plasma surfaces driven at relativistic intensity, Phys. Plasmas **3**, 3425 (1996)
25. N.A. Papadogiannis et al., Two XUV-photon ionization of He through a superposition of higher harmonics, Phys. Rev. Lett. **90**, 133902 (2003)
26. Y. Nabekawa et al., Production of doubly charged helium ions by two-photon absorption of an intense sub-10-fs soft X-ray pulse at 42 eV photon energy, Phys. Rev. Lett. **94**, 043001 (2005)
27. N. Miyamoto et al., Observation of Two-Photon Above-Threshold Ionization of Rare Gases by xuv Harmonic Photons, Phys. Rev. Lett. **93**, 083903 (2004)
28. E.P. Benis et al., Two-photon double ionization of rare gases by a superposition of harmonics, Phys. Rev. A **74**, 051402(R) (2006)
29. E.P. Benis et al., Frequency-resolved photoelectron spectra of two-photon ionization of He by an attosecond pulse train, New J. Phys. **8**, 92 (2006)
30. L.A.A. Nikolopoulos et al., Second order autocorrelation of an XUV attosecond pulse train, Phys. Rev. Lett. **94**, 113905 (2005)
31. M. Hentschel et al., Attosecond metrology, Nature **414**, 509 (2002)
32. R. Kienberger et al., Atomic transient recorder, Nature **427**, 817 (2004)
33. G. Sansone et al., Isolated single-cycle attosecond pulses, Science **314**, 443 (2006)
34. E. Goulielmakis et al., Single-Cycle Nonlinear Optics, Science **320**, 1614 (2008)
35. A. Baltuska et al., Attosecond control of electronic processes by intense light fields, Nature **421**, 611 (2003)
36. P. Tzallas et al., Generation of intense continuum extreme-ultraviolet radiation by many-cycle laser fields, Nat. Phys. **3**, 846 (2007)
37. T. Wittman et al., Single-shot carrier-envelope phase measurement of few-cycle laser pulses, Nat. Phys. **5**, 367 (2009)
38. G.G. Paulus et al., Absolute-phase phenomena in photoionization with few-cycle laser pulses, Nature **414**, 182 (2001)
39. P.B. Corkum, N.H. Burnett, M.Y. Ivanov, Subfemtosecond pulses, Opt. Lett. **19**, 1870 (1994)
40. I.J. Sola et al., Controlling attosecond electron dynamics by phase-stabilized polarization gating, Nat. Phys. **2**, 319 (2006)
41. H. Mashiko et al., Double Optical Gating of High-Order Harmonic Generation with Carrier-Envelope Phase Stabilized Lasers, Phys. Rev. Lett. **100**, 103906 (2008)

42. V.V. Strelkov, E. Mevel, E. Constant, Generation of isolated attosecond pulses by spatial shaping of a femtosecond laser beam, New J. Phys. **10**, 083040 (2008)
43. C. Altucci et al., Single isolated attosecond pulse from multicycle lasers, Opt. Lett. **33**, 2943 (2008)
44. E. Skantzakis et al., Coherent continuum extreme ultraviolet radiation in the sub-100-nJ range generated by a high-power many cycle laser field, Opt. Lett. **34**, 1732 (2009)
45. D. Charalambidis et al., Exploring intense attosecond pulses, New J. Phys **10**, 025018 (2008)
46. D. Oran et al., Efficient polarization gating of high-order harmonic generation by polarization-shaped ultrashort pulses, Phys. Rev. A **72**, 063816 (2006)
47. A. Peralta Conde et al., Realization of time-resolved two-vacuum-ultraviolet-photon ionization, Phys. Rev. A **79**, 061405(R) (2009)
48. Y. Kobayashi et al., 27-fs extreme ultraviolet pulse generation by high order harmonics, Opt. Lett. **23**, 64 (1998)
49. P. Antoine et al., Theory of high-order harmonic generation by an elliptically polarized laser field, Phys. Rev. A **53**, 1725 (1996)
50. K.S. Budil et al., Influence of ellipticity on harmonic generation, Phys. Rev. A **48**, R3437 (1993)
51. N.L. Manakov, V.D. Ovsyannikov, Zh. Eksp. Teor. Fiz. 79, 1769 (1980)
52. V. Strelkov et al., Single attosecond pulse production with an ellipticity-modulated driving IR pulse, J. Phys. B **38**, L161 (2005)
53. V. Strenkov et al., Generation of attosecond pulses with ellipticity-modulated fundamenta, Appl. Phys. B **78**, 879 (2004)
54. O. Tcherbakoff et al., Time-gated high-order harmonic generation, Phys. Rev. A **68**, 043804 (2003)
55. V.T. Platonenko, V.V. Strelkov, Single attosecond soft-x-ray pulse generated with a limited laser beam, J. Opt. Soc. Am. B **16**, 435 (1999)
56. S. Witte et al., A source of 2 terawatt, 2.7 cycle laser pulses based on noncollinear optical parametric chirped pulse amplification, Opt. Express **14**, 8168 (2007)
57. D. Herrmann et al., Generation of sub-three-cycle, 16 TW light pulses by using noncollinear optical parametric chirped-pulse amplification, Opt. Lett. **34**, 2459 (2009)
58. X. Feng et al., Generation of isolated attosecond pulses with 20 to 28 femtosecond lasers, Phys. Rev. Lett. **103**, 183901 (2009)
59. E.J. Takahashi et al., Infrared Two-Color Multicycle Laser Synthesis for Generating an Intense Attoseond Pulse, Phys. Rev. Lett. **104**, 233901 (2010)
60. S.G. Rykovanov et al., Intense single attosecond pulses from surface harmonics using the polarization gating, New J. Phys. **10**, 025025 (2008)
61. G. Sansone, Quantum path analysis of isolated attosecond pulse generation by polarization gating Phys. Rev. A **79**, 053410 (2009)
62. P. Antoine et al., Theory of high-order harmonic generation by an ellipticity polarized laser field, Phys. Rev. A **53**, 1725 (1996)
63. D.B. Milosevic, W. Becker, Role of long quantum orbits in high-order harmonic generation, Phys. Rev. A **66**, 063417 (2002)
64. M.B. Gaarde, K.J. Schafer, Space-Time Considerations in the Phase Locking of High Harmonics, Phys. Rev. Lett. **89**, 213901 (2002)
65. M. Belini et al., Temporal Coherence of Ultrashort High-Order Harmonic Pulses, Phys. Rev. Lett. **81**, 297 (1998)
66. G. Sansone et al., Observation of Carrier-Envelope Phase Phenomena in the Multi-Optical-Cycle Regime, Phys. Rev. Lett. **92**, 113904 (2004)
67. G. Wahlstrom et al., High-order harmonic generation in rare gases with an intense short-pulse laser, Phys. Rev. A **48**, 4709, (1993)
68. E. Constant et al., Optimizing high harmonic generation in absorbing gases: Model and experiment Phys. Rev. Lett. **82**, 1668 (1999)
69. E. Takahashi et al., Generation of highly coherent submicrojoule soft x rays by high-order harmonics Phys. Rev. A **66**, 021802(R) (2002)

70. T. Ditmire et al., Energy-yield and conversion efficiency measurements of high-order harmonic radiation Phys. Rev. A **51**, R902 (1995)
71. J.F. Hergott et al., Extreme-ultraviolet high-order harmonic pulses in the microjoule range Phys. Rev. A **66**, 021801(R) (2002)
72. F. Lukes, Oxidation of Si and GaAs in air at room temperature, Surf. Sci. **30**, 91 (1972)
73. E.J. Takahashi et al., High-throughput, high-damage-threshold broadband beam splitter for high-order harmonics in the extreme-ultraviolet region, Opt. Lett. **29**, 507 (2004)
74. V.T. Platonenko, V.V. Strelkov, Generation of a single attosecond x-ray pulse, Quantum Electron. **28**, 749 (1998)
75. G.V. Marr, J.B. West, Atomic Data Nucl. Data Tables **18**, 497, (1976)
76. T. Baeva, S. Gordienko, A. Pukhov, Relativistic plasma control for single attosecond x-ray burst generation, Phys. Rev. E **74**, 065401 (2006)
77. J. Seres et al., Coherent superposition of laser-driven soft-X-ray harmonics from successive sources, Nat. Phys. **3**, 878 (2007)
78. E. Skantzakis et al., Tracking autoionazing-wavepacket dynamics at 1-fs time scale Phys. Rev. Lett. **105**, 043902 (2010)
79. P. Tzallas et al., Extreme-ultraviolet pump-probe studies of one femtosecond scale electron dynamics arXiv:1103.0873 (2011)

Chapter 9
Relativistic Laser Plasmas for Electron Acceleration and Short Wavelength Radiation Generation

A. Pukhov, D. an der Brügge, and I. Kostyukov

Abstract We consider here a few options to use relativistic laser plasmas for novel sources of short wavelength radiation. Electrons accelerated in underdense plasmas in the bubble regime wiggle in an ion channel. This leads to broadband incoherent synchrotron-like radiation bursts, which are of femtosecond duration. The photon energies are in kilo electron volt (keV) to mega electron volt (MeV) energy range; however, this radiation is not coherent. To reach coherency, the electron bunch must have structure at the wavelength of the emitted x-rays. This can be achieved, in principle, by sending the laser-accelerated electron bunch through an external wiggler. However, to reach free electron lasing in the x-ray regime, the energy spread of the laser-accelerated electrons must be reduced dramatically. Another option is to use high harmonic generation at overdense plasma boundaries. The laser-driven plasma surface oscillates at relativistic velocities and severely alters the frequency of the reflected laser light. The high harmonics are emitted in coherent subfemtosecond flashes. The theory of harmonics generation in the relativistic regime predicts a power law energy spectrum with an exponent $-8/3$. However, for short laser pulses and high intensities, the electrons self-organize in nanobunches that lead to coherent synchrotron emission. The power law harmonic spectrum can become very flat in this case with the exponent as low as $-6/5$. This can make the high harmonics potentially the brightest laser-driven short wavelength sources with unique properties.

A. Pukhov (✉) and D. an der Brügge
Institute for theoretical physics I, University of Duesseldorf, 40225 Duesseldorf, Germany
e-mail: pukhov@tp1.uni-duesseldorf.de, dadb@tp1.uni-duesseldorf.de

I. Kostyukov
Institute for applied physics, Nizhni Novgorod, Russia
e-mail: kost@appl.sci-nnov.ru

K. Yamanouchi et al. (eds.), *Progress in Ultrafast Intense Laser Science VII*, Springer Series in Chemical Physics 100, DOI 10.1007/978-3-642-18327-0_9,

9.1 Introduction

One of the most important future applications for relativistic laser plasmas is the development of novel sources of short wavelength radiation. Because the driving laser pulses are of femtosecond duration, one may expect that the generated flashes of x-rays also can be potentially made extremely short. The conversion of the laser pulse energy into the short wavelength radiation in plasmas is always mediated by electrons. The ultra strong laser and plasma fields accelerate electrons to relativistic energies. Then, these electrons wiggle in the transverse fields and radiate. Depending on the size of the radiating electron bunch or layer and the relative phasing of the individual electrons, the radiation may have different coherency properties.

When a relativistically intense, $I\lambda^2 \gg 1.37\times10^{18}$ Wμm^2/cm^2, laser pulse propagates in underdense plasmas, its ponderomotive force is strong enough to expell all electrons from the filament occupied by the laser. Here, I is the laser intensity, and λ is its wavelength. As a consequence, an ion channel is formed that traps the accelerated electrons transversely.

When the laser pulse duration T is long in comparison with the plasma period $2\pi/\omega_p$ so that $\omega_p T \gg 1$, the accelerated electrons and the laser fields are overlapped in space. Here, $\omega_p = \sqrt{4\pi n e^2/m}$ is the background plasma frequency, and n is the electron density. Under these conditions, the betatron resonance between the transverse oscillation frequency and the Doppler-downshifted frequency of the electromagnetic field is reached [1]. Swinging around the betatron resonance, the electrons gain energy from the laser pulse and attain quasithermal spectra with multi-MeV temperatures. The high energy tail of the electron distribution will then radiate x-rays due to the transverse oscillatory motion. However, because of the low number of electrons in the high energy tail of the Boltzmann-like spectrum, the number of expected x-ray photons is small in this case.

When the intense laser pulse is short, so that $\omega_p T < 1$, a bubble is formed [2] that leads to quasimonoenergetic acceleration of electron bunches [3–5]. The energy conversion efficiency from the laser pulse to the electron beam can be as high as 20% [6]. The electrons propagate slightly out of axis. Thus, they oscillate transversely in the bubble fields and radiate x-rays [7–10]. The radiation is synchrotron-like, broadband, and incoherent. The incoherency of the radiation in this regime is caused by the absense of regular structure in the electron bunch at the x-ray wavelength.

One option to generate coherent x-rays and to increase the source brightness is to feed the laser-accelerated electron bunch into an external magnetic wiggler [11]. This approach has the potential to reduce the size of x-ray free electron lasers (x-FELs) dramatically. However, before this scheme can be realized, the energy spread of the laser-generated electron bunches still must be reduced dramatically.

Another path toward coherent x-ray radiation is to employ ultra-thin targets, such as nanoscale diamond foils [12]. In the case, the fine structure of the target could be preserved during the laser-driven acceleration, coherent short wavelength radiation will be emitted coherently. One of the main problems in this scheme is the fast Coulomb explosion of the target electron layer as soon as the laser pulse separates the electrons from heavier ions.

Finally, high harmonics generated at the surfaces of solid targets when irradiated by relativistically intense lasers are a potentially bright source of coherent XUV and x-ray radiation. Here, the coherency is defined by the sharp boundary between the laser pulse and the plasma. The resulting spectra have a power law decay. When the boundary remains step-like during the interaction, harmonics have the universal spectrum revealed by Baeva, Gordienko, and Pukhov [13]. The BGP spectrum is a power law with the exponent $-8/3$, smoothly rolling off into an exponential decay at some frequency scaling as $I^{3/2}$. This type of spectrum could be observed experimentally [28]. The roll-off frequency scaling was also confirmed [14].

However, the BGP spectrum is not the flattest one. Recently, it has been shown that when the driving laser is ultra-short, the plasma electrons can self-organize into nanobunches and lead to coherent synchrotron emission (CSE) with power law spectra as flat as $I_n \sim n^{-6/5}$ [15]. This CSE regime of relativistic harmonic generation might lead to a novel bright coherent source of short wavelength radiation.

Below, we shortly review all these mechanisms and compare their potential.

9.2 Synchrotron Emission from Relativistic Electrons in the Bubble Regime

The plasma-based accelerators support fields that are many orders of magnitude higher than those of the convensional accelerators. This allows to accelerate particles to high energies in extremely short distances. Especially efficient in electron acceleration is the so-called bubble regime [2]. This regime is reached when the relativistic amplitude of the laser pulse $a = eA/mc^2 \gg 1$, and the pulse duration is short, $\omega_p T < 1$. Here, A is the laser vector potential. In this case, the ponderomotive force of the laser creates a cavity – the bubble – free of background electrons. The electrons flow along the boundary of the bubble and some of them get trapped at the end of the cavity, Fig. 9.1a. The electrons are accelerated by the longitudinal field. At the same time, they wiggle in the transverse fields of the bubble and emit synchrotron radiation.

The electromagnetic fields inside the relativistic cavity are [16]

$$\begin{aligned} \frac{eE_x}{mc\omega_0} &= \frac{k_p\xi}{2}, \\ \frac{eE_y}{mc\omega_0} &= -\frac{eB_z}{mc\omega_0} = \frac{k_p y}{4}, \\ \frac{eE_z}{mc\omega_0} &= \frac{eB_y}{mc\omega_0} = \frac{k_p z}{4}, \end{aligned} \tag{9.1}$$

where ω_0 is the laser frequency, $k_p = \omega_p/c$ is the plasma wavenumber, and $\xi = x - ct$.

To estimate the x-ray emission from the bubble, we have to understand the trapping process. Recently, it has been shown that the trapping in the bubble is a

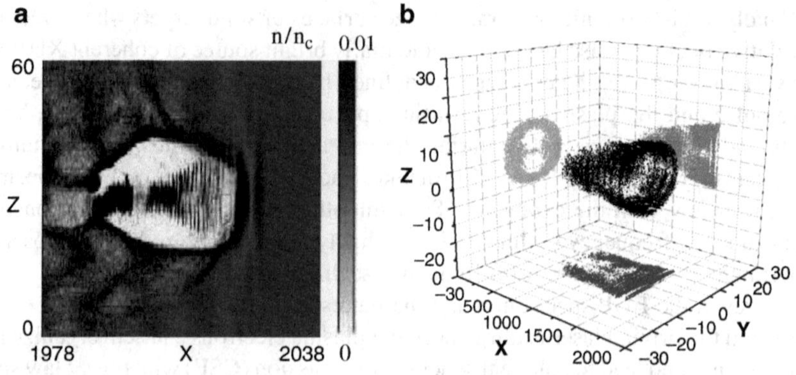

Fig. 9.1 (**a**) The distribution of the electron density in the plane $x - z$ as obtained from the 3D PIC simulation. (**b**) The distribution of initial positions of accelerated electrons with energy above 75 MeV at the moment $t = 2{,}000$. The x-axis is the direction of the laser propagation. The coordinates and time are given in the laser wavelength and laser period, the density is given in the critical plasma density, respectively

continuous process and is related to the dynamic development of the cavity [17–19]. In a 3D PIC simulation, we have recorded initial positions of the trapped electrons. It appears that the electrons are trapped out of a thin ring-like structure with the radius close to that of the bubble, Fig. 9.1b.

The simulation shown in Fig. 9.1 is done for a circularly polarized laser pulse with the envelope $a = a_0 \exp\left(-r^2/r_L^2\right) \cos\left(\pi t/T_L\right)$, where $T_L = 30\,\mathrm{fs}$ is the pulse duration, $r_L = 9\mu\mathrm{m}$ is the focused spot size and $a_0 \equiv eA/mc^2 = 1.5$ is the normalized vector potential, which corresponds to the laser intensity $I = 3 \times 10^{18}\,\mathrm{W/cm^2}$, $\lambda = 0.82\mu\mathrm{m}$ is the laser wavelength. The plasma density is $n_0 = 1.16 \times 10^{19}\,\mathrm{cm^{-3}}$. The electron density in Fig. 9.1a is shown when the laser pulse has passed the distance $2{,}000\lambda$ in plasma. Note that the bubble shape deviates from the ideal spherical one because of the large number of trapped particles.

The radiation spectrum of a single individual electron is the synchrotron one:

$$\frac{\mathrm{d}I}{\mathrm{d}\omega} = \frac{4\sqrt{3}e^2}{c} N_\beta K \gamma F\left(\frac{\omega}{\omega_c}\right), \tag{9.2}$$

where N_β is the number of betatron oscillations, $K = \sqrt{\gamma/2}k_p r_b$ is the bubble wiggler strength for an electron propagating with the relativistic gamma factor γ and betatron oscillation amplitude r_b in the radial direction. The function $F(\xi) = \xi \int_\xi^{\inf} K_{5/3}(\xi')\mathrm{d}\xi'$ is the standard synchrotron emission function [20] and

$$\omega_c = 0.75\omega_p \gamma^2 k_p r_b, \tag{9.3}$$

is the critical synchrotron frequency.

Because the spectrum radiated by each individual electron is synchrotron-like and is characterized by the single parameter – the critical frequency ω_c, we can easily estimate the number of photons N_{ph} emitted:

$$N_{\text{ph}} \approx \frac{4\pi}{9}\frac{e^2}{\hbar c} N_\beta K = \frac{1}{9}\frac{e^2}{\hbar c}(k_p L)(k_p r_b). \tag{9.4}$$

Here, L is the propagation length. Now, we have to estimate the radial excursion length r_b. To do this, we mention that the trapped electrons originate their motion in the bubble at the ring of radius $k_p R \approx \sqrt{a_0}$. As they are accelerated, the oscillation amplitude decreases and becomes according to [16]:

$$k_p r_b \approx \frac{k_p R}{\gamma^{1/4}} \approx \left(\frac{a_0^2}{\gamma}\right)^{1/4}. \tag{9.5}$$

However, if the electron beam propagates long enough in the cavity and reaches the laser pulse, its betatron oscillations can become resonant with the laser pulse field and grow until the radial excursion r_b reaches the bubble radius R. This increases the number of emitted photons by the factor $\gamma^{1/4}$.

The energy conversion efficiency from the accelerated electron into the x-rays is thus

$$\frac{N_{\text{ph}}\hbar\omega_c}{\gamma mc^2} \approx \frac{1}{12}\frac{e^2}{\hbar c}\frac{\hbar\omega_p}{mc^2} k_p L (k_p r_b)^2 \gamma = \frac{1}{12}\frac{e^2}{\hbar c}\frac{\hbar\omega_p}{mc^2} k_p L a_0 \sqrt{\gamma}. \tag{9.6}$$

In the case the electrons overlap with the laser, the conversion efficiency increases by the factor $\sqrt{\gamma}$ to

$$\frac{N_{\text{ph}}\hbar\omega_c}{\gamma mc^2} = \frac{1}{12}\frac{e^2}{\hbar c}\frac{\hbar\omega_p}{mc^2} k_p L a_0 \gamma. \tag{9.7}$$

The conversion efficiency (9.6)–(9.7) can be done significantly high for large acceleration distances $k_p L$ high laser amplitudes a_0.

It is relatively easy to extend the synchrotron photon spectrum to quite high energies as it follows from the formula (9.3). However, the spectrum is very broad and the radiation is incoherent.

A possible solution is to feed an external undulator with the quasimonoenergetic electron beam accelerated in the bubble regime [11]. This would be the regime of a table top x-ray free electron laser. However, to achieve the generation, the electron bunch must be dense and simultaneously have a very low energy spread, at the level of 0.1%. The feasibility of such high quality electron beams from laser plasmas must yet be demonstrated.

9.3 High Harmonics Generation in the Relativistic Regime

High harmonics generated at plasma surfaces in the relativistic regime [13, 14, 21–33, 35] provide an extremely bright coherent source of short wavelength radiation and attosecond pulses, perfectly suited for the study of ultra-fast processes in atoms, molecules, and solids at intensities significantly higher than those obtained from strong field laser–atom interactions [38–42].

For the first time, this spectacular phenomenon was observed with nanosecond pulses of long wavelength (10.6 μm) CO_2 laser light [43]. Short after the experimental observation in 1981, Bezzerides et al. studied the problem of harmonic light emission theoretically [44]. Their approach based on nonrelativistic equations of motion and hydrodynamic approximation for the plasma predicted a cutoff of the harmonic spectrum at the plasma frequency.

Ten years later, in 1993, a new approach to the interaction of an ultrashort, relativistically strong laser pulse with overdense plasma was proposed by Bulanov et al. [45]. They "interpreted the harmonic generation as due to the Doppler effect produced by a reflecting charge sheet, formed in a narrow region at the plasma boundary, oscillating under the action of the laser pulse" [45].

At the beginning of 1996, numerical results of particle-in-cell simulations of the harmonic generation by femtosecond laser–solid interaction were presented by P. Gibbon [46]. He demonstrated numerically that the high harmonic spectrum goes well beyond the cutoff predicted in [44] and also presented a numerical fit for the spectrum, which approximated the intensity of the n-th harmonic as $I_n \propto n^{-5}$. At about the same time, the laser-overdense plasma interaction was also studied by Lichters et al. [47].

The same year the analytical work by von der Linde and Rzazewski [48] appeared. The authors used the "oscillating mirror" model and approximated the oscillatory motion of the mirror as a sin-function of time without analysis of the applicability of this approximation. With the explicit form of the mirror motion, an analytical formula for the harmonic spectrum was obtained.

In the weakly and moderately relativistic regime, when the laser intensity $I\lambda^2 \leq 10^{18}\,\mathrm{W\,\mu m^2/cm^2}$, the dynamics of the reflecting surface can be also influenced by harmonic resonances [49, 50].

The physics understanding and the analytic theory leading to the universal power-law spectrum has been developed by Baeva, Gordienko, and Pukhov [13].

The plasma harmonics are generated in the relativistic regime, when the laser pulse intensity $I \gg 10^{18}\,\mathrm{W/cm^2}$. At these intensities, electrons are driven up to relativistic velocities and move along curved trajectories. Their motion is due to combined fields resulting from complicated nonlinear laser-plasma dynamics. Yet, the presence of a plasma surface leads to translational symmetry along it and thus to conservation of the generalized momentum tangential component:

$$\mathbf{P}_\tau = \mathbf{p}_\tau - \frac{e}{c}\mathbf{A}_\tau, \tag{9.8}$$

where $\mathbf{p} = \gamma m \mathbf{v}$ is the kinetic momentum, $\mathbf{v}$ is the electron velocity, $\gamma = (1 - v^2/c^2)^{-1/2}$ is the relativistic factor, and $\mathbf{A}$ is the laser vector potential.

This conservation means that all the plasma surface electrons are driven by the laser coherently and emit x-rays in the form of very short pulses just when their momenta point toward the observer. This coherent x-ray emission we observe is the high harmonic emission [13, 35]. Due to the physical mechanism, the harmonics are all phase-locked and appear in the from of ultra-short, (sub-)attosecond pulses [13, 32]. The theory of high harmonic generation in relativistic regime has been developed in the work of Baeva, Gordienko, and Pukhov (BGP) [13]. The theory is based on the dynamics of the plasma surface and results in universal high harmonic spectra.

In this work, we limit ourselves to the one-dimensional (1D) theory and simulations. This greatly simplifies the theoretical understanding as well as make the simulations more handy. Even in this 1D limit, we can incorporate the obliquely incident laser pulses due to the Lorentz frame transformation [36]. This is acceptable as long as the focal spot size is big compared to the laser wavelength [37].

9.3.1 *Physical Picture of HHG at Overdense Plasma Boundary*

In this section, we state the problem of high harmonic generation at the boundary of overdense plasma and qualitatively describe its main features, which will find their analytical and numerical confirmation in what follows.

Let us consider a laser pulse of ultra-relativistic intensity, interacting with the sharp surface of an overdense plasma slab (see Fig. 9.2).

We assume that the incident laser pulse is short, so that we can neglect the slow ion dynamics and consider the electron motion only. The electrons are driven by the laser light pressure, a restoring electrostatic force comes from the ions. As a consequence, the plasma surface oscillates and the electrons gain a normal momentum component.

Since the plasma is overdense, the incident electromagnetic wave is not able to penetrate it. This means that there is an electric current along the plasma surface. For this reason, the momenta of electrons in the skin layer have, apart from the components normal to the plasma surface, also tangential components.

According to the relativistic similarity theory [6], both the normal and the tangential components are of the order of the dimensionless electromagnetic potential a_0.

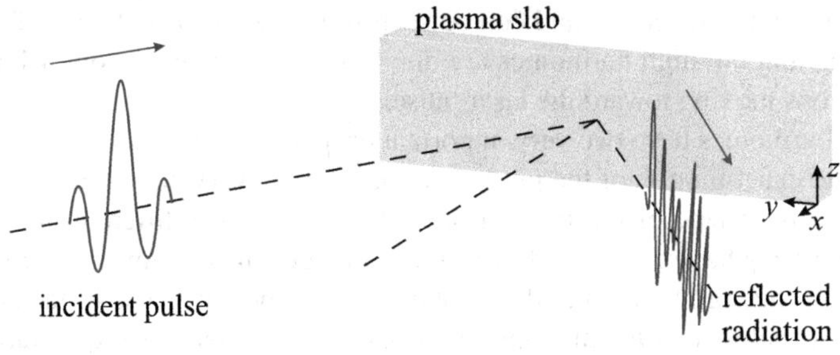

Fig. 9.2 Geometry of the problem. The laser pulse is moving toward the overdense plasma slab, x is perpendicular to the surface, y and z are parallel to it

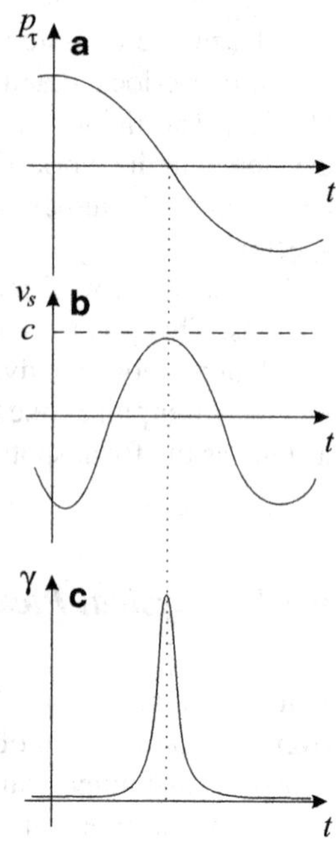

Fig. 9.3 (**a**) Electron momentum component parallel to the surface as a function of time (**b**) The velocity of the plasma surface v_s is a smooth function of time, unlike the γ-factor of the surface (**c**)

Consequently, the actual electron momenta make a finite angle with the plasma surface for most of the time.

Since we consider a laser pulse of ultra-relativistic intensity, the motion of the electrons is ultra-relativistic. In other words, their velocities are approximately c. Although the motion of the plasma surface is qualitatively different: its velocity v_s is not ultra-relativistic for most of the times but smoothly approaches c only when the tangential electron momentum vanishes (see Fig. 9.3b).

The γ-factor of the surface γ_s also shows specific behavior. It has sharp peaks at those times when the velocity of the surface approaches c (see Fig. 9.3c). Thus, while the velocity function v_s is characterized by its smoothness, the distinctive features of γ_s are its quasi-singularities.

When v_s reaches its maximum and γ_s has a sharp peak, high harmonics of the incident wave are generated and can be seen in the reflected radiation. Physically, this means that the high harmonics are due to the collective motion of bunches of fast electrons moving toward the laser pulse.

These harmonics have two very important properties. First, their spectrum is universal. The exact motion of the plasma surface can be very complicated, since it is affected by the shape of the laser pulse and can differ for different plasmas. Yet, the qualitative behavior of v_s and γ_s is universal, and since it governs the HHG, the spectrum of the high harmonics does not depend on the particular surface motion.

We show below that the high harmonic spectrum contains two qualitatively different parts: power law decay and exponential decay (see Fig. 9.4). In the power law part, the spectrum decays as

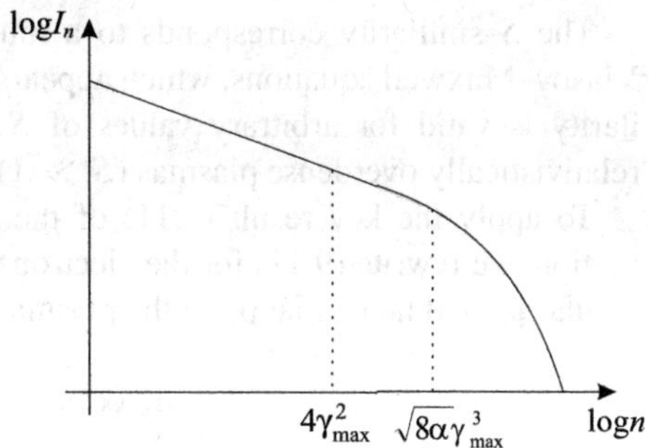

Fig. 9.4 The universal high harmonic spectrum contains power law decay and exponential decay (plotted in log–log scale)

$$I_n \propto 1/n^{8/3}, \tag{9.9}$$

up to a critical harmonics number that scales as γ^3_{max}, where I_n is the intensity of the nth harmonic (see Sect. 9.3.4). Here, γ_{max} is the maximal γ-factor of the point, where the component of the electric field tangential to the surface vanishes (see Sect. 9.3.3).

The second important feature of the high harmonics is that they are phase locked. This observation is of particular value, since it allows for the generation of attosecond and even subattosecond pulses [32].

9.3.2 *Ultra-Relativistic Similarity and the Plasma Surface Motion*

The analytical theory presented in this work is based on the similarity theory developed in [6] for collisionless ultra-relativistic laser-plasma regime and is valid both for under- and overdense plasmas.

The ultra-relativistic similarity theory states that when the dimensionless laser vector potential $\mathbf{a}_0 = e\mathbf{A}_0/mc^2$ is large ($a_0^2 \gg 1$) the plasma electron dynamics does not depend on a_0 and the plasma electron density N_e separately. Instead, they merge in the single dimensionless similarity parameter S defined by

$$S = \frac{N_e}{a_0 N_c}, \tag{9.10}$$

where $N_c = \omega_0^2 m/4\pi e^2$ is the critical electron density for the incident laser pulse with amplitude a_0 and carrier frequency ω_0.

In other words, when the plasma density N_e and the laser amplitude a_0 change simultaneously, so that $S = N_e/a_0 N_c = const$, the laser-plasma dynamics remains similar. In particular, this basic ultra-relativistic similarity means that for different interactions with the same $S = const$, the plasma electrons move along similar trajectories while their momenta $\mathbf{p}$ scale as

$$\mathbf{p} \propto a_0. \tag{9.11}$$

The S-similarity corresponds to a multiplicative transformation group of the Vlasov–Maxwell equations, which appears in the ultra-relativistic regime. The similarity is valid for arbitrary values of S. Physically, the S-parameter separates relativistically overdense plasmas ($S \gg 1$) from underdense ones ($S \ll 1$).

To apply the key result (9.11) of the similarity theory to the plasma surface motion, we rewrite (9.11) for the electron momentum components that are perpendicular $\mathbf{p}_n$ and tangential $\mathbf{p}_\tau$ to the plasma surface:

$$\mathbf{p}_n \propto a_0, \quad \mathbf{p}_\tau \propto a_0. \tag{9.12}$$

This result is significant. It shows that when we increase the dimensionless vector potential a_0 of the incident wave while keeping the plasma overdense, so that $S = const$, both $\mathbf{p}_n$ and $\mathbf{p}_\tau$ grow as a_0. In other words, the velocities of the skin layer electrons

$$v = c\sqrt{\frac{\mathbf{p}_n^2 + \mathbf{p}_\tau^2}{m_e^2c^2 + \mathbf{p}_n^2 + \mathbf{p}_\tau^2}} = c(1 - O(a_0^{-2})), \tag{9.13}$$

are about the speed of light almost at all times. Yet, the relativistic γ-factor of the plasma surface $\gamma_s(t')$ and its velocity $\beta_s(t')$ behave in a quite different way. To realize this key fact let us consider the electrons at the very boundary of the plasma. The scalings (9.12) state that the momenta of these electrons can be represented as

$$\begin{aligned} \mathbf{p}_n(t') &= a_0 \mathbf{P}_n(S, \omega_0 t') \\ \mathbf{p}_\tau(t') &= a_0 \mathbf{P}_\tau(S, \omega_0 t'), \end{aligned} \tag{9.14}$$

where $\mathbf{P}_n$ and $\mathbf{P}_\tau$ are universal functions, which depend on the pulse shape and the S-parameter rather than on a_0 or N_e separately. Consequently, for $\beta_s(t')$ and $\gamma_s(t')$, one obtains

$$\beta_s(t') = \frac{p_n(t')}{\sqrt{m_e^2c^2 + \mathbf{p}_n^2(t') + \mathbf{p}_\tau^2(t')}} = \frac{P_n(t')}{\sqrt{\mathbf{P}_n^2(t') + \mathbf{P}_\tau^2(t')}} - O(a_0^{-2}), \tag{9.15}$$

$$\gamma_s(t') = \frac{1}{\sqrt{1 - \beta_s^2(t')}} = \sqrt{1 + \frac{\mathbf{P}_n^2(t')}{\mathbf{P}_\tau^2(t')}} + O(a_0^{-2}). \tag{9.16}$$

One sees from (9.13) that when a_0 gets large, the relativistic γ-factor of the electrons becomes too large and their velocities approach the velocity of light. However, the dynamics of the plasma boundary is significantly different. For large a_0s, the plasma boundary motion does not enter the ultra-relativistic regime and its relativistic γ-factor $\gamma_s(t')$ is generally of the order of unity. Yet, there is one exception: if at the moment t'_g it happens that $\mathbf{P}_\tau(S, t'_g) = 0$, i.e.

$$\mathbf{p}_\tau(S, t'_g) = 0, \tag{9.17}$$

we have

$$\gamma_s = \frac{1}{\sqrt{1-\beta_s^2}} = \sqrt{\frac{\mathbf{p}_n^2 + m_e^2 c^2}{m_e^2 c^2}} \propto a_0. \tag{9.18}$$

So the relativistic γ-factor of the boundary jumps to $\gamma_s(t'_g) \propto a_0$ and the duration of the relativistic γ-factor spike can be estimated as

$$\Delta t' \propto 1/(a_0 \omega_0). \tag{9.19}$$

For the velocity of the plasma boundary, one finds analogously that it smoothly approaches the velocity of light as $\beta_s(t'_g) = (1 - O(a^{-2}))$. Figure 9.3 represents schematically this behavior.

As we will see later, the γ_s spikes cause the generation of high harmonics in the form of ultra short pulses.

9.3.3 Boundary Condition: Energy Conservation and the Apparent Reflection Point

In this section, we introduce the boundary condition describing the laser-overdense plasma interaction appealing to physical arguments, just as it was previously done in [32, 51]. However, for the purposes of this work it is sufficient to treat this problem on a more intuitive basis [32].

To derive the correct boundary condition, let us consider the tangential vector potential components of a laser pulse normally incident onto a overdense plasma slab. These components satisfy the equation

$$\frac{1}{c^2}\frac{\partial^2 \mathbf{A}_\tau(t,x)}{\partial t^2} - \frac{\partial^2 \mathbf{A}_\tau(t,x)}{\partial x^2} = \frac{4\pi}{c}\mathbf{j}(t,x), \tag{9.20}$$

where $\mathbf{A}_\tau(t, x = -\infty) = 0$ and $\mathbf{j}$ is the tangential plasma current density. Equation (9.20) yields

$$\mathbf{A}_\tau(t,x) = 2\pi \int_{-\infty}^{+\infty} \mathbf{J}\left(t,x,t',x'\right) \mathrm{d}t'\mathrm{d}x'. \tag{9.21}$$

Here, $\mathbf{J}(t,x,t',x') = \mathbf{j}(t',x')(\Theta_- - \Theta_+)$, where we have defined $\Theta_- = \Theta(t - t' - |x - x'|/c)$ and $\Theta_+ = \Theta(t - t' + (x - x')/c)$, using the Heaviside step-function $\Theta(t)$. Due to this choice of $\mathbf{J}$, the vector potential $\mathbf{A}_\tau(t,x)$ satisfies both (9.20) and the boundary condition at $x = -\infty$ since $\mathbf{J}(t, x = -\infty, t', x') = 0$. The tangential electric field is $\mathbf{E}_\tau = -(1/c)\partial_t \mathbf{A}_\tau(t,x)$. If we denote the position of the electron fluid surface by $X(t)$, we have

$$\mathbf{E}_\tau(t, X(t)) = \frac{2\pi}{c} \sum_{\alpha=-1}^{\alpha=+1} \alpha \int_0^{-\infty} \mathbf{j}(t + \alpha\xi/c, X(t) + \xi)\, \mathrm{d}\xi, \tag{9.22}$$

where $\xi = x' - X(t)$.

Now one has to estimate the parameters characterizing the skin layer, i.e., the characteristic time τ_s of skin layer evolution (in the co-moving reference frame) and the skin layer thickness δ. Since the plasma is driven by the light pressure, one expects that $\tau_s \propto 1/\omega_0$. The estimation of δ is more subtle. From the ultra-relativistic similarity theory follows that $\delta \propto (c/\omega_0)S^{\Delta}$, where $S \gg 1$ for strongly overdense plasmas and Δ is an exponent that has to be found analytically. In this work, we do not discuss the exact value of Δ, but notice that this quantity does not depend on neither S nor a_0. On the other hand, the denser the plasma, the less the value of δ. This condition demands that $\Delta < 0$ and we get $c/\omega_0 \gg \delta$ for $S \gg 1$.

If the characteristic time τ_s of the skin layer evolution is long ($c\tau_s \gg \delta$), then we can use the Taylor expansion $\mathbf{j}\,(t \pm \xi/c, x' = X(t) + \xi) \approx \mathbf{j}(t, x') \pm \epsilon$, where $\epsilon = (\xi/c)\partial_t \mathbf{j}(t, x')$, and substitute this expression into (9.22). The zero-order terms cancel each other and we get $\mathbf{E}_\tau(t, X(t)) \propto J_p(\delta/c\tau) \ll E_l$, where $J_p \propto cE_l$ is the maximum plasma surface current. Thus, as long as the skin-layer is thin and the plasma surface current is limited, we can use the Leontovich boundary condition [52]

$$\mathbf{e}_n \times \mathbf{E}(t, X(t)) = 0. \tag{9.23}$$

This condition has a straightforward relation to energy conservation. Indeed, if we consider the Poynting vector,

$$\mathbf{S} = \frac{c}{4\pi}\mathbf{E} \times \mathbf{B}. \tag{9.24}$$

we notice that the boundary condition (9.23) represents balance between the incoming and reflected electromagnetic energy flux at the boundary $X(t)$.

The boundary condition (9.23) allows for another interpretation. An external observer sees that the electromagnetic radiation gets reflected at the point $x_{\mathrm{ARP}}(t)$, where the normal component of the Poynting vector $\mathbf{S}_n = c\mathbf{E}_\tau \times \mathbf{B}_\tau/4\pi = 0$, implied by $\mathbf{E}_\tau(x_{\mathrm{ARP}}) = 0$. We call the point $x_{\mathrm{ARP}}(t)$ *the apparent reflection point* (ARP).

If the reflected field is a phase modulation of the incident one, the actual location of the ARP can be easily found from the electromagnetic field distribution in front of the plasma surface. The incident laser field in vacuum runs in the negative x-direction, $\mathbf{E}^i(x, t) = \mathbf{E}^i(x + ct)$, while the reflected field is translated backward: $\mathbf{E}^r(x, t) = \mathbf{E}^r(x - ct)$. The tangential components of these fields interfere destructively at the ARP position $x_{\mathrm{ARP}}(t)$, so that the implicit equation for the apparent reflection point $x_{\mathrm{ARP}}(t)$ is

$$\mathbf{E}_\tau^i(x_{\mathrm{ARP}} + ct) + \mathbf{E}_\tau^r(x_{\mathrm{ARP}} - ct) = 0. \tag{9.25}$$

We want to emphasize that (9.25) contains the electromagnetic fields in vacuum. That is why the reflection point x_{ARP} is *apparent*. The real interaction within the plasma skin layer can be very complex. Yet, an external observer, who has information about the radiation in vacuum only, sees that $\mathbf{E}_\tau = 0$ at x_{ARP}. The ARP is located within the skin layer at the electron fluid surface, which is much shorter than the laser wavelength for overdense plasmas, for which the similarity parameter is $S \gg 1$. In this sense, the ARP is attached to the oscillating plasma surface.

9.3.4 High Harmonic Universal Spectrum

According to (9.23), the electric field of the reflected wave at the plasma surface is

$$\mathbf{E}_r(t', X(t')) = -\mathbf{E}_i(t', X(t')), \tag{9.26}$$

where $\mathbf{E}_i(t', X(t')) = -(1/c)\partial_{t'}\mathbf{A}_i(t', X(t'))$ is the incident laser field, t' is the reflection time. The one-dimensional wave equation translates signals in vacuum without change. Thus, the reflected wave field at the observer position x and time t is $\mathbf{E}_r(t, x) = -\mathbf{E}_i(t', X(t'))$. Setting $x = 0$ at the observer position, we find that the Fourier spectrum of the electric field $\mathbf{E}_r(t, x = 0)$ is

$$\mathbf{E}_r(\Omega) = \frac{m_e c\omega}{e\sqrt{2\pi}} \int\limits_{-\infty}^{+\infty} \mathrm{Re}\left[i\mathbf{a}\left(\left(ct' + X(t')\right)/c\tau_0\right)\exp(-\mathrm{i}\omega_0 t' - \mathrm{i}\omega_0 X(t')/c)\right] \times \exp(-\mathrm{i}\Omega t)\,\mathrm{d}t, \tag{9.27}$$

where

$$t' - X(t')/c = t, \tag{9.28}$$

is the retardation relation [47].

The fine structure of the spectrum of $\mathbf{E}_r(t)$ depends on the particular surface motion $X(t)$, which is defined by the complex laser–plasma interaction at the plasma surface. Previous theoretical works on high-order harmonic generation from plasma surfaces [46–48, 51] tried to approximate the function $X(t)$ to evaluate the harmonic spectrum. For the first time, analytical description of the high harmonic intensity spectrum and the concept of universality were presented in [32]. This work has shown for the first time that the most important features of the high harmonic spectrum do not depend on the detailed structure of $X(t)$. The relativistic similarity theory which was developed later [6] not only simplifies the physical picture of HHG, but, as we will see below, also influences the saddle point technique [53] in this regime.

To find the spectrum, we notice that the investigation of $\mathbf{E}_r(\Omega)$ (9.27) is equivalent to the investigation of the function

$$f(n) = f_+(n) + f_-(n), \tag{9.29}$$

where

$$f_\pm = \pm \int_{-\infty}^{+\infty} \mathbf{g}(\tau' + x(\tau')) \exp(\pm i(\tau' + x(\tau')) - in\tau)\, \mathrm{d}\tau. \tag{9.30}$$

Here, $\tau = \omega_0 t$, $\tau' = \omega_0 t'$, $n = \Omega/\omega_0$, $x(\tau') = (\omega_0/c)X(t')$ and $\mathbf{g}$ is a slowly varying function ($|dg(\tau')/\mathrm{d}\tau'| \ll 1$), which is trivially related to $\mathbf{a}$ as

$$\mathbf{g}(\tau' + x(\tau')) = \frac{-im_e c}{2e\sqrt{2\pi}} \mathbf{a}((ct' + X(t'))/c\tau_0). \tag{9.31}$$

Making use of (9.28), we re-write (9.30) as

$$f_\pm = \pm \int_{-\infty}^{+\infty} g(\tau' + x(\tau')) \exp(i\tau'(-n \pm 1) + ix(\tau')(n \pm 1)) \left(1 - x'\left(\tau'\right)\right) \mathrm{d}\tau'. \tag{9.32}$$

We wish to examine the integral (9.32) for very large n. For this purpose, we notice that the derivative of the phase

$$\Theta(\tau') = \tau'(-n \pm 1) + x(\tau')(n \pm 1), \tag{9.33}$$

is negative everywhere except in the vicinity of $\tau'_g = \omega_0 t'_g$ for which $x'(\tau'_g) \approx 1$ (see Fig. 9.5a).

The physical meaning of τ'_g and the behavior of $x(\tau')$ in the vicinity of these times is explained by (9.17). Since the time derivative of $\Theta(\tau')$ is negative for all τs that are not too close to one of the τ'_g, we can shift the path over which we integrate to the lower half of the complex plane everywhere except in the neighborhoods of τ'_g (see Fig. 9.5b). The contributions of the parts remote from the real axis are

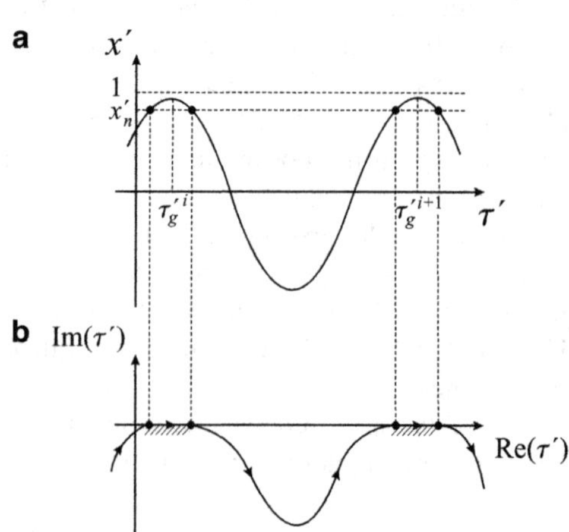

Fig. 9.5 Surface dynamics and path integration in (9.32). (**a**) Velocity $x'(\tau')$ of the plasma surface; $x'_n = (n-1)/(n+1)$ are the saddle points corresponding to $\mathrm{d}\Theta/\mathrm{d}\tau' = 0$. (**b**) The integration path can be shifted below the real axis everywhere except in the neighborhoods of τ'_g (*dashed regions*)

exponentially small. We can shift the path to the complex plane till the derivative equals zero or we find a singularity of the phase Θ.

To calculate the contributions of τ_g's neighborhoods, we can expand $x'(\tau')$ near each of its maxima at τ_g'. Since every smooth function resembles a parabola near its extrema, the expansion of $x'(\tau')$ is a quadratic function of $(\tau' - \tau_g')$. Simple integration leads to the following expression for $x(\tau')$

$$x(\tau') = x(\tau_g') + v_0(\tau_g')(\tau' - \tau_g') - \frac{\alpha(\tau_g')}{3}(\tau' - \tau_g')^3. \tag{9.34}$$

The Taylor expansion given by (9.34) has three important properties related to its dependence on S and a_0: 1) for $S = const$ and $a_0 \to +\infty$ one finds that $v_0 \to c$; 2) for $a_0 \to +\infty$, α depends only on the parameter S; 3) the expansion (9.34) is a good approximation for $|\tau' - \tau_g'| \ll (2\pi/\omega_0) f_1(S)$, where the function f_1 does not depend on a_0. These three properties are mathematical statements of the physical picture described in Sect. 9.3.1 combined with the similarity theory developed in Sect. 9.3.2. In other words, the properties of the expansion (9.34) just mentioned are direct consequences of the physical picture presented in Fig. 9.3.

Substitution of (9.34) into $f_\pm(n)$ yields

$$f_\pm(n) = \sum_{\tau_g'} f_\pm(\tau_g', n), \tag{9.35}$$

where the sum is over all times τ_g',

$$f_+(\tau_g', n) = g\left(\tau_g' + x(\tau_g')\right) \exp(\mathrm{i}\Theta_+(\tau_g', n)) F(\tau_g', n), \tag{9.36}$$

$$f_-(\tau_g', n) = -g\left(\tau_g' + x(\tau_g')\right) \exp(\mathrm{i}\Theta_-(\tau_g', n)) F(\tau_g', -n), \tag{9.37}$$

$$F(\tau_g', n) = \frac{4\sqrt{\pi}}{\left(\sqrt[4]{\alpha(\tau_g')n}\right)^{4/3}} \mathrm{Ai}\left(\frac{2}{n_{\mathrm{cr}}(\tau_g')} \frac{n - n_{\mathrm{cr}}(\tau_g')}{(\alpha(\tau_g')n)^{1/3}}\right), \tag{9.38}$$

$$\Theta_\pm = \pm(\tau_g' + x(\tau_g')) + n(x(\tau_g') - \tau_g'), \tag{9.39}$$

and $n_{\mathrm{cr}} = 2/(1 - v_0)$. In (9.38) Ai is the well-known Airy-function, defined as

$$\mathrm{Ai}(x) = \frac{1}{\sqrt{\pi}} \int_0^{+\infty} \cos\left(ux + \frac{1}{3}u^3\right) \mathrm{d}u. \tag{9.40}$$

Note that if $x(\tau' + \pi) = x(\tau')$ and $g(\tau') = g(\tau' + \pi)$, then $f_\pm(2n) = 0$.

Using equations (9.35)–(9.39), we can show analytically that the spectrum of radiation generated by the plasma is described by a universal formula.

For the intensity of the nth harmonic, we obtain

$$I_n \propto \left|\exp\left(\mathrm{i}\Theta_+(n)\right) F\left(n\right) - \exp\left(\mathrm{i}\Theta_-(n)\right) F\left(-n\right)\right|^2, \tag{9.41}$$

where

$$F(n) = \frac{4\sqrt{\pi}}{(\sqrt[4]{\alpha n})^{4/3}}\mathrm{Ai}\left(\frac{2}{n_{\mathrm{cr}}}\frac{n-n_{\mathrm{cr}}}{(\alpha n)^{1/3}}\right), \tag{9.42}$$

$$\Theta_{\pm}(n) = \pm\Theta_0 - n\Theta_1 \tag{9.43}$$

with the Airy function $Ai(x)$ defined in (9.40) and the critical harmonic number n_{cr} satisfying $n_{\mathrm{cr}} = 4\gamma_{\max}^2$, where $\gamma_{\max}$ is the largest relativistic factor of the plasma boundary.

Equation (9.41) gives an exact formula for the high harmonic spectrum, which includes both power law and exponential decay parts. Now we want to use different asymptotic representations of the Airy function to demonstrate explicitly these two quite different laws of high harmonic intensity decay.

For $n < \sqrt{\alpha/8}n_{\mathrm{cr}}^{3/2}$ $(2|1-n/n_{\mathrm{cr}}| \ll (\alpha n)^{1/3})$, we can substitute the value of the Airy function at $x = 0$ ($Ai(0) = \sqrt{\pi}/(3^{2/3}\Gamma(2/3)) = 0.629$, $Ai'(0) = -3^{1/6}\Gamma(2/3)/(2\sqrt{\pi}) = -0.459$) in (9.41), and obtain

$$I_n \propto \frac{1}{n^{8/3}}\left|\sin\Theta_0 + \frac{Ai'(0)}{Ai(0)}B(n,\Theta_0)\right|^2, \tag{9.44}$$

where

$$B(n,\Theta_0) = \frac{2\sin\Theta_0}{(\alpha n_{\mathrm{cr}})^{1/3}}\left(\frac{n}{n_{\mathrm{cr}}}\right)^{2/3} + \frac{2i\cos\Theta_0}{(\alpha n)^{1/3}}. \tag{9.45}$$

This means that the universal spectrum

$$I_n \propto 1/n^{8/3}, \tag{9.46}$$

is observed everywhere except for $\sin\Theta_0 \approx 0$, when the dominant term in the expansion is zero.

The power-law spectrum (9.46) continues upto a cutoff – that is rather a smooth rollover to an exponential decay – at the harmonic number n_{cutoff}

$$n_{\mathrm{cutoff}} = \sqrt{8\alpha}\gamma_{\max}^3. \tag{9.47}$$

This is a much stronger scaling than one would get applying just the naive Doppler shift of $4\gamma_{\max}^2$. The difference is due to the sharp spike in the surface γ-factor that broadens the spectrum significantly.

For $n > \sqrt{\alpha/8}n_{\mathrm{cr}}^{3/2}$ $(2|1-n/n_{\mathrm{cr}}| \gg (\alpha n)^{1/3})$, (9.38) can be rewritten as

$$I_n \propto \frac{n_{\mathrm{cr}}^{1/2}}{n^3}\exp\left(-\frac{16\sqrt{2}}{3\alpha^{1/2}}\frac{n-n_{\mathrm{cr}}}{n_{\mathrm{cr}}^{3/2}}\right). \tag{9.48}$$

It is interesting to notice that the approximation used in [32] also gives (9.48) for this area.

9.3.5 Cutoff and the Structure of Filtered Pulses

As we have seen, the relativistic plasma harmonics are phase locked and can be used to generate ultra short pulses. However, to extract these ultra short pulses, one has to remove the lower harmonics. The high energy cutoff (9.47) of the power law spectrum defines the shortest pulse duration that can be achieved this way.

Let us apply a high-frequency filter that suppresses all harmonics with frequencies below Ω_f and study how the relative position of the Ω_f and the spectrum cutoff affects the duration of the resulting (sub)attosecond pulses.

According to (9.41), the electric field of the pulse after the filtration is

$$E \propto Re \int_{\Omega_f/\omega_0}^{+\infty} \left(\exp\left(i\Theta_+(n)\right) F(n) - \exp\left(i\Theta_-(n)\right) F(-n)\right) \exp(int)\, dn. \quad (9.49)$$

The structure of the filtered pulses depends on where we set the filter threshold Ω_f. In the case $1 \ll \Omega_f/\omega_0 \ll \sqrt{\alpha/8n_{cr}^{3/2}}$, we use (9.46) and rewrite (9.49) as

$$E \propto Re \int_{\Omega_f/\omega_0}^{+\infty} \frac{\exp(int)}{n^{4/3}}\, dn = \left(\frac{\omega_0}{\Omega_f}\right)^{1/3} Re \exp(i\Omega_f t - i\Theta_1) P(\Omega_f t), \quad (9.50)$$

where the function P

$$P(x) = \int_1^{+\infty} \frac{\exp(iyx)}{y^{4/3}}\, dy, \quad (9.51)$$

gives the slow envelope of the pulse.

It follows from the expression (9.50) that the electric field of the filtered pulse decreases very slowly with the filter threshold as $\Omega_f^{-1/3}$. The pulse duration decreases as $1/\Omega_f$. At the same time, the fundamental frequency of the pulse is Ω_f. Therefore, the pulse is hollow when $\Omega_f/\omega_0 \ll \sqrt{8\alpha}\gamma_{max}^3$, i.e. its envelope is not filled with electric field oscillations. One possible application of these pulses is to study atom excitation by means of a single strong kick.

The pulses structure changes differ significantly when the filter threshold is placed above the spectrum cutoff. For $\Omega_f/\omega_0 \gg \sqrt{8\alpha}\gamma_{max}^3$, we use (9.48) and (9.49) to obtain

$$E \propto \left(\frac{\omega_0}{\Omega_f}\right)^{3/2} \exp\left(-\frac{8\sqrt{2}\Omega_f}{3\sqrt{\alpha}\omega_0 n_{cr}^{3/2}}\right) Re \frac{\exp(i\Omega_f t - i\Theta_1)}{8\sqrt{2}/\left(3\sqrt{\alpha} n_{cr}^{3/2}\right) + i\omega_0 t}. \quad (9.52)$$

The amplitude of these pulses decreases fast when Ω_f grows. However, the pulse duration $\propto 1/\sqrt{\alpha}\gamma_{max}^3$ does not depend on Ω_f. Since the fundamental frequency of the pulse grows as Ω_f, the pulses obtained with an above-cutoff filter are filled with electric field oscillations. Therefore, these pulses are suitable to study the resonance excitation of ion and atom levels.

The minimal duration of the pulse obtained by cutting-off low order harmonics is defined by the spectrum cutoff $\propto \sqrt{\alpha 8 \gamma_{\max}^3}$. Physically, this result is the consequence of the ultra-relativistic spikes in the plasma surface γ-factor.

9.3.6 Numerical Results

To check our analytical results, we have performed a number of 1d PIC simulations with the 1d particle-in-cell code VLPL [54]. In all simulations, a laser pulse with the Gaussian envelope $a = a_0 \exp\left(-t^2/\tau_L^2\right) \cos\left(\omega_0 t\right)$, duration $\omega_0 \tau_L = 4\pi$ and dimensionless vector potential $a_0 = 20$ was incident onto a plasma layer with a step density profile.

9.3.6.1 Apparent Reflection Point

First, we study the oscillatory motion of the plasma and the dynamics of the apparent reflection point defined by the boundary condition (9.23). The plasma slab is initially positioned between $x_R = -1.5\lambda$ and $x_L = -3.9\lambda$, where $\lambda = 2\pi/\omega_0$ is the laser wavelength. The laser pulse has the amplitude $a_0 = 20$. The plasma density is $N_e/N_c = 90$ $(S = 4.5)$.

At every time step, the incident and the reflected fields are recorded at $x = 0$ (the position of the "external observer"). Being solutions of the wave equation in vacuum, these fields can be easily chased to arbitrary x and t. To find the ARP position x_{ARP}, we solve numerically equation (9.25). The trajectory of $x_{\mathrm{ARP}}(t)$ obtained in this simulation is presented in Fig. 9.6a. One can clearly see the oscillatory motion of the point $x_{\mathrm{ARP}}(t)$. The equilibrium position is displaced from the initial plasma boundary position x_R due to the mean laser light pressure.

Since only the ARP motion toward the laser pulse is of importance for the high harmonic generation, we cut out the negative ARP velocities $v_{\mathrm{ARP}}(t) = dx_{\mathrm{ARP}}(t)/dt$ and calculate only the positive ones (Fig. 9.6b). The corresponding γ-factor $\gamma_{\mathrm{ARP}}(t) = 1/\sqrt{1 - v_{\mathrm{ARP}}(t)^2/c^2}$ is presented in Fig. 9.6c. Notice that the ARP velocity is a smooth function. At the same time, the γ-factor $\gamma_{\mathrm{ARP}}(t)$ contains sharp spikes, which coincide with the velocity extrema. These numerical results confirm the predictions of the ultra-relativistic similarity theory, which were presented in Sect. 9.3.2.

9.3.6.2 High Harmonic Spectrum

For the same laser-plasma parameters ($a_0 = 20$, $N_e = 90N_c$), the spectrum of high harmonic radiation is presented in Fig. 9.7. The maximum γ-factor of the apparent

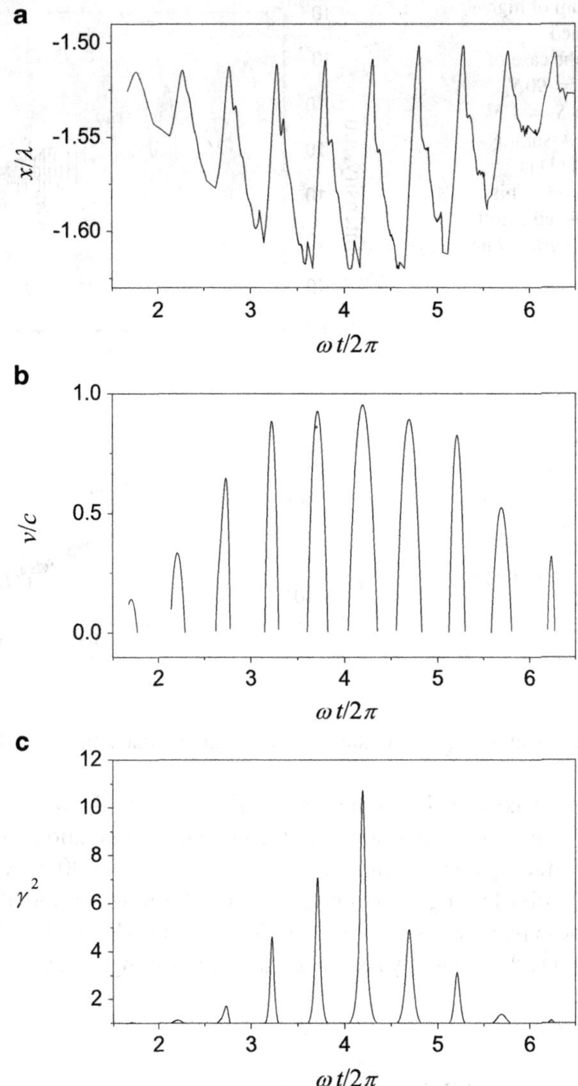

Fig. 9.6 1D PIC simulation results for the parameters $a_0 = 20$ and $N_e = 90N_c$. (**a**) Oscillatory motion of the point $x_{\mathrm{ARP}}(t)$ where $\mathbf{E}_\tau(x(t)) = 0$. (**b**) Velocity $v_{\mathrm{ARP}}(t) = \mathrm{d}x_{\mathrm{ARP}}(t)/\mathrm{d}t$; only the positive velocities are shown, since they correspond to motion towards the laser pulse in the geometry of this simulation. Notice that the ARP velocity is a smooth function. (**c**) The corresponding γ-factor $\gamma_{\mathrm{ARP}}(t) = 1/\sqrt{1 - v_{\mathrm{ARP}}(t)^2/c^2}$ contains sharp spikes, which coincide with the velocity extrema

reflection point in this numerical simulation is $\gamma_{\max} \approx 3.3$ (compare with Fig. 9.6). Consequently, the maximal harmonic number predicted by the "oscillating mirror" model lies at $4\gamma_{\max}^2 \approx 40$, while the harmonic cutoff predicted by the relativistic spikes is about 100. Figure 9.7 clearly demonstrates that there is no change of the

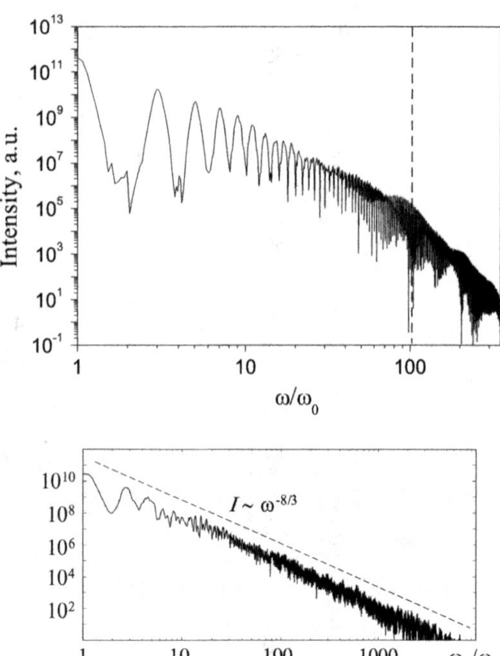

Fig. 9.7 Spectrum of high harmonics obtained numerically for the case of $a_0 = 20$ and $N_e = 90N_c$, corresponding to $S = 4.5$ and $\gamma_{\max} \approx 3.3$. Assuming $\alpha \approx 1$, the cutoff (1) is expected at $n \approx 100$. This analytically predicted cutoff is marked by the *dashed line*

Fig. 9.8 Spectra of the reflected radiation for the laser amplitude $a_0 = 20$ and the plasma density $N_e = 30N_{\text{cr}}$. The *broken line* marks the universal scaling $I \propto \omega^{-8/3}$

spectrum behavior at $4\gamma_{\max}^2$, while steeper decay takes place above 100, as predicted by our theory.

To be able to make a real statement about the power in the power law decay of the spectrum, we need more harmonics in the numerical simulation. For this reason, we made the simulation with parameters $a_0 = 20$ and $N_e = 30N_c$, which roughly corresponds to solid hydrogen or liquid helium. The reflected radiation spectrum obtained for these parameters is shown in Fig. 9.8 in log–log scale. The power law spectrum $I_n \propto 1/n^{8/3}$ is clearly seen here, thus confirming the analytical results of Sect. 9.3.4.

9.3.6.3 Subattosecond Pulses

Let us take a closer look at Fig. 9.8. The power law spectrum extends at least till the harmonic number 2000, and zeptosecond (1zs $= 10^{-21}$s) pulses can be generated. The temporal profile of the reflected radiation is shown in Fig. 9.9. When no spectral filter is applied, Fig. 9.9a, a train of attosecond pulses is observed [51]. However, when we apply a spectral filter selecting harmonics above $n = 300$, a train of much shorter pulses is obtained, Fig. 9.9b. Figure 9.9c zooms in to one of these pulses. Its full width at half maximum is about 300 zs. At the same time, its intensity normalized to the laser frequency is huge $(eE_{\text{zs}}/mc\omega)^2 \approx 14$. This corresponds to the intensity $I_{\text{zs}} \approx 2 \times 10^{19}$ W/cm^2.

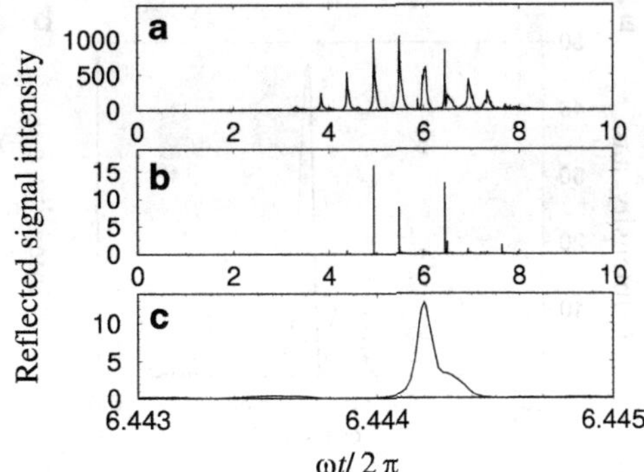

Fig. 9.9 Zeptosecond pulse train: (**a**) temporal structure of the reflected radiation; (**b**) zeptosecond pulse train seen after spectral filtering; (**c**) one of the zeptosecond pulses zoomed, its FWHM duration is about 300 zs

9.3.6.4 Filter Threshold and the Attosecond Pulse Structure

The dependence of the short pulses on the position of the filter also can be studied numerically. We apply a filter with the filter function $f(\omega) = 1 + \tanh((\omega - \Omega_f)/\Delta\omega)$. It passes through frequencies above Ω_f and suppresses lower frequencies. We choose the simulation case of laser vector potential $a_0 = 20$ and plasma density $N_e = 90N_c$. The spectrum of high harmonics is given in Fig. 9.7. We zoom in to one of the pulses in the pulse train obtained and study how the shape of this one pulse changes with Ω_f. Figure 9.10 represents the pulse behavior for four different positions of Ω_f. We measure to what degree the pulse is filled by the number of field oscillations within the FWHM. One clearly sees that for filter threshold below the cutoff frequency, Fig. 9.10a, b, the pulse is hollow. Notice that the case of Fig. 9.10b corresponds to the cutoff frequency predicted by the "oscillating mirror" model. Only for filter threshold positions above the spectrum cutoff given by (9.47) the pulse becomes filled, Fig. 9.10c,d. These results confirm once again the real position of the harmonic cutoff.

9.3.7 Efficiency of the BGP Spectrum

If we are interested in a keV x-ray source based on the plasma surface harmonics and take the BGP spectrum (9.46) around the harmonic number $n = 1{,}000$, we will need a laser pulse with the relativistic amplitude $a_0 = 10$ at least. The energy conversion efficiency η in a narrow spectral range $\Delta n/n$ is then

$$\eta \propto \frac{\Delta n}{n}\frac{1}{n^{5/3}}. \tag{9.53}$$

Assuming the spectral range to be $\Delta n/n = 0.01$, we obtain the energy conversion efficiency from the laser pulse to the keV x-rays $\eta \approx 10^{-7}$. Thus, a 10 J laser pulse would generate about 1 μJ, or some 10^{10} keV photons. Even though the BGP spectrum is a power law, the exponent is $-8/3$ and the efficiency still drops fast with the harmonic number.

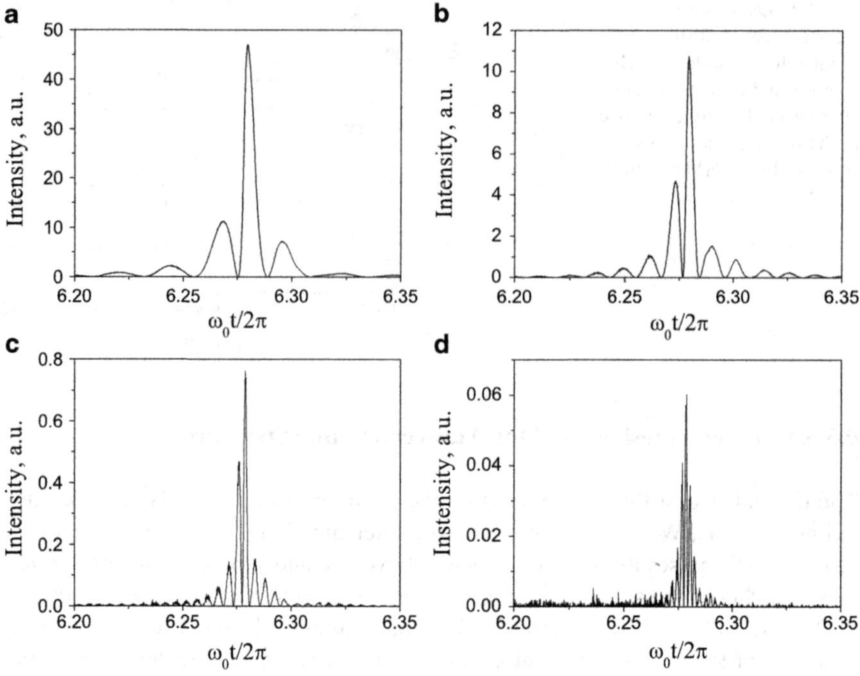

Fig. 9.10 Dependence of the pulses filling on the position of the sharp filter boundary for $a_0 = 20$ and $N_e = 90N_c$ and filter positions: (**a**) $\Omega_f = 20\omega_0$, $\Delta\omega = 2\omega_0$; (**b**) $\Omega_f = 40\omega_0$, $\Delta\omega = 2\omega_0$; (**c**) $\Omega_f = 100\omega_0$, $\Delta\omega = 2\omega_0$; (**d**) $\Omega_f = 200\omega_0$, $\Delta\omega = 2\omega_0$

9.4 Electron Nanobunching and High Harmonics in the Coherent Synchrotron Emission Regime

The BGP harmonic spectrum relies essentially on the boundary condition (9.25). In this case, it is clear from (9.25) that the reflected field is nothing but a phase modulation of the negative of the incident field. This can easily be checked inside a PIC simulation. Figure 9.11a shows an exemplary result. For this simulation, the laser was normally incident on a sharp-edged plasma density profile. Here, (9.25) is fulfilled to a good approximation as can be verified from Fig. 9.11a: Apart from minor deviations, the extremal values and the sequence of the monotonic intervals agree, the reflected field is approximately a phase modulation of the incident one. In general, we find in our PIC simulations that this condition works well when the sharp plasma boundary remains more or less step-like during the interaction. This is usually the case. However, one can find special regimes, when the boundary condition (9.25) breaks.

Let us now have a look at Fig. 9.11c, which shows the result of another PIC simulation. In this simulation, the plasma density profile was extended over a few fractions of a laser wavelength, and the p-polarized laser was incident at an angle of $\alpha = 63°$. The laser pulse used was the same as in the first simulation.

It is evident that the maximum of the reflected field reaches out about an order of magnitude above the amplitude of the incident laser. From the boundary condition (9.25) it follows that the reflected field E_r must be a phase modulation of the incident field E_i. Here, the reflected radiation can clearly *not* be obtained from the incident one just by phase modulation. By this we conclude that the boundary condition (9.25) fails here. Consequently, the spectrum deviates from the $-8/3$-power law, compare Fig. 9.11d. Indeed, we see that the efficiency of harmonic generation is much higher than estimated by the BGP calculations: about two orders of magnitude at the hundredth harmonic. Also, we can securely exclude "coherent wake emission" (CWE, see [55]) as the responsible mechanism, since this would request a cutoff around $\omega_p \approx 10\omega_0$. From this, we conclude that the radiation cannot be attributed to any of the known mechanisms.

To get a picture of the physics behind, let us have a look at the motion of the plasma electrons that generate the radiation. Figure 9.12 shows the evolution of the electron density in both our sample cases. In addition to the density, contour lines of the spectrally filtered reflected radiation are plotted. These lines illustrate where the main part of the high frequency radiation emerges. We observe that in both cases the main part of the harmonics is generated at the point, when the electrons move toward the observer. This shows again that in both cases the radiation does not stem from CWE. For CWE harmonics, the radiation is generated inside the plasma, at the instant when the Brunel electrons re-enter the plasma [55]. Apart from that mutuality, the two presented cases appear to be very different.

Figure 9.12a corresponds to the BGP case. It can be seen that the density profile remains roughly step-like during the whole interaction process and the plasma skin layer radiates as a whole. This explains why the BGP theory works well here, as we have seen before in Fig. 9.11a and b.

However, Fig. 9.12b looks clearly different. The density distribution at the moment of harmonics generation is far from being step-like, but possesses a highly dense (up to $\sim$10,000 N_c density) and very narrow δ-like peak, with a width of only a few nanometers. This electron "nanobunch" emits synchrotron radiation coherently.

The radiation is emitted by a highly compressed electron bunch moving *away* from the plasma. However, the electrons first become compressed by the relativistic ponderomotive force of the laser that is directed into the plasma, compare the blue lines in Fig. 9.13. During that phase, the longitudinal electric field component grows until the electrostatic force turns around the bunch, compare the green lines in Fig. 9.13. Normally, the bunch will lose its compression in that instant, but in some cases, as in the one considered here, the fields and the bunch current match in a way that the bunch maintains or even increases its compression. The final stage is depicted by the red lines in Fig. 9.13.

We emphasize that such extreme nanobunching does not occur in every case of p-polarized oblique incidence of a highly relativistic laser on an overdense plasma surface. On the contrary, it turns out that the process is highly sensitive to changes in the plasma density profile, laser pulse amplitude, pulse duration, angle of incidence,

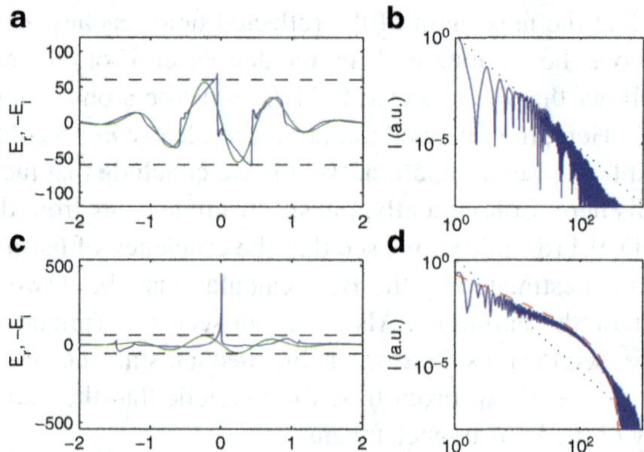

Fig. 9.11 Radiation in time [(**a**) and (**c**)] and spectral [(**b** and (**d**)] domain for two simulations. (**a**) and (**b**) correspond to the BGP case: normal incidence, plasma density $N_e = 250\,N_c$, sharp-edged profile; (**c**) and (**d**) correspond to the nanobunching case: plasma density ramp $\propto \exp(x/(0.33\,\lambda))$ up to a maximum density of $N_e = 95\,N_c$ (lab frame), oblique incidence at 63° angle (p-polarized). Laser field amplitude is $a_0 = 60$ in both cases. In all frames, the reflected field is represented by a blue line. In (**a**) and (**c**), the *green line* represents the field of the incident laser and the *black dashed lines* mark the maximum field of it. In (**b**) and (**d**), the *dotted black line* represents an 8/3 power law, the *red dashed line* corresponds to the 1D synchrotron spectrum (9.57) and (9.58), with $\omega_{\rm rs} = 800\,\omega_0$ and $\omega_{\rm rf} = 225\,\omega_0$

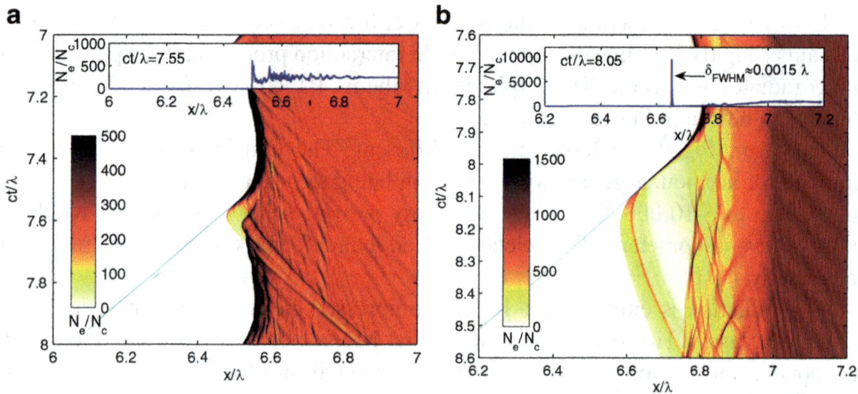

Fig. 9.12 The electron density and contour lines (*cyan, or light gray*) of the emitted harmonics radiation for $\omega/\omega_0 > 4.5$, in (**a**) the BGP and (**b**) the synchrotron (or nanobunching) regime. The small windows inside the main figures show the detailed density profile at the instant of harmonic generation. All magnitudes are taken in the simulation frame. The parameters used are the same as in the other figures

and even the carrier envelope phase of the laser. The parameters in the example were selected in a way to demonstrate the new effect unambiguously, i.e. the nanobunch is well formed and emits a spectrum that clearly differs from the BGP one. The dependence of the effect on some parameters is discussed in Sect. 9.4.3.

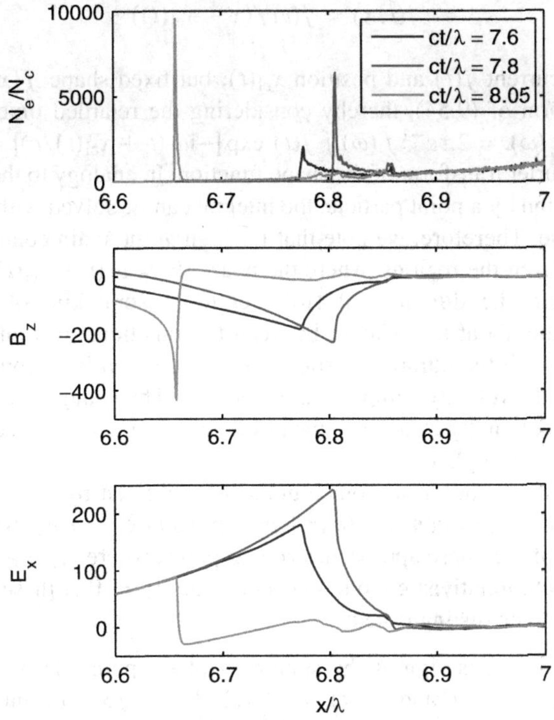

Fig. 9.13 Formation of the nanobunch in the simulation corresponding to Figs. 9.11c-d and 9.12b. We depict the electron density N_e in units of the critical density N_c, the transverse magnetic field component B_z and the longitudinal electric field component E_x in relativistically normalized units

Because of the one-dimensional slab geometry, the spectrum is not the same as the well-known synchrotron spectra [20] of a point particle. We now calculate the spectrum analytically.

9.4.1 Spectrum of 1D Coherent Synchrotron Emission

The radiation field generated to the left of a one-dimensional current distribution in a completely general way can be expressed as:

$$E_{\mathrm{sy}}(t,x) = \frac{2\pi}{c}\int_{-\infty}^{+\infty} j\left(t+\frac{x-x'}{c},x'\right)\mathrm{d}x'. \tag{9.54}$$

Optimal coherency for high frequencies will certainly be achieved, if the current layer is infinitely narrow: $j(t,x) = j(t)\delta(x - x_{\mathrm{el}}(t))$. To include more realistic cases, we allow in our calculations for a narrow, but finite electron distribution:

$$j(t,x) = j(t) f(x - x_{\mathrm{el}}(t)) \tag{9.55}$$

with variable current $j(t)$ and position $x_{\mathrm{el}}(t)$, but fixed shape $f(x)$. We take the Fourier transform of (9.54), thereby considering the retarded time, and arrive at the integral $\tilde{E}_{\mathrm{sy}}(\omega) = 2\pi c^{-1} \tilde{f}(\omega) \int j(t) \exp[-\mathrm{i}\omega (t + x_{\mathrm{el}}(t)/c)]\, \mathrm{d}t$. Here, $\tilde{f}(\omega)$ denotes the Fourier transform of the shape function. In analogy to the standard synchrotron radiation by a point particle, the integral can be solved with the method of stationary phase. Therefore, we note that for high ω the main contributions to the integral come from the regions, where the phase $\Phi = \omega (t + x_{\mathrm{el}}(t)/c)$ is approximately stationary, i.e. $\mathrm{d}\Phi/\mathrm{d}t \approx 0$. However, to get some kind of result we need some assumption about the relation between the functions $j(t)$ and $x_{\mathrm{el}}(t)$. Since we are dealing with the ultrarelativistic regime $a_0 \gg 1$, it is reasonable to assume that changes in the velocity components are governed by changes in the direction of movement rather than by changes in the absolute velocity, which is constantly very close to the speed of light c.

Let the electron motion in momentum space be given by $(p_x, p_y) = a_0 m_e c\, (\hat{p}_x, \hat{p}_y)$, so that $j(t) = ecn_e \hat{p}_y/(a_0^{-2} + \hat{p}_x^2 + \hat{p}_y^2)$ and $\dot{x}_{\mathrm{el}} = c\hat{p}_x/(a_0^{-2} + \hat{p}_x^2 + \hat{p}_y^2)$. The derivative of the phase approaches zero at points where $\dot{x}_{\mathrm{el}} \approx -c$, so from our assumption of ultrarelativistic motion we conclude $\hat{p}_y = 0$ at these instants. Now, two cases have to be distinguished:

1. The current changes sign at the stationary phase point. Then we can Taylor expand $j(t) = \alpha_0 t$ and $x_{\mathrm{el}}(t) = -v_0 t + \alpha_1 t^3/3$. The integral can now be expressed in terms of the well-known Airy-function, yielding $\tilde{E}_{\mathrm{sy}}(\omega) = C\, \tilde{f}(\omega)\, \omega^{-2/3} \mathrm{Ai}'\left(\frac{1-v_0}{\sqrt[3]{\alpha_1}} \omega^{2/3}\right)$, where Ai′ is the Airy function derivative and $C = \mathrm{i}(2\pi)^2 e n_e c^{-1} \alpha_0 \alpha_1^{-2/3}$ is a complex prefactor. We now find the spectral envelope

$$I(\omega) \propto |\tilde{f}(\omega)|^2\, \omega^{-4/3} \left[\mathrm{Ai}'\left(\left(\frac{\omega}{\omega_{\mathrm{rs}}}\right)^{2/3}\right)\right]^2 \tag{9.56}$$

with $\omega_{\mathrm{rs}} \approx 2^{3/2} \sqrt{\alpha_1} \gamma_0^3$, where $\gamma_0 = (1 - v_0^2)^{-1/2}$ is the relativistic γ-factor of the electron bunch at the instant when the bunch moves toward the observer. As in the BGP case, the spectral envelope (9.56) does not depend on all details of the electron bunch motion x_{el}, but only on its behavior close to the stationary points, i.e. the γ-spikes.
2. The current does not change sign at the stationary phase point. Because of the assumption of highly relativistic motion the changes in absolute velocity can again be neglected compared to the changes in direction, and it follows that in this case the third derivative of x_{el} is zero at the stationary phase point. Therefore, our Taylor expansions look like $j(t) = \alpha_0 t^2$ and $x_{\mathrm{el}}(t) = -v_0 t + \alpha_1 t^5/5$. This leads us to the spectral envelope

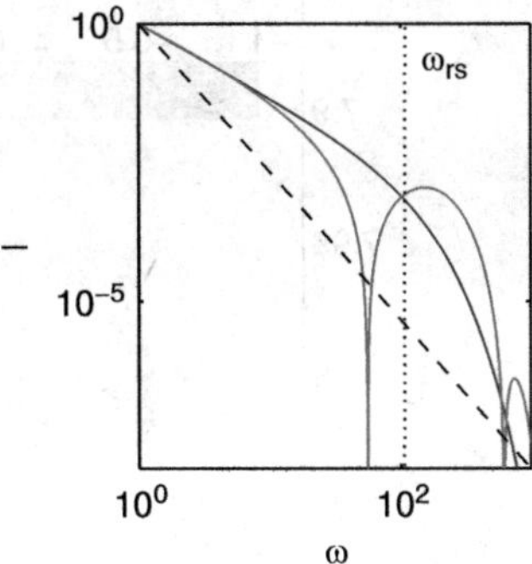

Fig. 9.14 Coherent 1D synchrotron spectra for an infinitely thin electron layer $\tilde{f}(\omega) \equiv 1$ and $\omega_{\mathrm{rs}} = 100$. The *blue line* corresponds to (9.56) and the *red line* to (9.57). For comparison, the *dashed black line* denotes the BGP 8/3-power law. The frequency ω is normalized to the laser fundamental frequency. The spectral intensity I of the reflected radiation is normalized to the intensity of the fundamental harmonic

$$I(\omega) \propto |\tilde{f}(\omega)|^2 \, \omega^{-6/5} \left[\mathrm{S}'' \left(\left(\frac{\omega}{\omega_{\mathrm{rs}}} \right)^{4/5} \right) \right]^2 \tag{9.57}$$

with S'' being the second derivative of $\mathrm{S}(x) \equiv (2\pi)^{-1} \int \exp\left[\mathrm{i}\left(xt + t^5/5\right)\right] \mathrm{d}t$, a special case of the canonical swallowtail integral [57]. For the characteristic frequency ω_{rs}, we now obtain $\omega_{\mathrm{rs}} \approx 2^{5/4} \sqrt[4]{\alpha_1} \gamma_0^{2.5}$. Because now even the derivative of $\ddot{x}_{\mathrm{el}}$ is zero at the stationary phase point, the influence of acceleration on the spectrum decreases and the characteristic frequency scaling is closer to the γ^2-scaling for a mirror moving with constant velocity.

In Fig. 9.14, the CSE spectra of the synchrotron radiation from the electron sheets are depicted. Comparing them to the 8/3-power law from the BGP-case, we notice that because of the smaller exponents of their power law part, the CSE spectra are much flatter, around the 100th harmonic we win more than two orders of magnitude. Another intriguing property are the side maxima found in the spectrum (9.57). This might provide an explanation for modulations that are occasionally observed in harmonics spectra, compare e.g. [56].

To compare with the PIC results, the finite size of the electron bunch must be taken into account. Therefore, we assume a Gaussian density profile which leads us to

$$|f(\omega)|^2 = \exp\left[-\left(\frac{\omega}{\omega_{\mathrm{rf}}} \right)^2 \right]. \tag{9.58}$$

Thus, the spectral cutoff is determined either by ω_{rs}, corresponding to the relativistic γ-factor of the electrons, or by ω_{rf}, corresponding to the bunch width. A look at the motion of the electron nanobunch in the PIC simulation (Fig. 9.15) tells us that there is no change in sign of the transverse velocity at the stationary phase

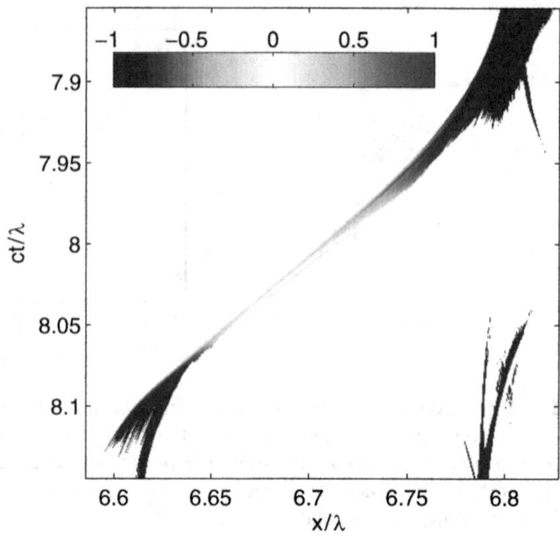

Fig. 9.15 Normalized transverse fluid velocity v_y/c of the electron nanobunch. Parts of the plasma with a density below 500 N_c are filtered out

point, consequently we use (9.57). We choose $\omega_{\mathrm{rf}} = 225\,\omega_0$ and $\omega_{\mathrm{rs}} = 800\,\omega_0$ to fit the PIC spectrum. ω_{rf} corresponds to a Gaussian electron bunch with a width of $\delta = 10^{-3}\lambda$, which matches reasonably well to the $\delta_{\mathrm{FWHM}} = 0.0015\,\lambda$ (see Fig. 9.12b) measured in the simulation, corresponding to a Gaussian electron bunch $f(x) = \exp\left[-(x/\delta)^2\right]$ with a width of $\delta = 10^{-3}\lambda$ and an energy of $\gamma \sim 10$. This matches well with the measured electron bunch width $\delta_{\mathrm{FWHM}} = 0.0015\,\lambda$ (see Fig. 9.12b) and the laser amplitude $a_0 = 60$, since we expect γ to be smaller but in the same order of magnitude as a_0. In this case $\omega_{\mathrm{rf}} < \omega_{\mathrm{rs}}$, so the cutoff is dominated by the finite bunch width. Still, both values are in the same order of magnitude, so that the factor coming from the Swallowtail-function cannot be neglected and actually contributes to the shape of the cutoff. The modulations that appear in Fig. 9.14 for frequencies around ω_{rs} and above cannot be seen in the spectra, because it is suppressed by the Gauss-function (9.58). The analytical synchrotron spectrum agrees excellently with the PIC result, as the reader may verify in Fig. 9.11d.

9.4.2 Properties of the CSE Radiation

As we see in Fig. 9.11c, the CSE radiation is emitted in the form of a single attosecond pulse whose amplitude is significantly higher than that of the incident pulse. This pulse has an FWHM duration of 0.003 laser periods, i.e. 9 as for a laser wavelength of 800 nm. This is very different from emission of the ROM harmonics, which need to undergo diffraction [37] or spectral filtering [13] before they take on the shape of attosecond pulses.

When we apply spectral a spectral filter in a frequency range $(\omega_{\mathrm{low}}, \omega_{\mathrm{high}})$ to a power-law harmonic spectrum with an exponent q, so that $I(\omega) = I_0(\omega_0/\omega)^q$, the energy efficiency of the resulting attosecond pulse generation process is

$$\eta_{\text{atto}} = \int_{\omega_{\text{low}}}^{\omega_{\text{high}}} I(\omega)\, d\omega \tag{9.59}$$
$$= \frac{I_0\omega_0}{q-1}\left[\left(\frac{\omega_0}{\omega_{\text{low}}}\right)^{q-1} - \left(\frac{\omega_0}{\omega_{\text{high}}}\right)^{q-1}\right].$$

The scaling (9.59) gives $\eta_{\text{atto}}^{\text{ROM}} \sim (\omega_0/\omega_{\text{low}})^{5/3}$ for the BGP spectrum with $q = 8/3$. For unfiltered CSE harmonics with the spectrum $q = 4/3$, the efficiency is close to $\eta_{\text{atto}}^{\text{CSE}} = 1$. This means that almost the whole energy of the original optical cycle is concentrated in the attosecond pulse. Note that absorption is very small in the PIC simulations shown; it amounts to 5% in the run corresponding to Fig. 9.11c-d and is even less in the run corresponding to Fig. 9.11a-b.

The ROM harmonics can be considered as a perturbation in the reflected signal as most of the pulse energy remains in the fundamental. On the contrary, the CSE harmonics consume most of the laser pulse energy. This is nicely seen in the spectral intensity of the reflected fundamental for the both cases (compare Fig. 9.11b and d). As the absorption is negligible, the energy losses at the fundamental frequency can be explained solely by the energy transfer to high harmonics. We can roughly estimate this effect by $I_0^{\text{BGP}}/I_0^{\text{CSE}} \approx \int_1^\infty \omega^{-8/3}\, d\omega / \int_1^\infty \omega^{-4/3}\, d\omega = 5$. This value is quite close to the one from the PIC simulations: $I_0^{(\text{Fig. 1b})}/I_0^{(\text{Fig. 1d})} = 3.7$.

Further, we can estimate amplitude of the CSE attosecond pulse analytically from the spectrum. Since the harmonic phases are locked, for an arbitrary power law spectrum $I(\omega) \propto \omega^{-q}$ and a spectral filter $(\omega_{\text{low}}, \omega_{\text{high}})$ we integrate the amplitude spectrum and obtain:

$$E_{\text{atto}} \approx \frac{2\sqrt{I|_{\omega=\omega_1}}}{q-2}\left[\left(\frac{\omega_0}{\omega_{\text{low}}}\right)^{\frac{q}{2}-1} - \left(\frac{\omega_0}{\omega_{\text{high}}}\right)^{\frac{q}{2}-1}\right]. \tag{9.60}$$

Apparently, when the harmonic spectrum is steep, i.e. $q > 2$, the radiation is dominated by the lower harmonics ω_{low}. This is the case of the BGP spectrum $q = 8/3$. That is why one needs a spectral filter to extract the attosecond pulses here. The situation changes drastically for slowly decaying spectra with $q < 2$ like the CSE spectrum with $q = 4/3$. In this case, the radiation is dominated by the high harmonics ω_{high}. Even without any spectral filtering the radiation takes on the shape of an attosecond pulse. As a rule of thumb formula for the attosecond peak field of the unfiltered CSE radiation, we can write:

$$E_{\text{atto}}^{\text{CSE}} \approx \sqrt{3}\left(m_c^{1/3} - 1\right) E_0. \tag{9.61}$$

Using $m_c = \omega_c/\omega_0 = 225$, the lower of the two cutoff harmonic numbers used for comparison with the PIC spectrum in Fig. 9.11d, we obtain $E_{\text{peak}} = 8.8\, E_0$. This is in nice agreement with Fig. 9.11c.

9.4.3 Parametrical Dependence of Harmonics Radiation

Now, we have a look at the dependence of the harmonics radiation in and close to the CSE regime on the laser and plasma parameters. We are going to examine two dimensionless key quantities: The intensity boost $\eta \equiv \max(E_r^2)/\max(E_i^2)$ and the pulse compression $\Gamma \equiv (\omega_0 \tau)^{-1}$. It is straightforward to extract both magnitudes from the PIC data, and both are quite telling. The intensity boost η is a sign of the mechanism of harmonics generation. If the ARP boundary condition (9.25) is approximately valid, we must of course have $\eta \approx 1$. Then again, if the radiation is generated by nanobunches, we expect to have strongly pronounced attosecond peaks (compare (9.61)) in the reflected radiation and therefore $\eta \gg 1$. The pulse compression Γ is defined as the inverse of the attosecond pulse duration. In the nanobunching regime, we expect it to be roughly proportional to η, as the total efficiency of the attosecond pulse generation remains $\eta_{\text{atto}} \leq 1$, compare (9.59). In the BGP regime, there are no attosecond pulses observed without spectral filtering. So the FWHM of the intensity peak is on the order of a quarter laser period, and we expect $\Gamma \sim 1$.

In Fig. 9.16, the two parameters η and Γ are shown in dependence of a_0. Except for the variation of a_0, the parameters chosen are the same as in Figs. 9.11c–d, 9.12b and 9.15.

First of all we notice that for all simulations in this series with $a_0 \gg 1$, we find $\eta \gg 1$. Thus, (9.25) is violated in all cases. Since we also notice $\Gamma \gg 1$ and $\Gamma \sim \eta$, we know that the radiation is emitted in the shape of attosecond peaks with an efficiency of the order 1. This indicates that we can describe the radiation as CSE. The perhaps most intriguing feature of Fig. 9.16 is the strongly pronounced peak of both curves around $a_0 = 55$. We think that because of some very special phase matching between the turning point of the electron bunch and of the electromagnetic wave, the electron bunch experiences an unusually high compression at this

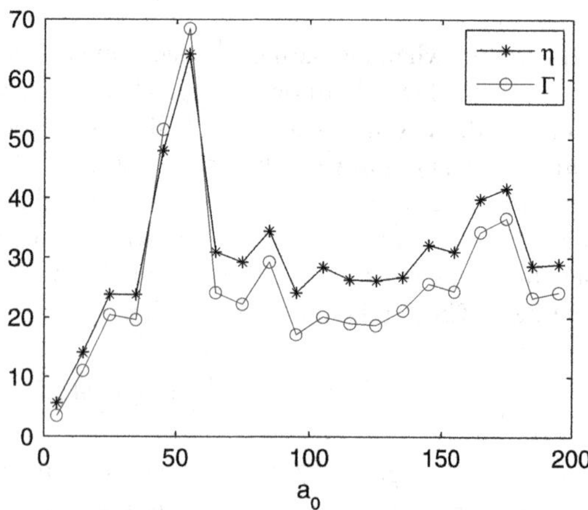

Fig. 9.16 Dependence of the intensity boost $\eta = \max(E_r^2)/\max(E_i^2)$ and the pulse compression $\Gamma = (\omega_0 \tau)^{-1}$, where τ is the FWHM width of the attosecond intensity peak in the reflected radiation, on a_0. The laser amplitude a_0 is varied between 5 and 195 in steps of 10. Other parameters are the same as in Fig. 9.12b

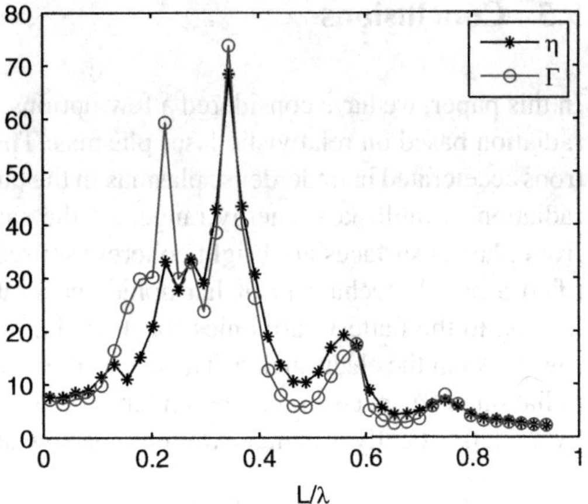

Fig. 9.17 Dependence of the intensity boost $\eta = \max(E_r^2)/\max(E_i^2)$ and the pulse compression $\Gamma = (\omega_0\tau)^{-1}$, on the plasma scale length in units of the laser wavelength L/λ in the lab frame. Except for the plasma scale length, parameters are the same as in Fig. 9.12b. The plasma ramp is again an exponential one $\propto \exp(x/L)$

parameter settings. This is the case that was discussed in Sect. 9.4. The thorough understanding of this effect requires detailed parametric studies that would result in a vast computational effort. Yet, the physics seems to be the trajectory crossing of Brunel electrons just at the right moment that results in a kind of wave breaking and thus enormous local density of the electron nanobunch, as seen in Fig. 9.12.

Figure 9.17 shows the two parameters η and Γ as functions of the plasma gradient scale length L. It is seen that both functions possess several local maxima. Further, η and Γ behave similar apart from one runaway value at $L = 0.225\lambda$, where the FWHM peak duration is extremely short, but the intensity boost is not as high. A look at the actual field data tells us that in this case the foot of the attosecond peak is broader, consuming most of the energy. This deviation might e.g. be caused by a different, non-gaussian shape of the electron nanobunch.

The maximum of both functions lies around $L = 0.33\lambda$, the parameter setting analyzed in detail before. In the limit of extremely small scale lengths $L \leq 0.1\lambda$, η and Γ become smaller, but they remain clearly bigger than one. Thus, the reflection in this parameter range can still not very well be described by the ARP boundary condition. For longer scale lengths $L > 0.8\lambda$, both key values approach 1, so the ARP boundary condition can be applied here. This is a possible explanation, why the BGP spectrum (9.46) could experimentally measured at oblique incidence [28].

The CSE spectrum (9.57) is significantly flatter than the universal BGP spectrum. However, we could observe it in our PIC simulations for ultra-short, sub-10 fs, laser pulses only. The spectrum is also sensitive to the carrier-envelope phase of the laser pulse. Further, the CSE spectrum cutoff is limited not only by the characteristic frequency ω_{rs}, but also by the thickness of the electron nanobunch [15].

Like the BGP-spectrum harmonics, the CSE harmonics are phase locked and appear in the form of attosecond pulses. In contrast to the BGP case, the attosecond pulses are visible immediately and have amplitudes much higher than the incident laser pulse. Therefore, no energy needs to be wasted in spectral filtering and the energy efficiency of attosecond pulse generation is close to 100%.

9.5 Conclusions

In this paper, we have considered a few options for novel sources of x-rays and xuv radiation based on relativistic laser plasmas. The synchrotron radiation out of electrons accelerated in underdense plasmas in the bubble regime can provide incoherent radiation in multi-keV energy range. At the same time, the relativistic harmonics from plasma surfaces are bright coherent sources of xuv and x-rays. We also identified a novel mechanism of harmonics generation at overdense plasma surfaces, leading to the flattest harmonics spectrum known so far. Extremely dense and narrow peaks in the electron density, we call them nanobunches, are responsible for the radiation. The spectrum can be understood as a 1D synchrotron spectrum emitted by a relativistically moving, extremely narrow and dense electron layer.

Acknowledgements The work has been supported in parts by DFG Transregio SFB TR18, DFG Graduiertenkolleg GRK 2103, EU FP7 project LAPTECH and by MES RF, project nr. 02.740.11.5168.

References

1. A. Pukhov, Z. M. Sheng, J. Meyer-ter-Vehn, Phys. Plasmas. **6**, 2847 (1999)
2. A. Pukhov, J. Meyer-ter-Vehn, Appl. Phys. B **74**, 355 (2002)
3. S.P.D. Mangles et al., Nature **431**, 535 (2004)
4. C.G.R. Gedder, et al., Nature **431**, 538 (2004)
5. J. Faure et al., Nature **431**, 541 (2004)
6. S. Gordienko, A. Pukhov, Phys. Plasmas. **12**, 043109 (2005)
7. S. Kiselev, A. Pukhov, I. Kostyukov, Phys. Rev. Lett. **93**, 135004 (2004)
8. A. Rousse, K.T. Phuoc, R. Shah et al., Phys. Rev. Lett **93**, 135005 (2005)
9. S.P.D. Mangles, G. Genoud, S. Kneip et al., Appl. Phys. Lett. **95**, 181106 (2009)
10. S. Kneip et al., *A Bright Spatially-Coherent Compact X-ray Synchrotron Source* arXiv:0912.1812v1 (2010)
11. F.Gruner et al., Appl. Phys. B **86**, 431 (2007)
12. J. Meyer-ter-Vehn, H.C. Wu, Eur. Phys. J. D **55**, 433 (2009)
13. T. Baeva, S. Gordienko, A. Pukhov, Phys. Rev. E **74**, 046404 (2006)
14. B. Dromey, S. Kar, C. Bellei et al., Phys. Rev. Lett. **99**, 085001 (2007)
15. D. an der Brugge, A. Pukhov, Phys. Plasmas. **17**, 033110 (2010)
16. I. Kostyukov, A. Pukhov, S. Kiselev, Phys. Plasmas. **11**, 5256 (2004)
17. I. Kostyukov, E. Nerush, A. Pukhov et al., New J. Phys. **12**, 045009 (2010)
18. I. Kostyukov, E. Nerush, A. Pukhov, et al. Phys. Rev. Lett. **103**, 175003 (2009)
19. S. Kalmykov, et al., Phys. Rev. Lett. **103**, 135004 (2009)
20. J.D. Jackson, *Classical Electrodynamics* (Wiley, New York, 1999)
21. S. Kohlweyer et al., Opt. Commun. **117**, 431 (1995)
22. D. von der Linde et al., Phys. Rev. A **52**, R25 (1995)
23. P.A. Norreys et al., Phys. Rev. Lett **76**, 1832 (1996)
24. M. Zepf, G.D. Tsakiris et al., Phys. Rev. E **58**, R5253 (1998)
25. U. Teubner et al., Phys. Rev. A 01381 (2003)
26. I. Watts et al., Phys. Rev. Lett. **88**, 155001-1 (2002)
27. K. Eidmann et al., Phys. Rev. E **72**, 036413 (2005)

28. B. Dromey, M. Zepf, A. Gopal, K. Lancaster, M.S. Wei, K. Krushelnick, M. Tatarakis, N. Vakakis, S. Moustaizis, R. Kodama, M. Tampo, C. Stoeckl, R. Clarke, H. Habara, D. Neely, S. Karsch, P. Norreys, Nature Phys. **2**, 456 (2006)
29. B. Dromey, D. Adams, R. Horlein, Y. Nomura, S.G. Rykovanov, D.C. Carroll, P.S. Foster, S. Kar, K. Markey, P. McKenna, D. Neely, M. Geissler, G.D. Tsakiris, M. Zepf, Nature Phys. **5**, 146 (2009)
30. C. Thaury, F. Quere, J.P. Geindre et al., Nature Phys. **3**, 424 (2007)
31. C. Thaury, H. George, F. Quere et al., Nature Phys. **4**, 631 (2008)
32. S. Gordienko, A. Pukhov, O. Shorokhov, T. Baeva, Phys. Rev. Lett. **93**, 115002 (2004)
33. G.D. Tsakiris et al., New J. Phys. **8**, 19 (2006)
34. B. Dromey, M. Zepf, A. Gopal et al., Nature Phys. **2**, 456 (2006)
35. A. Pukhov, Nature Phys. **2**, 439 (2006)
36. A. Bourdier, Phys. Fluid. **26**, 1804 (1983)
37. D. An der Brügge, A. Pukhov, Phys. Plasma. **14**, 093104 (2007)
38. R. Kienberger, F. Krausz, Topics Appl. Phys. **95**, 343 (2004)
39. E. Goulielmakis et al., Science **305**, 1267 (2004)
40. R. Kienberger et al., Nature **427**, 817 (2004)
41. J. Itatani et al., Nature **432**, 867 (2004)
42. H. Niikura et al., Nature **417** 917 (2002)
43. R.L. Carman et al., Phys. Rev. Lett. **46**, 29 (1981)
44. B. Bezzerides et al., Phys. Rev. Lett. **49**, 202 (1982)
45. S.V. Bulanov et al., Phys. Plasma. **1**, 745 (1993)
46. P. Gibbon, Phys. Rev. Lett. **76**, 50 (1996)
47. R. Lichters et al., Phys. Plasma. **3**, 3425 (1996)
48. D. von der Linde, K. Rzazewski, Appl. Phys. B **63**, 499 (1996)
49. R. Ondarza, Phys. Rev. E **67**, 066401 (2003)
50. K. Eidmann et al., Phys. Rev. E **72**, 036413 (2005)
51. L. Plaja et al., J. Opt. Soc. Am. B **7**, 1904 (1998)
52. L.D. Landau, E.M. Lifshitz, L.P. Pitaevskii *Electrodynamics of Continuous Media* (Pergamon Press, Oxford, 1984)
53. N.G. de Bruijn *Asymptotic Methods in Analysis* (Dover, New York, 1981)
54. A. Pukhov, J. Plasma Phys. **61**, 425 (1999)
55. F. Quere, C. Thaury, P. Monot, S. Dobosz, P. Martin, J.-P. Geindre, P. Audebert, Phys. Rev. Lett. **96**, 125004 (2006)
56. T.J.M. Boyd, R. Ondarza-Rovira, Phys. Rev. Lett. **101**, 125004 (2008)
57. J.N.L. Connor, P.R. Curtis, D. Farrelly, J. Phys. A Math. General, **17**, 283 (1984)

Chapter 10
Ion Acceleration in Subcritical Density Plasma via Interaction of Intense Laser Pulse with Cluster-Gas Target

Y. Fukuda, A.Ya. Faenov, M. Tampo, T.A. Pikuz, T. Nakamura, M. Kando, Y. Hayashi, A. Yogo, H. Sakaki, T. Kameshima, K. Kawase, A.S. Pirozhkov, K. Ogura, M. Mori, T. Zh. Esirkepov, J. Koga, A.S. Boldarev, V.A. Gasilov, A.I. Magunov, T. Yamauchi, R. Kodama, P.R. Bolton, K. Kondo, S. Kawanishi, Y. Kato, T. Tajima, H. Daido, and S.V. Bulanov

Abstract We present substantial enhancement of the accelerated ion energies up to 10–20 MeV per nucleon by utilizing the unique properties of the cluster-gas target irradiated with 40-fs laser pulses of only 150 mJ energy, corresponding to

---

Y. Fukuda (✉), A.Ya. Faenov, M. Tampo, T.A. Pikuz, T. Nakamura, M. Kando, Y. Hayashi, A. Yogo, H. Sakaki, T. Kameshima, K. Kawase, A.S. Pirozhkov, K. Ogura, M. Mori, T. Zh. Esirkepov, J. Koga, P.R. Bolton, K. Kondo, S. Kawanishi, Y. Kato, T. Tajima, H. Daido, and S.V. Bulanov
Kansai Photon Science Institute (KPSI) and Photo-Medical Research Center (PMRC), Atomic Energy Agency (JAEA), Kyoto, 615–0215 Japan
e-mail: fukuda.yuji@jaea.go.jp, tampo.motonobu@jaea.go.jp, nakamura.tatsufumi@jaea.go.jp, kando.masaki@jaea.go.jp, hayashi.yukio@jaea.go.jp, yogo.akifumi@jaea.go.jp, sakaki.hironao@jaea.go.jp, kameshima@spring8.or.jp, kawase@sanken.osaka-u.ac.jp, Pirozhkov.Alexander@jaea.go.jp, ogura.koichi@jaea.go.jp, mori.michiaki@jaea.go.jp, Timur.Esirkepov@jaea.go.jp, koga.james@jaea.go.jp, bolton.paul@jaea.go.jp, kondo.kiminori@jaea.go.jp kawanishi.shunichi@jaea.go.jp, y.kato@gpi.ac.jp, Toshiki.Tajima@physik.uni-muenchen.de, daido.hiroyuki@jaea.go.jp

A.Ya. Faenov and T.A. Pikuz
Joint Institute of High Temperatures, Russian Academy of Sciences, Moscow 125412 Russia
e-mail: anatolyf@hotmail.com, tapikuz@yahoo.com

A.S. Boldarev and V.A. Gasilov
Keldysh Institute of Applied Mathematics, Russian Academy of Sciences (RAS), Moscow, 125047 Russia
e-mail: boldar@imamod.ru, vgasilov@yandex.ru

A.I. Magunov and S.V. Bulanov
A.M. Prokhorov Institute of General Physics, Russian Academy of Sciences (RAS), Moscow, 119991 Russia
e-mail: mai@mail.ru, bulanov.sergei@jaea.go.jp

T. Yamauchi
Graduate School of Maritime Sciences, Kobe University, Kobe 658–0022, Japan
e-mail: yamauchi@maritime.kobe-u.ac.jp

R. Kodama
Faculty of Engineering and Institute of Laser Engineering, Osaka University, Osaka, 565–0871 Japan
e-mail: kodama@eie.eng.osaka-u.ac.jp

K. Yamanouchi et al. (eds.), *Progress in Ultrafast Intense Laser Science VII*, Springer Series in Chemical Physics 100, DOI 10.1007/978-3-642-18327-0_10,

approximately tenfold increase in the ion energies compared to previous experiments using thin foil targets. A particle-in-cell simulation infers that the high energy ions are generated at the rear side of the target due to the formation of a strong dipole vortex structure in subcritical density plasmas. The demonstrated method can be important in the development of efficient laser ion accelerators for hadron therapy and other applications.

10.1 Introduction

The laser-driven ion acceleration via the interaction of short, intense laser pulses with matter, known as laser–plasma acceleration, is featured by its high accelerating electric fields and short pulse length compared to the conventional radiofrequency (RF) accelerators. It has been one of the most active areas of research during the last several years [1–3]. The laser-driven ion beams can be employed in a broad range of applications in cancer therapy [4–6], isotope preparation for medical applications [7], proton radiography [8] and controlled thermonuclear fusion [9].

In the case of multi-MeV ion acceleration in "overdense plasmas" via laser-thin foil target interactions, widely used as a model for the study of ion acceleration, a number of experiments [10–12] have been so far carried out where protons have been accelerated from the rear surface of thin foil targets by a strong electrostatic field set up by relativistic electron beam, i.e. the process known as target normal sheath acceleration (TNSA) [13] (see for details [14–18]). In this realm, quasimonoenergetic ions can be generated using microstructured targets [4, 19–21]. The divergence and the energy spread of ions can be controlled with a plasma microlens [22] or magnet systems [23–26]. Recent improvement of the contrast ratio of laser pulses allowed the ion energy increase by decreasing the target thickness [27–30]. The effect of the self-induced transparency and the laser radiation pressure is dominant in [28] and [29], respectively. Increase in the laser peak intensity using a compact plasma-based focusing optics also allowed the ion energy increase [31]. These advances establish a firm basis for realizing various applications of laser-driven ions.

On another front, the multi-MeV ion acceleration in "underdense plasmas" has been demonstrated in experiments where the lower-density plasma has been produced by direct ionization of gas targets [32–36] or by evaporation of thin solid targets by the laser prepulse [37, 38]. Ions are accelerated radially in [32–34] by a collisionless shock acceleration mechanism, while in the case of [35, 36] and [37, 38], an electric field is generated at the plasma–vacuum interface through a combination of charge separation and a quasistatic magnetic field produced by the fast electron current which can accelerate ions in the forward direction and collimate them radially [39–42]. The effect of the quasistatic magnetic field is dominant for the shorter laser pulse lengths in [37] and [38].

When a cluster-gas target, which consists of solid-density clusters embedded in a background gas, is irradiated by high intensity laser light, it renders ion acceleration in underdense plasmas a truly unique property [43]. Because a cluster and a background gas in the cluster-gas target produce overdense plasma and underdense plasma, respectively, the target can produce subcritical density plasma, through which a laser pulse can penetrate and effectively induces various nonlinear effects. In fact, efficient plasma waveguide formation in cluster-gas targets even at non-relativistic laser intensity has been reported [44, 45], where, in addition to usual self-focusing due to the Kerr effect, a positive and concave shaped refractive index structure created during the early part of the laser–cluster interaction induces additional self-focusing. Generation of a large quantity of high energy electrons during the intense laser pulse interaction with the cluster-gas target has been reported [46], where the massive electrons are injected when they are expelled from the clusters by the strong laser pulse fields. These properties are quite favorable for enhancement of the ion energy to the range usable in cancer therapy, because the self-focusing can lead to intensification of the laser pulse and transport the laser energy to the rear side of the target and the high electron current can create the strong electromagnetic structures. Such electromagnetic structures can remain for a time much longer than the laser pulse duration. Accordingly, the cluster-gas target is expected to accelerate ions to higher energies with relatively low input laser energy. Furthermore, the replenishable cluster-gas target enables high repetition rate of high energy ion generation, free of plasma debris. These properties are quite advantageous for practical applications of laser-driven ion beam. We note that this approach is different from previous works on ion acceleration due to the Columb explosion of clusters [47–49].

Here we present details of ion acceleration with tenfold enhancement of accelerated ion energies in the forward direction up to 10–20 MeV per nucleon in a mixture of submicron-sized CO_2 clusters and He gas irradiated by laser pulses of moderate energy. Ions are detected by using CR39 nuclear track detectors and by using the time-of-flight (TOF) method. Through a shadowgraphy, we found a 5-mm long stable channel formation demonstrating the highly nonlinear laser-plasma coupling, inducing the strong long-lasting electromagnetic structures, from which the high energy ions are accelerated in the laser propagation direction.

10.2 Experimental Methods

10.2.1 Laser System

The experiment has been conducted using the JLITE-X Ti:sapphire laser at JAEA-KPSI [50]. A schematic of experimental setup is shown in Fig. 10.1a. The laser delivers pulses with a central wavelength of 820 nm, a pulse duration of 40 fs, a pulse energy of 150 mJ, and a contrast ratio of 10^{-6} at a repetition rate of 1 Hz. The laser pulse was split into a main pulse and a lower energy probe pulse. The main laser pulse was focused to a $30-\mu$m diameter spot ($1/e^2$ intensity) with an off-axis parabola having an effective focal length of 646 mm. This yields a peak vacuum intensity of 7×10^{17} W/cm^2.

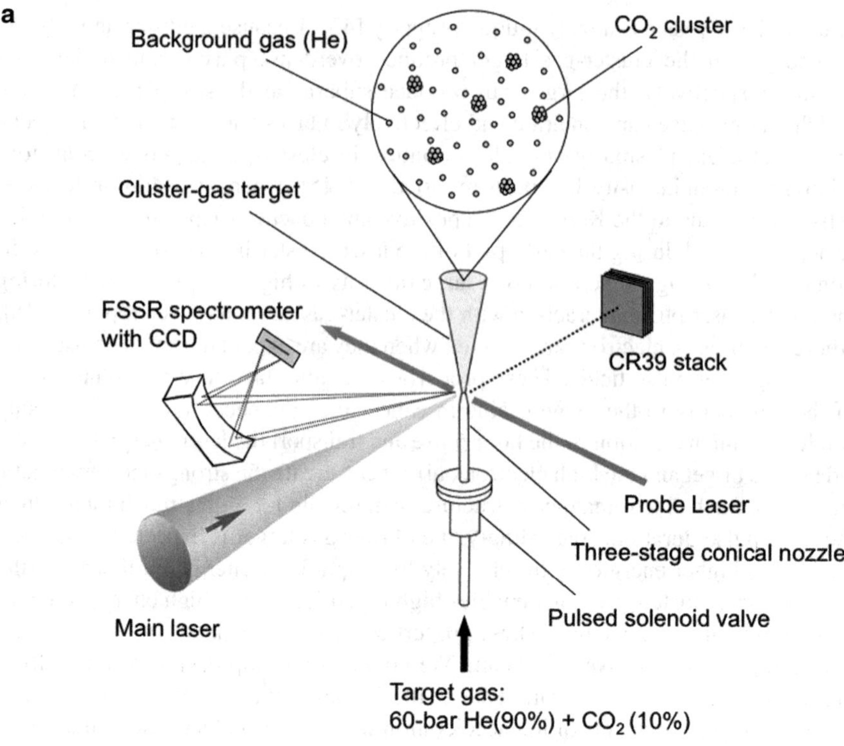

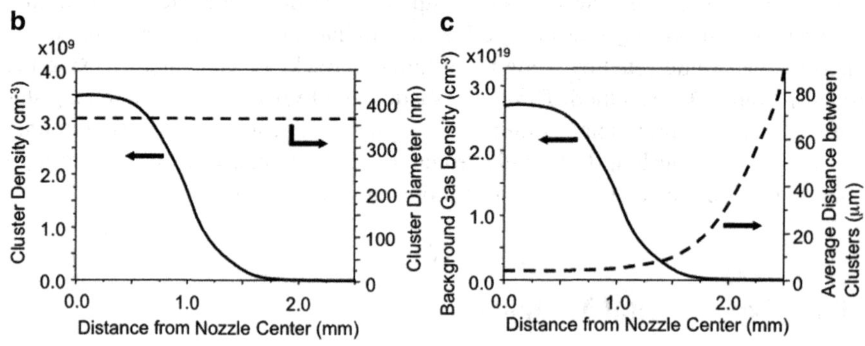

Fig. 10.1 (**a**) Schematic of the experimental setup for ion measurements with a stack of CR-39 detectors. (**b**) The cluster density and the average distance between clusters, (**c**) the background gas density and average cluster diameter, at the distance of 1.5 mm from the nozzle orifice, calculation [51]

10.2.2 Cluster-Gas Target

A pulsed solenoid valve connected to a specially designed circular nozzle having a three-stage conical structure with an orifice diameter of 2 mm was used to produce submicron-size CO_2 clusters embedded in He gas. With the aid of a numerical model [51], we find that a 60-bar gas consisting of 90% He and 10% CO_2 is optimal for the production of submicron-size CO_2 clusters. The reliability of the model has been verified by the experiment with the use of Rayleigh scattering of the light on clusters [52]. To accelerate ions, the laser beam was focused near the rear side of the gas jet, 1.5 mm above the nozzle orifice, where, using the model described in [51], we calculated that solid density CO_2 clusters with an average diameter of $\sim 0.4\,\mu$m (a root mean square deviation of $\sim 0.1\,\mu$m), containing $\sim 5 \times 10^8$ molecules each, are embedded in the He gas of density $\sim 2 \times 10^{19}\,\text{cm}^{-3}$ which corresponds to $\sim 0.02 n_c$, where $n_c = m_e \omega^2/4\pi e^2$ is the critical density. Here, a high cluster density ($\sim 3 \times 10^9\,\text{cm}^{-3}$) and a small intercluster distance ($\sim 5\,\mu$m) are achieved (see Fig. 10.1b–c).

10.2.3 Plasma Conditions

Our target is a cloud of solid-density submicron-size clusters embedded in a background gas. Under the action of the prepulse accompanying the main laser pulse, the clusters are expected to be evaporated forming a subcritical inhomogeneous plasma with its profile determined by the initial density distribution of the cluster-gas target.

The plasma condition is monitored by measuring soft X-ray spectra of the He_β (665.7 eV) and Ly_α (653.7 eV) lines of oxygen using a focusing spectrometer with two-dimensional spatial resolution [53] equipped with a spherically bent mica crystal and a back illuminated CCD camera, Fig. 10.1a. Plasma parameter diagnostics using comparison of relative intensities and shapes of X-ray spectral lines with numerical simulation [54] indicates the generation of subcritical density plasma (see Fig. 10.2a). It gives the electron density of $1.0 \pm 0.2 \times 10^{20}\,\text{cm}^{-3}$, which is about $0.1 n_c$.

It is expected that at least four times enhancement of the peak laser intensity occurs during its propagation through the target, because the peak power of the main laser pulse is well above the critical power for relativistic self-focusing. The laser propagation in the target was monitored by shadowgraph images with the probe laser. The shadowgraph image shown in Fig. 10.2b reveals the formation of a channel of approximately 5 mm in length, substantially longer than the nozzle orifice diameter (2 mm) and the Rayleigh length (900 μm).

Regarding the ionization levels of clusters, in addition to field ionization such as barrier suppression ionization having a timescale of femtoseconds, it is well known that collisional ionization having a timescale of picoseconds by energetic electrons play an important role in ionizing clusters to produce unexpectedly high charge states [55–57]. Even without considering the self-focusing effect, the ponderomotive potential (effective quiver energy acquired by an oscillating electron) of the laser field for $7 \times 10^{17}\,\text{W/cm}^2$ is about 40 keV. Since the ionization potential for C^{5+} and O^{7+} are 490 and 871 eV, respectively, it is highly plausible to produce fully stripped ions via collisional ionization by energetic electrons. In fact, we have observed soft X-ray emissions from the Rydberg states of O^{7+}, i.e. He_β (665.7 eV) and Ly_α (653.7 eV) lines of oxygen. This is the strong evidence for the production of completely ionized plasma under our experimental conditions.

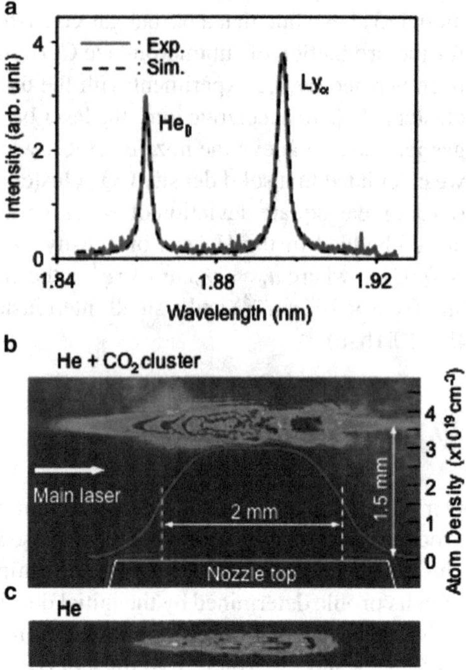

Fig. 10.2 (**a**) Measured (*solid curve*) and calculated (*dotted curve*) soft X-ray spectra for the subcritical density plasma of the order of 10^{20} cm^{-3}, which is about $0.1n_c$. (**b**) The shadowgraph image for a mixture of He gas and CO_2 clusters. The *solid curve* shows the initial atom density profile. (**c**) The shadowgraph image for a pure He gas target

10.3 Experimental Results

10.3.1 Characterization of High Energy Ions Using a Stack of CR-39

To detect high energy ions accelerated in the forward direction (the laser propagation direction), we placed a stack of solid state nuclear track detectors (SSNTDs) on the laser propagation axis at a distance 200 mm from the laser focal plane. The SSNTD stack consists of ten sheets of 10-μm thick polycarbonate and twelve sheets of 100-μm thick CR-39 (HARZLAS TNF-1, Nagase-Landauer) with a transverse size of 40 × 40 mm. One layer of 6-μm thick Al foil was placed ahead of these to protect the plastic nuclear detectors from damage induced by the transmitted portion of the main femtosecond pulses. The CR-39 is a polymer of optical quality, which suffers a microscopic defect when hit by an energetic particle. This defect can be enhanced into a visible track when the CR-39 sample is etched in concentrated KOH solution.

We accumulated ion signals for about six thousand laser shots. The pits were established following 9 h of etching in a solution of 6N-KOH at a temperature of 70°C. Figure 10.3a–d shows typical images of etched pits registered in the same position of the 1st, 8th, 11th, and 12th layers of CR39, respectively, observed with a differential interference microscope. We note that ions penetrate through several

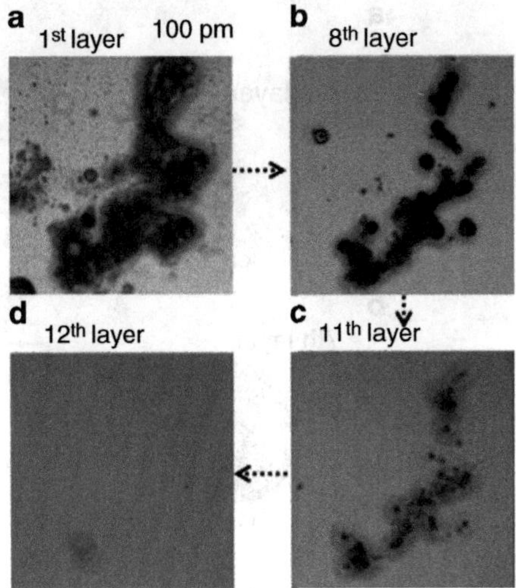

Fig. 10.3 Typical microscope images of the etched pits registered in the (**a**) 1st, (**b**) 8th, (**c**) 11th, and (**d**) 12th layers of CR-39, observed with a differential interference microscope

successive CR39 layers at exactly the same lateral and vertical positions, and vanish at some layers which correspond to the depth of the Bragg peak for ions in the CR39 stack.

Figures 10.4a–d show detailed images of the ion pits registered on the front surface of the 6th, 7th, 8th, and 9th layers of CR39, respectively, observed with a confocal laser scanning microscope. The line profiles of the front and rear surfaces of these layers are shown in the right side of Fig. 10.4a–d. The etched pits have a conical shape and clearly indicate that high energy ions penetrate through successive CR39 layers and vanish. For example, the track of ion 1 vanishes at the rear surface of the 8th layer, while the track of ion 3 appears at the rear surface of the 7th layer. The track diameter of ion 2 takes its maximum at the rear surface of 6th layer and decreases thereafter. These observations are consistent with the Bragg peak phenomenon associated with ion stopping power, i.e. the well localized maximum of ion energy loss in matter.

We note that the diameter of the etched pits has two-humped distribution. Since the cluster-gas target is a mixture of He and CO_2 clusters, highly charged ions of helium, carbon, and oxygen are the possible candidates for the accelerated ions registered in the CR39. Since the track registration sensitivity depends strongly on the ion charge [58], we ascribe that the smaller pits are due to the helium ions and the larger pits to the carbon or oxygen ions.

The energy range of ions is determined quantitatively from the extent of these tracks made in the CR-39 stack by calculating their stopping ranges using the SRIM code [59]. We observe the ion tracks in CR39 up to the 11th layer and none in the 12th layer, which corresponds to maximum ion energies of 10, 17, and 20 MeV per nucleon for helium, carbon, and oxygen, respectively. The number of ions detected on the 11th layer of the CR39 stack is about 5/MeV/Sr/shot. The track images on the CR39 show that these high energy ions are well collimated with a divergence (full angle) of 3.4° in the forward direction.

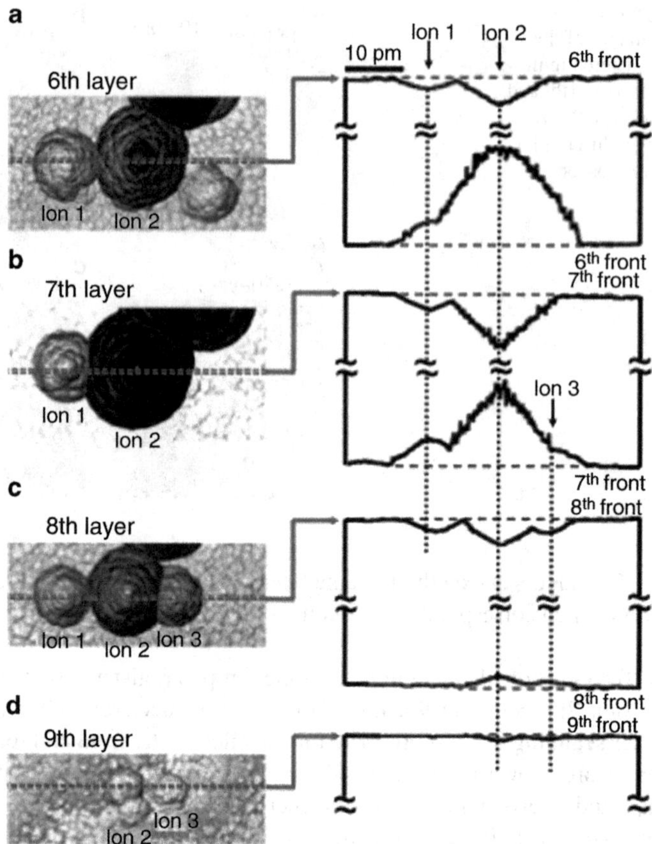

Fig. 10.4 Detailed microscope images of the etched pits registered in the front surface of the (**a**) 6th, (**b**) 7th, (**c**) 8th, and (**d**) 9th layers of CR-39, observed with a confocal laser scanning microscope. The line profiles of the front and rear surfaces of these layers are shown in the right side

10.3.2 Characterization of High Energy Ions Using a TOF Method

Further evidence for high energy ion production is obtained from TOF measurements made using a microchannel plate (MCP) detector having an effective area of 14.5 mm in diameter (F4655–12, Hamamatsu), that was placed 930 mm from the gas nozzle along the laser propagation axis in place of the SSNTD stack in a separate data acquisition run of the same series of experiment (see Fig. 10.5). The acceptance angle of MCP (1.0°) is smaller than the divergence angle (3.4°) of high energy ions. The MCP registers a real-time signal from each repetitive laser shot. Since the MCP is sensitive not only to ions, but also to electrons and X rays produced in the laser–cluster interaction, an electromagnet (0.15 T, 40 mm in diameter) was

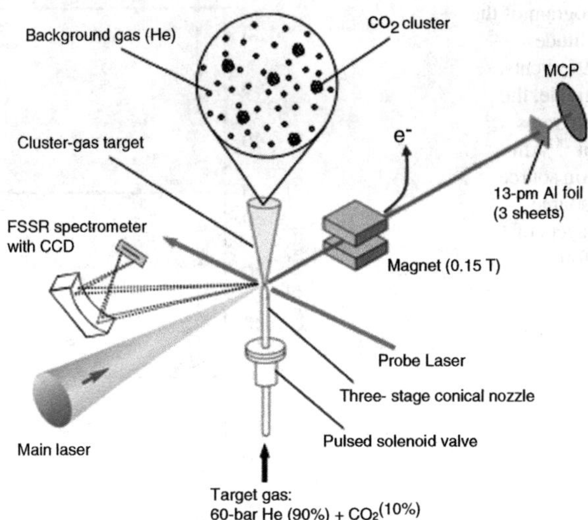

Fig. 10.5 Schematic of the experimental setup for ion measurements with a time-of-flight method

placed between the focal plane for the main laser pulse and the MCP detector. Consequently, electrons with energies below 20 MeV were deflected and did not reach the MCP detector. In addition, three layers of 13-μm thick Al foil were inserted to block the transmitted portion of the main femtosecond pulses and to reduce the incident X-ray irradiation. Thus helium, carbon, and oxygen ions with energy greater than 1.9, 3.2, and 3.6 MeV per nucleon, respectively, can pass through three Al foils and reach the MCP.

In these measurements, signals of MCP output greater than 20 mV are attributed to ions, while those smaller than 20 mV can be electrons and/or noise signal created by MCP itself. This discrimination level is based on calibration measurements of the MCP's output intensity by using an ^{241}Am source, which mainly emits 5.5-MeV alpha particles and 60-keV γ rays. Figure 10.6 represents a histogram of the MCP signal amplitude measured over 297 events. The upper (green) shows a histogram measured without ^{241}Am source, which shows that the noise level of MCP output is smaller than 20 mV. The middle (red) shows a histogram measured with ^{241}Am source, which shows that MCP outputs greater than 20 mV are due to 5.5-MeV alpha particles and 60-keV γ rays. The bottom (blue) shows a histogram measured with ^{241}Am source and two layers of 13-μm thick Al foil, where 5.5-MeV alpha particles are stopped by two layers of 13-μm thick Al foil (stopping range for 5.5-MeV alpha is 25 μm) and thus the small number of MCP outputs greater than 20 mV can be 60-keV γ rays. Therefore, we conclude that MCP outputs greater than 20 mV are attributed to high energy ions.

Figure 10.7a shows a typical TOF spectrum obtained in one laser shot, which registers one 15-MeV/u ion signal. A saturated peak around the flight time $t = 5$ is caused by hard X-rays emitted from the laser–cluster interaction region and pass through three Al foils. Figure 10.7b shows an ion energy spectrum thus obtained

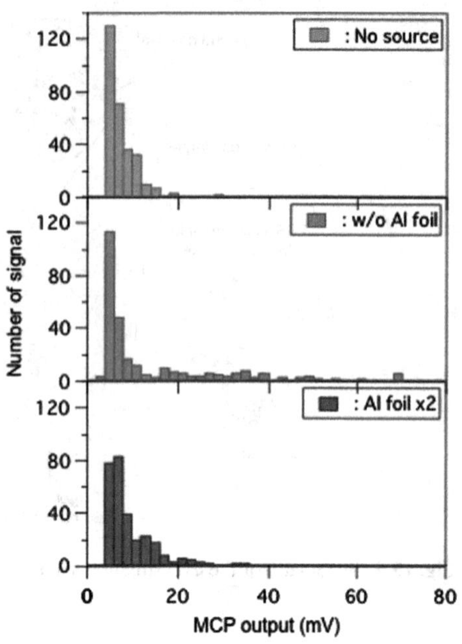

Fig. 10.6 A histogram of the MCP signal amplitude measured over 297 events. The upper, the middle, the bottom show histograms measured without ^{241}Am source, with ^{241}Am source, and with ^{241}Am source blocked by two layers of 13-μm thick Al foil, respectively

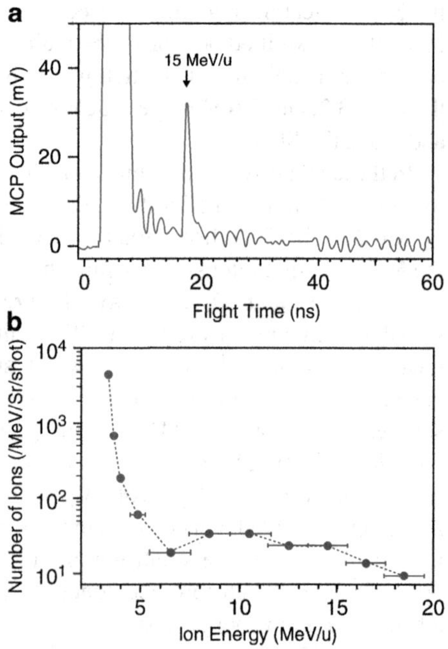

Fig. 10.7 (**a**) A typical TOF spectrum obtained in one laser shot, which registers $15\,\mathrm{MeV}/u$ ion signal. A saturated signal around the flight time $t = 5$ is caused by hard X-rays emitted from the laser–cluster interaction region. (**b**) The ion energy spectrum obtained by the TOF method

from the TOF measurements over 285 consecutive laser shots. Energy scale of abscissa axis is calculated assuming that the observed ion signals are carbon ions. The maximum ion energy is measured to be 18.5 ± 1 MeV per nucleon, which is consistent with the energy observed with the CR-39 track detectors. We note that even if we calculate the energy scale of abscissa axis of Fig. 10.7b assuming that the observed ion signals are helium or oxygen ions, the maximum ion energy of 18.5 ± 1 MeV per nucleon does not change within an experimental error regardless of ion species.

The number of ions detected by the TOF method is about 9/MeV/Sr/shot in the case for 18.5 ± 1-MeV/u ions. This agrees well with that obtained by using the CR39 detector.

10.3.3 Comparison of Our Experiment with Others

In our experiment, we have produced high energy ions up to 10–20 MeV per nucleon. This is approximately tenfold increase in accelerated ion energy compared to previous experiments with micron-thick solid targets, where 1.3–1.5 MeV protons were produced using ultrashort laser pulses (<100 fs) with energies of 120–200 mJ [60, 61].

We note that 40 MeV (10 MeV per nucleon) He^{2+} ions were produced using 1-ps laser pulses with a laser energy of 340 J in underdense helium plasma ($\sim 0.04 n_c$) [35, 36]. In our experiment, we have produced high energy ions with energy comparable to [35, 36] using ultrashort laser pulses with two thousand times lower energy. Concerning the number of ions, it should be noted that the laser energy in our experiment is more than two thousand times smaller than that of [35, 36]. Taking into account the nonlinear nature of the ion acceleration processes, there is no surprise that the number of ions is smaller in our experiment. Instead, there is a surprise that the ion energy is nearly the same.

10.3.4 Effect of Laser Prepulse on Ion Acceleration

When a larger prepulse with the contrast ratio of 10^{-4} was intentionally introduced to destroy the clusters well before the arrival of the main pulse, neither a long channel nor ion acceleration was observed. Furthermore, when the laser was focused onto a 60-bar He gas jet of density $\sim 0.02 n_c$ without clusters using the same nozzle, the rear part of the channel structure as well as the fringe pattern disappeared, Fig. 10.2c, and no high energy ions were observed. These results clearly show that the clusters are necessary for retaining the plasma density and its profile to make them favorable for the long channel formation and the ion acceleration.

10.4 Simulations

10.4.1 Simulation Methods

To elucidate the ion acceleration process, we conducted two-dimensional particle-in-cell (PIC) simulations, where we assume that the laser intensity is $1 \times 10^{19}\,\mathrm{W/cm^2}$, which includes the effect of intensity enhancement due to self-focusing, and the plasma parameters correspond to those of the experiment, i.e. a homogeneous plasma with the maximum density of $0.1 n_c$ is assumed as follows.

The plasma is composed of electrons and ions with the atomic number to the nucleon number ratio of $Z/A = 1/2$, i.e., He^{2+}, C^{6+} and O^{8+}, which can be produced via field ionization and/or collisional ionization. As mentioned above, high energy ion generation has been observed when the laser pulse is focused near the rear side of the gas jet. This region of plasma is simulated in the interval $2 < x < 112\,\mu\mathrm{m}$. As shown in Fig. 10.8, the density is equal to $0.02 n_c$ for $2 < x < 13\,\mu\mathrm{m}$, then linearly increases to $0.1\,n_c$ for $13 < x < 20\,\mu\mathrm{m}$, then remains constant for $20 < x < 65\,\mu\mathrm{m}$, then linearly decreases to $0.02\,n_c$ for $65 < x < 82\,\mu\mathrm{m}$, and then remains constant again for $82 < x < 112\,\mu\mathrm{m}$. The plasma slab is surrounded by vacuum regions. The density gradient employed for the PIC simulations is derived from the cluster-gas target model calculation [51].

10.4.2 Simulation Results

The linearly polarized (y-direction) laser pulse propagates along the x-axis from the left to the right. The laser pulse forms a channel in the plasma and undergoes relativistic self-focusing, Fig. 10.9a. The energy transmittance of about 10% is in agreement with the experimental observation of about 8%, i.e., the laser pulse has deposited almost all its energy into the plasma. Inside the plasma channel, the quasistatic magnetic field is generated. In the region $70 < x < 90\,\mu\mathrm{m}$, corresponding to the rear plasma slope, fast electrons accelerated in the channel form a dipole vortex [10, 39, 41, 42], associated with the strong quasistatic bipolar magnetic field with a maximum intensity of about 35 MG as seen in Fig. 10.9b. The magnetic field

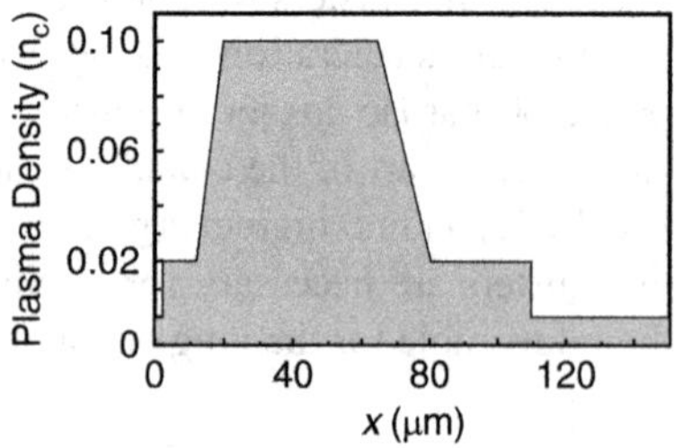

Fig. 10.8 Initial plasma density profile used for two-dimensional particle-in-cell (PIC) simulations

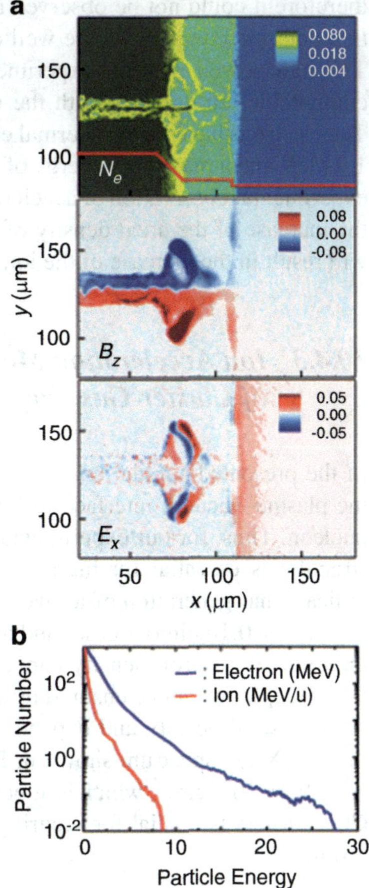

Fig. 10.9 (**a**) The electron density N_e (*upper*) normalized by n_c. The red line shows the initial density profile. The magnetic field B_z (*middle*) normalized by the laser field, and the longitudinal electric field E_x (*bottom*) normalized by the laser field, at $t = 900$ fs. (**b**) The calculated spectra of the ions in units of MeV/u and the electrons in MeV at $t = 6$ ps

presses out the cold plasma electrons forming the low density regions surrounded by dense thin shells. We note that the density profile in Fig. 10.9a resembles a bilobed structure at the end of the channel seen in the shadowgraph image, Fig. 10.2b. In addition to this, the fast electrons produce a quasistatic electric field at the plasma–vacuum interface (Fig. 10.9c, region $x > 112\,\mu$m), which is associated with the acceleration mechanism known as TNSA. However, we find that a much stronger electric field is generated at the shells wrapping the dipole vortex (Fig. 10.9c, region $70 < x < 90\,\mu$m).

According to the simulations and [39–42], ions are accelerated along the laser propagation axis in a time-dependent electric field generated during the magnetic field annihilation. This process is similar to the ion acceleration in plasma pinch discharges [62]. In the experiment, we have observed the ions accelerated in the forward direction. In addition, ions are accelerated perpendicular to the shell surface formed by the inhomogeneous magnetic field pressure. The second component of fast ions is expected to be seen in the direction at about ±45° from the laser axis and therefore it could not be observed in the experiment. In the simulations, we found that both ion components are well collimated with a divergence (full angle) of 5°. This agrees well with the experimental observation (3.4°). Both components give comparable ion energies with the maximum of 8.5 MeV per nucleon, Fig. 10.9d. The electrons have a quasithermal energy spectrum with an effective temperature of 1.9 MeV and a maximum energy of $\sim$ 20 MeV, Fig. 10.4d. This well agrees with the experimental observation of accelerated electrons [63]. Our simulations suggest that the increase of the areal density of the target, *nl*, and the energy of the laser pulse will result in the increase of the ion energy.

10.4.3 Ion Acceleration Mechanism and Role of Cluster-Gas Target

In the presented simulations, the contribution of the TNSA mechanism acting at the plasma–vacuum interface to ion energies is estimated to be about 2 MeV per nucleon. Thus, for our experimental conditions, the formation of the dipole vortex structure is essential for high energy ion generation. Our computer simulations indicate that generation of a favorable magnetic field requires an optimal electron density ($\sim 0.1n_c$ in our case) and an optimal slope-step profile. In this experiment, the optimal electron density cannot be provided only by the background He gas ($\sim 0.02n_c$), but can be ensured by a contribution from CO_2 clusters, which are initially at solid density and expand during the laser irradiation. In fact, analysis of the soft X-ray spectrum shown in Fig. 10.2a determines the electron density to be $1.0\pm0.2\times10^{20}$ cm^{-3} which is about $0.1n_c$. Thus, the use of a mixture of He gas and CO_2 clusters is crucial for securing the proper electron density and the slope-step profile.

10.5 Conclusion

We have demonstrated efficient generation of high energy ions with energies up to 10–20 MeV per nucleon and with a small full-angle divergence of 3.4° by irradiating the replenishable cluster-gas target with 40-fs laser pulses of only 150 mJ energy at 1 Hz repetition rate. This corresponds to approximately tenfold increase in the ion energies compared to previous experiments using thin foil targets [60, 61].

This high energy ion yield with modest laser intensity can be important in the development of more efficient laser ion accelerators for hadron therapy in the oncology, which can be achieved with smaller energy, high repetitive laser compared to the conventional laser-driven acceleration techniques.

Acknowledgements This work was supported by the Special Coordination Funds (SCF) for Promoting Science and Technology commissioned by MEXT of Japan.

References

1. M. Borghesi, et al., Fusion Sci. Technol. **49**, 412 (2006)
2. J. Fuchs, et al., Nature Phys. **2**, 48 (2006)
3. L. Robson, Nature Physics **3**, 58 (2007)
4. S.V. Bulanov, V.S. Khoroshkov, Plasma Phys. Rep. **28**, 453 (2002)
5. T. Tajima, D. Habs, X. Yan, Rev. Accel. Sci. Tech. **2**, 201 (2009)
6. A. Yogo et al., Appl. Phys. Lett. **94**, 181502 (2009).
7. I. Spencer, et al., Nucl. Instrum. Methods Phys. Res., Sect. B **183**, 449 (2001)
8. M. Borghesi et al., Phys. Plasmas **9**, 2214 (2002)
9. M. Roth et al., Phys. Rev. Lett. **86**, 436 (2001)
10. E.L. Clark et al., Phys. Rev. Lett. **84**, 670 (2000)
11. A. Maksimchuk et al., Phys. Rev. Lett. **84**, 4108 (2000)
12. R.A. Snavely et al., Phys. Rev. Lett. **85**, 2945 (2000)
13. S.P. Hatchett et al., Phys. Plasmas **7**, 2076 (2000)
14. A.V. Gurevich et al., Sov. Phys. JETP **22**, 449 (1966)
15. Y. Kishimoto et al., Phys. Fluids **26**, 2308 (1983)
16. F. Mako et al. Phys. Fluids **27**, 1815 (1984)
17. P. Mora, Phys. Rev. Lett. **90**, 185002 (2003)
18. M. Passoni et al., Phys. Rev. Lett. **101**, 115001 (2008)
19. T.Zh. Esirkepov et al., Phys. Rev. Lett. **89**, 175003 (2002)
20. B.M. Hegelich et al., Nature (London) **439**, 441 (2006)
21. H. Schwoerer et al., Nature (London) **439**, 445 (2006)
22. T. Toncian, et al., Science **312**, 410 (2006)
23. W. Luo et al., Med. Phys. **32**, 794 (2005)
24. S. Ter-Avetisyan et al., Laser Part. Beams **26**, 637 (2008)
25. M. Schollmeier et al., Phys. Rev. Lett. **101**, 055004 (2008)
26. M. Nishiuchi et al., Appl. Phys. Lett. **94**, 061107 (2009)
27. T. Ceccotti et al., Phys. Rev. Lett. **99**, 185002 (2007)
28. A. Henig et al., Phys. Rev. Lett. **103**, 045002 (2009)
29. A. Henig et al., Phys. Rev. Lett. **103**, 245003 (2009)
30. M. Nishiuchi et al., submitted to Proc. SPIE (2011)
31. M. Nakatsutsumi et al., Opt. Lett. **35**, 2314 (2010)
32. G.S. Sarkisov et al., Phys. Rev. E **59**, 7042 (1999)
33. K. Krushelnick et al., Phys. Rev. Lett. **83**, 737 (1999)
34. M.S. Wei et al., Phys. Rev. Lett. **93**, 155003 (2004)
35. L. Willingale et al., Phys. Rev. Lett. **96**, 245002 (2006)
36. L. Willingale et al., Phys. Rev. Lett. **98**, 049504 (2007)
37. K. Matsukado et al., Phys. Rev. Lett. **91**, 215001 (2003)
38. A. Yogo et al., Phys. Rev. E **77**, 016401 (2008)
39. A.V. Kuznetsov et al., Plasma Phys. Rep. **27**, 211 (2001)
40. H. Amitani et al., AIP Conf. Proc. 611, 340 (2002)
41. S.V. Bulanov et al., Plasma Phys. Rep. **31**, 369 (2005)
42. S.V. Bulanov, T.Zh. Esirkepov, Phys. Rev. Lett. **98**, 049503 (2007)
43. Y. Fukuda et al., Phys. Rev. Lett. **103**, 165002 (2009)
44. I. Alexeev, et al., Phys. Rev. Lett. **90**, 1034021 (2003)
45. H.M. Milchberg, et al., Phil. Trans. R. Soc. A **364**, 647 (2006)
46. Y. Fukuda, et al., Phys. Lett. A **363**, 130 (2007)
47. E. Springate et al., Phys. Rev. A **61**, 063201 (2000)
48. M. Eloy et al., Phys. Plasmas **8**, 1084 (2001)
49. K. Nishihara et al., Nucl. Instrum. Methods Phys. Res., Sect. A **464**, 98 (2001)
50. M. Mori, et al., Laser Phys. 16, 1092 (2006)
51. A.S. Boldarev, et al., Rev. Sci. Inst. 77, 083112 (2006)
52. F. Dorchies et al., Phys. Rev. A **68**, 023201 (2003)

53. A.Ya. Faenov et al., Phys. Scr. **50**, 333 (1994)
54. A.I. Magunov, et al., Laser and Particle Beam **21**, 73 (2003)
55. A. McPherson et al., Nature **370**, 631 (1994)
56. T. Ditmire et al., Phys. Rev. A **57**, 369 (1998)
57. Y. Fukuda et al., Phys. Rev. A **73**, 031201 (2006)
58. B. Dorschel et al., Radiat. Meas. **35**, 287 (2002)
59. J.F. Ziegler, J.P. Biersack, M.D. Ziegler, SRIM-The stopping and Range of Ions in Matter. SRIM Co. (2008)
60. I. Spencer et al., Phys. Rev. E **67**, 046402 (2003)
61. Y. Oishi et al., Phys. Plasmas **12**, 073102 (2005)
62. N.V. Filippov, JETP Lett. **31**, 120 (1980)
63. K. Kawase et al., in preparation

Index

K. Yamanouchi et al. (eds.), *Progress in Ultrafast Intense Laser Science VII*, Springer Series in Chemical Physics 100, DOI 10.1007/978-3-642-18327-0,

9783642183263